The Reinvention of Science

Slaying the Dragons of Dogma and Ignorance

The Reinvention of Science

Slaying the Dragons of Dogma and Ignorance

Bernard J. T. Jones
University of Groningen, The Netherlands

Vicent J. Martínez
University of Valencia, Spain

Virginia L. Trimble
University of California, Irvine, USA

 **World Scientific**

NEW JERSEY · LONDON · SINGAPORE · BEIJING · SHANGHAI · HONG KONG · TAIPEI · CHENNAI · TOKYO

Published by

World Scientific Publishing Europe Ltd.

57 Shelton Street, Covent Garden, London WC2H 9HE

Head office: 5 Toh Tuck Link, Singapore 596224

USA office: 27 Warren Street, Suite 401-402, Hackensack, NJ 07601

Library of Congress Cataloging-in-Publication Data
Names: Jones, B. J. T. (Bernard Jean Trefor), 1946– author. |
    Martinez, V. J. (Vicent J.), 1962– author. | Trimble, Virginia, author.
Title: The reinvention of science : slaying the dragons of dogma and ignorance /
    Bernard J. T. Jones (University of Groningen, The Netherlands), Vicent J. Martínez
    (University of Valencia, Spain), Virginia L. Trimble (University of California, Irvine, USA).
Description: New Jersey : World Scientific, 2024. | Includes bibliographical references and index.
Identifiers: LCCN 2022052458 | ISBN 9781800613362 (hardcover) |
    ISBN 9781800613607 (paperback) | ISBN 9781800613379 (ebook for institutions) |
    ISBN 9781800613386 (ebook for individuals)
Subjects: LCSH: Errors, Scientific--History--Miscellanea. | Science--History--Miscellanea.
Classification: LCC Q172.5.E77 J66 2024 | DDC 001.9/6--dc23/eng20230117
LC record available at https://lccn.loc.gov/2022052458

British Library Cataloguing-in-Publication Data
A catalogue record for this book is available from the British Library.

For any available supplementary material, please visit
https://www.worldscientific.com/worldscibooks/10.1142/Q0394#t=suppl

Desk Editors: Soundararajan Raghuraman/Adam Binnie/Shi Ying Koe

Typeset by Stallion Press
Email: enquiries@stallionpress.com

Preface

Fairy tales are more than true: not because they tell us that dragons exist but because they tell us that dragons can be beaten.

Neil Gaiman, Coraline

The Reinvention of Science: Slaying the Dragons of Dogma and Ignorance

The volume you now hold in your (real or virtual) hands is the product of a long and somewhat untidy process. It began as an essay by authors Martínez and Trimble. Under the title *Erasing the Dragons from the Maps of Science*, we looked at a handful of entities that had once seemed essential to explain the real world. These included phlogiston to account for fire, the luminiferous ether for propagation of radiation, the homunculus to provide for heredity and crystalline spheres to carry the wandering planets around the Earth.

In due course, the essay was declared ineligible for the prize competition at which it had been directed. Gradually, the idea arose that this viewpoint on the history of science was important enough to be a whole book, not just an essay. At this point, Bernard Jones joined the team as the leader (not just alphabetically), and we all reluctantly admitted that only a very few medieval maps actually said *here be dragons* in any language. Thus, we had to find another title, which you now see on the cover. In addition, the decision was made

v

to adopt a roughly chronological approach to the history of science, though some division by disciplines remains, and you will find each of our original "dragons" somewhere.

Many gaps therefore had to be filled in to describe what we think has happened, why and the extent to which groups of scientists or proto-scientists (the word dates only from the mid-1800s) clinging to some ancient idea or entity had retarded progress towards what each discipline now regards as the right way of looking at its territory.

What is the right way of looking at things? We, like the majority of our colleagues rejoice that "science is true, whether or not you believe in it!", as Neil deGrasse Tyson, director of the Rose Planetarium in New York, has said on a television program and on Twitter.

But "true" is significantly different for us than for our friends in mathematics, who (among other impressive tasks) establish theorems that MUST follow from a set of postulates. Our "true" is just the ideas (theories, hypotheses, models and scenarios) that agree most closely with the results of observations and experiments. In fact, as Naomi Oreskes, professor of History of Science at Harvard University, replied to Neil's tweet, "Scientists may reject things they later accept as true." And yes, progressing from one set of ideas to another promising better agreement can be retarded by hanging on to some of those dragons. Indeed, we probably cherish a few ourselves.

Nevertheless, assembling the pages that follow has led your authors to a greatly increased appreciation for the creativity and determination of their predecessors and contemporaries. And we hope that reading these same pages might bring the same gift to you.

Contents

Part I
Breaking the Dragon's Egg
Science Is Born

Chapter 1

Ether and Atoms

Be not the first by whom the new are tried,
Nor yet the last to leave the old aside.

Alexander Pope, An Essay on Criticism

1.1 The Most Famous Failed Experiment

It was July 1887. Everything was arranged in the basement of the dormitory building at Western Reserve University in Cleveland (Ohio) to carry out a physics experiment that, much later, would be considered one of the most important ever. Two people entered the basement that warm July morning: Albert A. Michelson (1852–1931) and Edward W. Morley (1838–1923). Five years earlier, Michelson had accepted an offer to be a professor of physics at Case School of Applied Science. The founders of Case had decided to hire a highly qualified expert in experimental science. He was the perfect choice to improve this industrial city's scientific culture. In the fall of 1886, a fire at Case destroyed the laboratory that Michelson had been setting up for 4 years. Morley, who was a Chemistry professor at the neighboring Western Reserve University, offered to install instruments that had been rescued from the fire in the basement of one of the university buildings. Morley was recognized as a skilled experimenter, and his contribution to setting up the experiment was crucial.

Albert Michelson made three large splashes in the waters of science from 1878 until his death. In two cases, his work confirmed

3

what his colleagues had been expecting. In the third, definitely not, and it took many iterations by him and others before everyone was convinced of his correctness.

First, he measured the speed of light from his student days at the United States Naval Academy (Annapolis Maryland) right up to the time of his death, adding several significant figures to a number that had been known approximately since the work of the Danish astronomer Ole Rømer (1644–1710) in 1676.

Second, he turned the 100-inch Hooker telescope at Mt. Wilson Observatory into an astronomical interferometer, proving (with Francis G. Pease (1881–1938)) that giant stars really are BIG, hundreds of times the diameter of our own Sun.

Third was the Michelson–Morley experiment. This is the one that used what is now called a Michelson interferometer. Interferometers of this type are used in the Laser Interferometric Gravitational-Wave Observatory (LIGO), which started detecting bursts in 2015.

Michelson had built such an interferometer while he was at Potsdam (Germany) as a postgraduate in 1880. It was designed to detect the movement of the Earth through the luminiferous (this is a fancy word for light-bearing) medium called the ether, a substance whose theoretical existence had been postulated many centuries ago and which for the majority of physicists of the 19th century was absolutely necessary for the transmission of light, as air is for sound.

The results of the first Michelson experiment seem to demonstrate that light is not like other things that move as waves (sound or ripples on water) or as particles (trains or race horses). For those, the speed you measure depends on how you and the source are moving, relative to each other or relative to some stationary background of water, air, or ground. But no matter who you are or what you are doing, if you measure the speed of light (in a vacuum or tenuous medium like air), you will always get the same answer, that c is about 300,000 km/sec that he had been measuring all along.

Michelson complained about the little attention his Potsdam experiment had received in the scientific community, but as a result of a letter received from Lord Rayleigh (1842–1919) encouraging him to repeat the test, he replied[1]: "Your letter has however once more fired my enthusiasm and it has decided my to begin to work at once".

What had everyone expected? Suppose you are a swimmer living near a river and swimming at a constant speed relative to water.

Swimming a mile with the flow is the fastest, swimming a mile against the flow is the slowest, and swimming a mile straight across comes in between.

Most of Michelson's contemporaries supposed that light moved in a medium called ether, in a way similar to the swimmer moving in water, or sound in air, only of course much faster.

If so, then if you set up an experiment in which you can compare how long it takes light to travel a mile parallel to Earth's motion through the ether vs perpendicular (or against Earth's travel), you should get different times, and a beam of light split in half, made to take the parallel and perpendicular paths, and then coming together should show the effect of the different times, by getting out of phase and producing an interference pattern.

The Michelson–Morley experiment was set up exactly in that way: swimmers in the river analogy were pulses of light in the interferometer (see Figure 1.1). The supposed ether wind played the role of the flow of water. But the result was, as Michelson wrote to Lord Rayleigh, "decidedly negative": both half pulses always arrived at the same time, independent of their direction with respect to the supposed ether wind.

Albert Michelson won the Nobel Prize in Physics in 1907. He was the first American who won it. Nevertheless, the experiment showing no Earth's drift through the postulated ether was not even mentioned in the citation for the prize. It was his measurement of the meter in terms of the wavelength of cadmium light that most impressed the Swedish Academy. That result was obtained also in collaboration with Morley and used the same interferometer.

The Michelson–Morley experiment is correctly described by Albert Einstein's special theory of relativity and might even have partly inspired it (though years after, he said that wasn't quite sure he had heard, before 1905, of that first 1887 result). In any case, there is no doubt that 1887 was an important year for the advancement of science. At the end of the very same year, Arthur Conan Doyle (1859–1930) published his first novel on Sherlock Holmes entitled *A Study in Scarlet*. In one of the insightful dialogs held with Dr. Watson, the famous detective says, "One's ideas must be as broad as Nature if they are to interpret Nature". Clearly, the Michelson–Morley experiment opened the door to broad ideas in science.

THE MICHELSON-MORLEY EXPERIMENT

1 The experiment was designed to show how the ether wind due to the Earth's motion changed the light speed in the same way that the water stream changes the swimmer's speed.

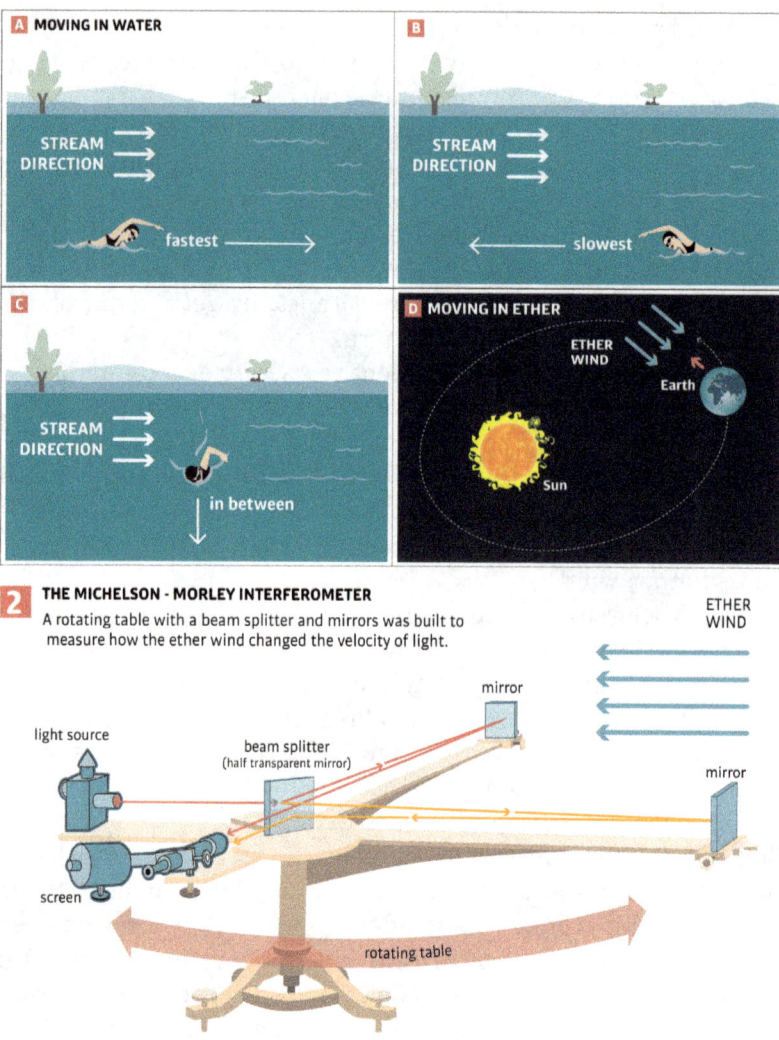

2 THE MICHELSON - MORLEY INTERFEROMETER

A rotating table with a beam splitter and mirrors was built to measure how the ether wind changed the velocity of light.

Figure 1.1 The analogy of the Michelson–Morley experiment with a swimmer in a river (top). Modern diagram of the experiment (bottom). Credit: Infographic sketched by the authors and elaborated by Javier Pérez Belmonte.

Over the years, lots of other physicists carried out similar experiments, and some even won prizes for NOT confirming Michelson and Morley. Not needing ether was not quite the same as not allowing for it. The experiments conducted by Thomas Young (1773–1829), which indicated that light took the form of a wave, revived the notion of "luminiferous ether", described by Agnes M. Clerke (1842–1907) in 1902 as "the ethereal vehicle of the vibrations of light", just as sound required a medium, air, for its propagation. She ignored the experiment conducted by Michelson and Morley some years earlier that called into question the existence of ether. It seems clear that Clerke along with many other scientists of that time knew the negative results of the Michelson–Morley test to detect the ether drift. Nevertheless, as the professor of History of Science Albert E. Moyer pointed out,[2] "Awareness of an experiment, however, differs from appreciation. Widespread appreciation of the test's significance come only after Michelson's Nobel Prize, when the physics community grasped the full implications of Albert Einstein's special theory of relativity".

Ether is without a doubt the "postulated but unseen entity" that has most frequently appeared throughout discussions of the history of science, from Aristotle to the present day.

The story begins in the 5th century BCE when a group of philosophers put together and wrote about their view of how the world was built. At the time, and for over 2,000 years thereafter, the dominant view was that Earth, Air, Fire and Water were the basic indivisible elements of the world. All matter was made up of combinations of these four elements. The idea is attributed to Empedocles (c. 494–434 BCE), in the 5th century BCE and was later strongly supported by Aristotle (384–322 BCE) who added a fifth element: the Quintessence. Similar themes arose, some at even earlier times, in different parts of the world.[3]

For Aristotle and many of his predecessors and contemporaries, the main function of ether was to prevent voids, thought to be logically impossible, but also to make up the heavenly bodies. Anaxagoras had somewhat heftier ether, friction with which heated the hot stones that were the Sun, Moon, and stars, although it also carried them around the Earth.

For most natural philosophers of later times, action-at-a-distance was as much anathema as a void, so ether was needed to connect causes and effects — to transmit forces in somewhat more modern language.

Throughout the history of science, different thinkers, philosophers and scientists postulated the existence of different entities that played a role similar to the one played by the ether for centuries. Entities that, in spite of not being visible or detectable in their time, or perhaps ever, were nevertheless necessary to maintain the stability of the cosmos as the philosophers saw and understood it. They could be entities whose existence explained, or at least justified, the observations or experiments of the moment. Over the course of this book, we will identify some of these, from Antiquity to the present. Those are collectively called in the subtitle of this book "Dragons". They were advocated on the maps of knowledge by scientists of all times. In many cases, the passage of time has revealed that these principles were wrong and that it was necessary to abandon the theories or beliefs on which they were based, as it happened with the ether. The experiment of Michelson–Morley erased this dragon from those maps of science forever (although at least two decades were necessary for a widespread recognition of the implications of the experiment). In other cases, the entities put forward were eventually discovered. Such findings were stellar moments in the history of science.

The fundamental questions are what picture of each particular aspect of nature was the proposer trying to explain, and would the suggested entity provide the required stability if it actually existed?

The planet Neptune is probably the object that best illustrates the category of entities that were initially proposed to explain observations of the moment and then subsequently discovered. In Chapter 6, we will see how, in the 19th century, the astronomers Urban Le Verrier (1811–1877) and John Couch Adams (1819–1892) tried to explain the anomalies observed in the orbit of the planet Uranus by proposing gravitational pull produced by a planet beyond Uranus' orbit as the cause. This is how Neptune was discovered in 1846. Throughout this book, we will also see other examples of objects (from stars to elementary particles) whose existence was conjectured at some point and which had not been observed when they were proposed but which were discovered later on. However, there are entities whose existence was postulated many years ago and are still needed to explain the cosmos as we understand it today but have not yet been detected. Readers' minds will surely have already turned to dark matter, which has been invoked by astronomers for 90 years to explain different cosmological observations. The nature of this possible dark matter remains a mystery, but few doubt its existence.[4]

A widespread early example of this kind was the need for there to be something to hold up the sky in order to make the room needed on Earth's surface for living creatures.

1.2 Separating the Heavens from Earth

In each era, the world has been seen in a particular way. Explanations of the structure and origin of the cosmos produce different civilizations' cosmology and cosmogony. The most ancient cultures had their own myths. Almost all ancient mythologies posited the existence of something or someone tasked with separating the heavens from Earth. Ancient mythologies and mythical thought obviously came about much earlier than scientific thought did. As Eudald Carbonell and Robert Sala explain in their book[5] *Encara no som Humans*, mythical thoughts must have appeared very long ago, even before our ice-age ancestors captured them in paintings on walls and ceilings of caves and in bone and stone carvings. We may need to look back as far as the species *Homo heidelbergensis*, in the Middle Pleistocene, 300,000 years ago, to find the earliest "abstract thought about human existence and the construction of a primitive cosmos".

For the Egyptians, Shu (air) kept Nut (the sky) above Geb (Earth). Nut and Geb were siblings and lovers. By separating them, their father, Shu, allowed life on Earth's surface to exist. Similar versions of this account exist in Chinese and Babylonian mythologies. For the Babylonians, Enlil separated Anu (Earth) from Ea (the waters of the heavens and the oceans). Babylonian astronomy was highly developed. It seems that they were able to predict lunar eclipses, planetary conjunctions and other astronomical events, and these things demonstrate an in-depth knowledge of the heavens, the stars and their movements. Nevertheless, it must be said that this cosmology remained in the mythological terrain described above, and the empirical knowledge that they gained regarding what was happening or was going to happen in the celestial sphere did not lead to a more rational approach to the cosmos. According to the Chinese, the deity Panku aided by a turtle and a set of mythical creatures (a phoenix, a dragon and a qilin) created a separation between Earth and the sky of about 43,000 km, at a rate of 3 meters per day for 18,000 years, as described by Helge Kragh in his *Conceptions of the Cosmos*. The turtle must have done most of the work, since the other three are mythical beasts (though the qilin must not be confused with the unicorn).

The book of Genesis in Judeo-Christian scripture contains no similar entity. Here, the Creator himself puts each thing in its place — including heaven and Earth — at the first attempt. However, a process of separation like that in the Eastern mythologies does appear in Genesis[6]1:4–9

> God saw that the light was good,
> and he separated the light from the darkness...
> And God said, "Let there be a vault between
> the waters to separate water from water..."
> And God said, "Let the water under the sky be gathered
> to one place, and let dry ground appear". And it was so.

The roughly contemporary account in the Theogony of Hesiod (8th century BCE, Greek) puts the in-principle unobservable deep abyss of the Tartaros as far below the (flat) Earth as heaven is above. How far is that? How far can a brass anvil fall in nine days and nights to arrive on the 10th? A little further than the distance to the Moon if Hesiod had been a Newtonian. One suspects, however, that he had probably never dropped an anvil more than 20 or 30 feet, yielding a speed of a couple of meters per second, or a total distance of a few thousand kilometers. We post-Galileans can also remark that it would all come out the same if the anvil had been made of iron rather than brass.

There is an aspect of Greek mythology that we have learned since childhood, which explains how the constellations arose from the relationship between different gods and also between the gods and humans. What is relevant in the context under consideration here is the role of Atlas, who was condemned by Zeus to "hold up the pillars of the heavens", as in the Farnese Atlas statue, though other more recent versions often show him holding up a spherical Earth, hanging on at places that Greeks rarely visited (but you should see in Figure 1.2 where Shu touches Nut!).

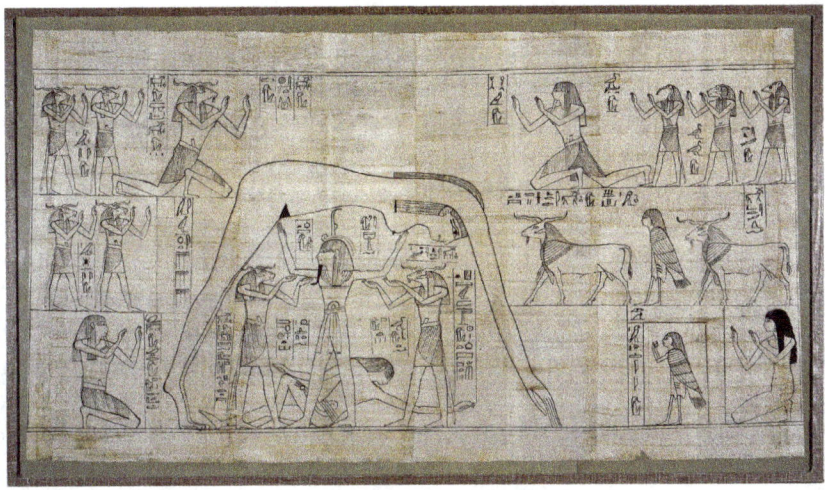

Figure 1.2 Details of the Greenfield Papyrus, Book of the Dead of Nestaneb-tasheru (sheet 87). The Egyptian god Shu (the air), holding Nut (the starry sky) apart from Geb (the Earth) allowing the life to exit. Credit: © The Trustees of the British Museum.

Myths came slowly to be abandoned, and more rational explanations for the cosmos began to be sought. However, it should be pointed out that this abandonment was gradual. Often, imagined entities were proposed, that were not strictly speaking mythical in the sense that they were not gods, goddesses or composite creatures. In many cases, knowledge subsequently gained revealed that these things did not exist. Nevertheless, they were perceived as necessary at the time when they were put forward in order to maintain the consistency of what was being posited. Included in this category of entities that were "postulated but not seen" were crystalline spheres, rotating transparent spheres made of ether carrying planets and stars, which were introduced in Greek astronomy to explain the movements of the celestial bodies and were used until the late 17th century (see Chapter 5).

1.3 Replacing Myths by Rational Thought

British evolutionary biologist Richard Dawkins (b. 1941) has said that science and literature are the human activities that justify the application of the specific name *Homo sapiens* to our species.[7] The precise moment in history when humans began to nurture these activities depends on the definitions that we attach to the words "literature" and "science". Some authors consider the birth of science to have taken place in the 16th century with the Copernican Revolution and the subsequent development of scientific method. However, in our opinion (and that of many others), this is too restrictive a view of science. We have to go much further back in time to find the origins of scientific thought. In his book *The Forgotten Revolution*, Italian physicist Lucio Russo (b. 1944) puts the date for its birth at 323 BCE. This is the beginning of the Hellenistic period according to the terminology introduced by German historian Johann Droysen (1808–1884). The year 323 BCE was when Alexander the Great (356–323 BCE) died. It was the era of Archimedes of Syracuse (ca. 287–212 BCE), Euclid of Alexandria (fl. 300 BCE), Herophilos of Chalcedon (ca. 335–280 BCE), Eratosthenes of Cyrene (ca. 276–195 BCE), Aristarchus of Samos (310–230 BCE), Apollonius of Perga (ca. 240–190 BCE) and Hipparchus of Nicaea (ca. 190–120 BCE), among others.

The most creative scientific activity occurred in the 3rd and 2nd centuries BCE, and a decline then began, following the conquest of Alexandrian lands by the Roman Empire, which ended in 30 CE with the annexation of Cleopatra's Egypt and the Romans' complete domination of the entire Mediterranean Basin. Formally speaking, this marked the end of the Hellenistic period, though there was still an impetus for scientific activity for some centuries in the former Alexandrian kingdoms under the *Pax Romana*. Greek culture and language were not ousted by the Roman Empire, and a clear division between the Latin West and the Greek East persisted. The technology and the economic activity of the Eastern part of the empire continued, allowing the continuity of scientific development, though without the initial strength and splendor. Alexandria would continue to be the nerve center of this activity. This was the era of Ptolemy (ca. 100–170 CE), Heron of Alexandria (ca. 10–70 CE), Galen of Pergamon (129–216 CE) and Diophantus of Alexandria (ca. 200–284 CE). The end of this period is usually set at the date when Hypatia of Alexandria (ca. 360–415 CE) was killed at the hands of Christian fanatics in 415 CE, a moment immortalized in Alejandro Amenábar's film *Agora*, in which Hypatia is played brilliantly by Rachel Weisz.

Russo believes that one of the key factors in the birth of scientific thought through the expansion of the Alexandrian Empire was the meeting of civilizations produced by that expansion, because the Greek hegemon, led by Alexander the Great, invaded Egypt and Mesopotamia, and thereby discovered civilizations with technology and economies that were superior to those of the conquerors themselves. To be sure, classical Greek culture had reached very high levels in aspects such as philosophy, theater, literature and politics (in the form of the birth of some aspects of democracy), but in technological terms, it was behind its eastern neighbors. As Russo states,

> The Greeks who moved to the new kingdoms that arose from Alexander's conquests had to administer and control those more advanced economies and technologies with which they were not familiar; their one crucial advantage and guide consisted in the sophisticated methods of rational analysis developed by the Greek cultural tradition during the preceding centuries. It is in this situation that science is born.

Technological development has always been linked to scientific advances. As Cambridge University cosmologist Malcolm S. Longair[8] argues, "Many of the pioneers of the technology of astronomy deserve their rightful places in the pantheon of the founders of modern cosmology". The impact that Galileo Galilei's *Sidereus Nuncius,* published in March 1609, had on 17th century scientific thought is a direct consequence of the astronomical observations that the Italian thinker undertook using a telescope. The Hellenistic era's forgotten scientific revolution was also accompanied by spectacular technological development. Unfortunately, few physical vestiges of those advances remain, but there is one, and its perfection must lead us to conclude that the level of sophistication in technology reached in the Hellenistic period was impressive and was not surpassed until many centuries later. This instrument is the Antikythera mechanism. Here is the story of its discovery and interpretation.

1.4 Analog Computers from the Distant Past

There was once an old boat that undertook a voyage, probably from the Greek coast towards Italy, with a hold filled with sculptures, ceramics and other riches of Hellenistic culture. But it was also transporting another item: a precision instrument that has turned out to be one of antiquity's most surprising objects. This item was a bronze analog astronomical calculator that was built in the 1st or 2nd century BCE. It could even have been built at the end of the 3rd century BCE. The date is uncertain. It showcases the level of technological sophistication and astronomical knowledge of its designers and makers. The ship sank near the small island of Antikythera, between the Peloponnese peninsula and the island of Crete. Its crew members perished, and the hull plus all its cargo ended up at the bottom of the Mediterranean Sea. And there it remained hidden for centuries. In 1900 at Easter, the Greek Elias Stadiatis donned a diving suit and descended about 45 meters in that precise place. He was looking for sea sponges but came across the remains of the old shipwreck. Stadiatis was traveling aboard a sponge fishing boat captained by Dimitrios Kondos. The captain reported the finding to the Greek authorities, and work to rescue the cargo began very soon afterward. The items retrieved were transferred to the National Museum of Archaeology in Athens.

Among such a wealth of eye-catching treasures, a rather corroded piece of bronze and wood initially went unnoticed until, on 17 May 1902, the Greek archaeologist Spyridon Stais noticed that there were cogwheels inside the contraption. Although its gears seemed surprising, it was not until much later, in the 1970s, that the Antikythera mechanism's extraordinary structure was revealed through the use of X-ray techniques. Earlier, the German philologist, Albert Rehm (1871–1949), in the period 1905–1906, was the first to understand that it was an astronomical calculating machine. British science historian Derek de Solla Price, Greek nuclear physicist Charalambos Karakalos and his wife, Emily, conducted the first detailed study of it. Price published their findings in 1974 in a lengthy article[9] entitled *Gears from the Greeks: The Antikythera Mechanism — A Calendar Computer from ca. 80 B.C.* A study of this curious object by Tony Freeth (University College London) and an interdisciplinary team appeared[10] in *Scientific Reports* in March, 2021. It carried the striking title: *A Model of the Cosmos in the ancient Greek Antikythera Mechanism.* In this work, the authors put forward a coherent story to explain the mechanism's different uses. They analyze the inscriptions that appear on both the wooden covers and the bronze pieces, and they also study the function of the different gears that make up the mechanism. They conclude that this planetarium, the oldest in the world, perfectly combines all the knowledge of Babylonian astronomy with the mathematical and astronomical expertise of ancient Greece.

Thirty gears from this piece of engineering have been partially preserved. Study and modeling work has led to the discovery that it could be used to predict astronomical phenomena, such as the phases of the Moon or the positions of the Sun and planets. It was at the same time a calculator, capable of predicting the dates of eclipses or the dates on which the different panhellenic games (the Olympic, Pythian, Nemean and Isthmian games) were to be held (see Figure 1.3).

The Antikythera mechanism is so sophisticated that it surely could not have been a one-off. It is very likely that other similar instruments had previously been built and allowed the necessary technology for the piece to be developed. Beyond the makers' understanding of the movement of the stars and the different cycles that were used in ancient times, their technical knowledge of gears, friction and strength of the materials used is surprising. Some have tried to link this instrument to similar pieces that are referred to in Roman texts. For example, in *De re Publica I*, Cicero (103–46 BCE) refers to a mechanical planetarium, probably an orrery, built by Archimedes.

Figure 1.3 Reconstruction of the Antikythera mechanism: Exploded model of Cosmos gearing. Credit: © 2020, Tony Freeth, Images First Ltd.

To be sure, the sage of Syracuse combined theoretical knowledge with a technological dimension to build different scientific instruments. According to Russo, Cicero relied on a now-lost text by Sulpicius Gallus in which the latter recounts having seen a planetarium in operation at the home of Consul Marcus Marcellus, the grandson of General Marcellus, who sacked Syracuse.

In any case, with the sinking of the Antikythera mechanism in the shipwreck and its 2,000-year stay on the bottom of the Mediterranean Sea, this technology was lost. Centuries would pass before similar instruments were developed. Similar instruments are mentioned in the literature in the centuries after the Antikythera Mechanism, though no physical instruments have been found. We find similar but much less sophisticated gears in a treatise on the astrolabe from the late 10th century written by Arabic mathematician al-Biruni (973–1048), in which he describes a lunisolar calendar. It was not until the analog clocks of the 14th century that examples of technology comparable to that of the Antikythera mechanism came along. As in other aspects of science and knowledge, the advances in technology of the Hellenistic revolution were also lost. The reinvention of science became essential, as researcher François Charette eloquently expresses in his article[11] "High tech from ancient Greece", published in the journal *Nature* in 2006: "The mind-boggling technological sophistication available in some parts of the Hellenistic and Graeco-Roman world was simply not transmitted further. The gear wheel had, in this case, to be reinvented. The Antikythera mechanism is a useful reminder that history seldom follows simple, linear paths".

1.5 Atoms and the Infinite

Why did Hellenistic culture and the nurturing of scientific thought disappear? Mainly because of the Roman invasion of different parts of the old Alexandrian Empire and the subsequent wars: the siege of Syracuse and the killing of Archimedes in 212 BCE, the destruction of Corinth and Carthage in 146 BCE, the defeat of Athens in 86 BCE, the pillage of Rhodes in 43 BCE and the final annexation of all of Egypt in 30 BCE. Libraries were broken up, and many Greek citizens who inhabited these lands and worked in their cultural centers were sent to Rome as slaves, tutors, copyists and scribes. The authority of Aristotle's philosophy, which prevailed from the 1st century CE, was another element that contributed to forgetting the Hellenistic thinkers, as did the subsequent adoption of Aristotelian thought by Christianity, especially by the Scholastics, represented by Thomas Aquinas (ca. 1224–1274).

The school of Plato and Aristotle was not the only school of thought in classical Greece. At least two other schools coexisted: the atomists and the stoics. The atomist school of philosophers envisaged an entire universe composed of invisible and indivisible small particles ("atoms") moving around in space (the void) out of which everything we see was built. In their view, the four elements of those ancient times, Earth, Air, Fire and Water, were themselves made from different organizations of atoms moving in the void. The idea of something more fundamental than the four elements is attributed to the 5th century BCE philosophers Leucippus and his pupil, Democritus (ca. 460–370 BCE). We have relatively little information about them. Not only were they sidelined by their opponents but also many of their manuscripts have been lost, destroyed or locked up in libraries that have since disappeared. Those who supported their view were known as "the atomists".

These atomists also supported the notion of a "void" in which the atoms moved freely. Nowadays we refer to this void as the "vacuum". Aristotle declined to think that such a concept should ever be seriously discussed. He took the not-unreasonable view that it is impossible to talk about "nothing". The concept of empty space has, of course, persisted into modern science as the vacuum, and even Aristotle's ether resembles modern-day "quintessence", whatever that might turn out to be.

The ideas of these leading classical figures on the constituents of matter and the introduction of the word "atom" were of fundamental importance to the atomist school. They spoke rationally about entities that they obviously could not observe, "dragons", but which were necessary for preserving their understanding of the cosmos. Almost 25 centuries later, atoms were introduced in modern physics, in a context of experimental and theoretical knowledge that was utterly different. To be sure, modern atoms have nothing in common, except for the name, with the classical version, but we must recognize that the modern concept is, at least, deeply rooted in the line of thought that began in classical Greece.

Different schools of thought during the Hellenistic era, holding opposing positions about the finiteness of the cosmos, coexisted with one another. The stoics for example, a school founded by Zeno of Citium (333–262 BCE) in Athens, argued, in contrast to the atomists, that there was a single, finite universe, although, they affirmed the existence of a non-physical infinite void surrounding the physical finite cosmos. This idea would be taken up by Johannes Kepler in the 17th century, and different astronomers supported it even in the 20th century — for instance, the North American Harlow Shapley (1885–1972), who was the director of the Harvard College Observatory for nearly 40 years. Modern cosmology does not have a definitive answer to whether the universe is finite or infinite, though we can state that it is very large compared to the part we can observe. Whether or not it is infinite depends on values of some cosmological parameters which have yet to be determined with accuracy.

Another example given by Russo in *The Forgotten Revolution* is that of the famous paradoxes of Zeno of Elea (ca. 490–430 BCE), and especially the paradox of Achilles and the tortoise.[12] We can clearly see in it the difference between the concepts introduced in the philosophical field in classical Greece. Later, in the Hellenistic period, these concepts would be addressed in a scientific way. These paradoxes involve ideas such as the continuity of physical magnitudes (for instance, time and space), though these were not part of his mathematical modeling. But they were in Euclid's, 150 years later, in the fifth book of his Elements. We must recognize, however, that Zeno's arguments are, at least, the forerunners of the basic ideas that underpin modern infinitesimal calculus developed in 1666 by Isaac Newton (1642–1727) and Gottfried Leibniz (1646–1716).

In short, although the birth of science can be placed at the beginning of the Hellenistic period, the contribution to rational thinking made by pre- and post-Socratic philosophers was essential in displacing cosmogonies based on ancient mythologies as the only way to explain the world. The transition would be smooth rather than abrupt, such that the particular elements of scientific thought grew with classical Greece's most impressive sequence of masters and disciples: Socrates, Plato and Aristotle. Appearing in Aristotle's natural philosophy, for example, are many of the elements needed to describe physical phenomena, such as the forces to explain movement.[13] But Aristotle works in a speculative manner, without making use of a scientific method, even a nascent one, as Archimedes would, decades later, to explain the same phenomena. We note that Aristotle was the teacher of Alexander the Great, with whom the Hellenistic period began.

To Aristotle, the four elements — Earth, Air, Fire and Water — were fundamental; everything in the sublunary region[14] was made from these four elements. To the atomists, there was a more fundamental indivisible building block, the "atom".

The atomist view was adopted in the 3rd century BCE by Epicurus of Samos who incorporated it within his ethical hedonistic philosophy in order to provide his views with a solid philosophical basis. He felt that life should be lived for the present and not to satisfy the whims of the gods or for the dream of an afterlife. For Epicurus, a good life was one that was free from pain.[15] He rejected the classical idea that our lives were controlled by the whims of self-serving and often disruptive gods: he was an atheist. Epicurus promoted the view that we should live our lives to the full because there was no afterlife. This one life was all we had.

Epicurus endowed the atoms with the additional property of weight and argued that their motion cannot be simple straight lines but rather that they should have a small random component of motion that enables them to interact. In the 3rd century BCE, randomness was a revolutionary concept. The atomists however believed everything was in motion and moreover that there was a randomness in those motions, leading to an indeterministic world. If there were no such random component, the laws of motion would be deterministic, and everything would depend on the starting conditions. Those would pre-determine human actions and there would be no free will.

However, the atomists' view did not go down well with the dominant Aristotelians. Epicurus and his followers were marginalized except in the popular writings of the day. It is, however, by Aristotle that we are told that Epicurus had endowed the atoms with hooks so that they could combine together to build other things. Despite this, the Epicurean view of life became quite popular in later Roman times, when Titus Lucretius Carus, "Lucretius" for short, wrote an epic 7500 line[16] poem in the 1st century BCE: *De Rerum Natura* presenting his version of Epicurus' philosophy. The poem is largely a detailed summary of the works of Epicurus with the addition of Lucretius' own reflections, written in six parts, beginning with

a description of matter and space in the first part and the movement and shapes of atoms in the second. It goes on to use atomism
as a basis for building and justifying the system of ethics and life
advocated by Epicurus. Bertrand Russell, one of the great philosophers of the 20th century, remarked that Lucretius' poem is our main
source about the ideas of Epicurus and the atomists. Curiously, the
poem was thought not to have been preserved after the decline of the
Roman Empire, until a manuscript that contained it was discovered
in 1417 in a European monastery. The existence and survival of this
poem is the central part of our story[17] in the next chapter.

Chapter 2

The Lost Poem

All of Nature as it is in itself consists of two things — bodies and a vacant space in which the bodies are situated and in which they move.

Lucretius, *De Rerum Natura*

2.1 On the Nature of Things

This is a story of the influence of the distant past on the present. The story tells of a schism about what the world was made of, and the ingredients of the tale include a hedonistic philosophy of life that sidelined the Gods, the manuscript of an epic poem that seemed to promote atheism and lascivious behavior, the loss of this manuscript, its rediscovery, and its translation into the English language by a Puritan lady who 20 years after she completed it regretted and repudiated her work. We are talking about the epic poem *De Rerum Natura (On the Nature of Things)*, by Lucretius (ca. 99–55 BCE).[1]

To Lucretius, atoms are always on the move, rebounding off one another and sometimes sticking together to build solids. He cites evidence that when looking at dust motes in a beam of sunlight, the motes are seen to move erratically within the light beam. He attributes that to collisions between the dust and the restless atoms and further remarks that the collection of motes, viewed as a whole, is not moving. He pre-discovers Brownian motion and identifies the central idea that the atoms swerve away, unpredictably, from their motion in straight lines, as a consequence of which the world made from atoms is not deterministic.[2] This is key because it is a statement of free will that denies the influence of the Gods — phenomena such as lightning strikes are not caused by angry Gods who must be appeased. This is indeed science and not introspective thinking — he cites evidence in describing his beliefs. He goes on to argue that, because of the Atoms, motion of the Universe must be infinite in both space and time, and because these atoms are indivisible and indestructible, no God was needed to create it.

Copies of the manuscript found their way to libraries, but little remained of the work of Epicurus, Lucretius and their followers after the 1st century CE other than through the writings of others.[3] Almost one and a half thousand years elapsed until 1417 when a manuscript hunter found a copy of the Lucretius work in a remote monastery. Copies were made and circulated, and copies of those copies. By the final decades of the 16th century and early 17th century, *De Rerum Natura* went viral and became part of the explosion of scientific thought during that period.

2.2   Rediscovery of Lucretius' Poem

The first decades of the 15th century, when this part of our story takes place, were not the best of times. Those famous icons of the Renaissance, Leonardo da Vinci, Erasmus, Copernicus and Michelangelo had not been born[4] and Gutenberg's movable-type printing had not yet been invented. This was during the last decades of the Hundred Years War: the Battle of Agincourt was fought in 1415 and Joan of Arc was burned at the stake in 1427. In 1400, the plague revisited Rome with a toll of 600–800 deaths per day leaving in excess of 10,000 deaths: the pilgrims traveling to and from Rome were the carriers. In 1405, a plague in the City of Padua claimed almost 20,000 lives. Cities started to employ mercenary bands of soldiers from the northern wars as private armies, and in 1423, Florence declared war on Milan. In 1410, three popes were at war with one another; the church was in disarray.

The 15th century was a period of dramatic change; it was the century of the Medici family in Florence, who spent their wealth on the arts and sciences. This was an age of wisdom: the birth of a great Artistic Renaissance and a lesser Scientific Renaissance.

The year 1417 was the last year of the Great Schism which in 1378 had split the Catholic Church into two warring factions, each with its own pope,[5] one in Avignon and the other in Rome. The Church was in chaos; the two-pope situation was unsustainable. Then, a group of Cardinals meeting in Pisa[6] in 1409 decided to create a third pope! Their first choice of pope, in fact an "anti-pope", was Peter of Candia, styled as Alexander V, but he died the following year to be replaced by the widely popular Baldassarre Cossa who was elected anti-Pope John XXIII in 1410. This had the support of the Pope in Rome, Gregory XII, but not of the Avignon pope, Benedict XIII.[7]

This third papacy established a Council to be held on the shores of Lake Constance in Germany, the Council of Constance. Importantly, this move had the support of the powerful King Sigismund of the Romans and Hungary, who would later become the Holy Roman Emperor and who had wanted all three popes to resign and thereby reboot the Church of Rome. The newly elected John XXIII was clearly unhappy about this strategy and did his best to oppose it — he presumably wanted to be the one Pope.

The Council started in 1414 and was attended by more than 400 delegates, abbots, bishops, archbishops and cardinals, each with his entourage, and, of course, lawyers. Among those delegates was Poggio Bracciolini[8] (1380–1459), who in 1410 had become Apostolic Secretary to the Council, having previously been the Apostolic Scribe (see Figure 2.1). Poggio was a dedicated searcher of ancient unknown manuscripts, and while he held the exalted post of Apostolic Secretary to John XXIII,[9] he wielded considerable power. Poggio lost the trappings of power and his source of income when his position ended with the formal deposing and impeachment of John XXIII as anti-Pope on 29 May 1415 by the 12th session of the Council.[10] He was once more just an ordinary well-educated citizen, just a scribe and in the spring of 1416 went to the town of Baden to take the waters.[11]

Between 1414 and 1417, Poggio and his friend, Bartholomew of Montepulciano, made excursions to abbeys and monasteries seeking out ancient manuscripts that were at the time unknown to Italian scholars. Poggio wanted to "rescue these precious relics from the hands of barbarians who were so little sensible of their value".[12] In 1414, he found Marcus Vitruvius' *Architectura* in the Abbey of

Figure 2.1 Engraving of Poggio Bracciolini. From *Icones quinquaginta virorum illustrium*, 1597, by Jean Jacques Boissard (1528–1602). Typ 520.97.225. Houghton Library, Harvard University.

St. Gallen. In 1416, by which time Poggio had lost his exalted job, they found a dust-covered copy of Marcus Quintilian's *Institutio Oratoria*. That same year, they also found Marcus Manilius' *Astronomica*, perhaps the earliest work on astrology, written sometime during the first 50 years CE in the form of a poem.[13] Until then, these were among the most important Roman period manuscripts which had been lost[14] with the passing of time.

Then, in the winter of 1417, they found a copy of Lucretius' *De Rerum Natura*. It is believed that the manuscript lay in the great library of the Abbey at Fulda, in what is now central Germany about

100 km northeast of Frankfurt above the river Fulda.[15] Their journey
to the abbey would have been bitterly cold and would have taken
weeks, if not months staying at hostels and monasteries on the way.
The scene is vividly reconstructed in the film *Name of the Rose* that
takes place in 1327 and revolves around finding a copy of Aristotle's
second book of Poetics: a master and his acolyte traveling to where
there is a library, rugged up to protect themselves from the cold,
making slow chilly progress on their donkeys.[16] The film is a much
simplified rendition of Umberto Eco's book, *Name of the Rose*. The
scene is also imagined in Stephen Greenblatt's Pulitzer Prize winning
The Swerve: How the World Became Modern.[17]

Once Poggio had found the manuscript, he sent it to be copied
and translated. He sent it to his friend Niccolò de'Niccoli in Florence,
one of the great copyists and his copy is the progenitor of lots of
other copies, many of which are inaccurate.[18] It was 17 years before
Poggio saw the manuscript again, although he repeatedly asked for
its return during those years. Niccoli's finest copy on parchment, his
apograph, was probably created in the mid-1430s and served as the
basis for all subsequent translations.[19] The copy found by Poggio in
1417 has now been lost. The two oldest full manuscripts we have
of the Lucretius poem are both kept at the University of Leiden in
the Netherlands.[20] One, from the early-9th century, is simply referred
to as "O" (Oblongus), and the other, from the mid-9th century, is
known as "Q" (Quadratus). Analysis[21] shows that a copy of "O" was
the basis for the manuscript found by Poggio.

2.3   The Italian Renaissance

Poggio Bracciolini was but one of the numerous manuscript hunters
of the Middle Ages, albeit a successful one who happened to be the
one who recovered the Lucretius manuscript. Manuscript hunting had
been the province of those who understood Latin and Greek, and perhaps Arabic: they were early 14th century writers like Petrarch[22] and
Boccaccio and they were scribes, like Poggio. Moveable-type printing had not yet been discovered and so possession of manuscripts
or copies was essential to academics all over Europe. Trade was evidently a good business for those hunters. Occasionally, an important
event would lead to a flood of new previously unseen manuscripts

becoming available as in the aftermath of the Fourth Crusade and the sack of Constantinople in 1204.[23] We should imagine scholars escaping from the war with whichever of their precious manuscripts they could carry. Some would have to sell their manuscripts in order to survive. This was a trigger for the early Renaissance, a period during which the scholars of the time had access to ancient works which have disappeared with the ravages of time and about which we know little or nothing today.

By the 15th century, Constantinople was in financial decline and in 1422 was unable to defend itself against the siege by the Ottoman army. Giovanni Aurispa had been in Constantinople since 1418 and managed to escape to Venice with a horde of 238 hitherto unknown documents which were copied and disseminated throughout southern Europe. This has been identified as one of the key events that kickstarted the Renaissance of the second half of the 15th century. The flow of documents from the East did not stop there. The final fall of Constantinople to the Ottoman army of the 21-year-old Sultan Mehmed II in 1453 led to an exodus of Byzantine scholars, mainly to Italy. It also led to another surge of newly available documents which continued into the following centuries. The fall of Constantinople in 1453 marked the end of one and a half thousand years of the Roman Empire and the end of the Medieval period of history.

One of the most influential documents to have come out of Byzantium during this period was the *Corpus Hermeticum*, a collection of 17 works ascribed to the mythical figure Hermes Trismegistus.[24] The *Hermetica*, as they are collectively known, went from Byzantium to Florence where in 1460 they landed on the desk of Marsilio Ficino, the head of Cosimo de Medici's Platonic Academy, with the instruction to prioritize its translation from the Greek. The translation was finished in 1464. It was a poor translation largely because Ficino did not understand what he was translating. Nevertheless, it was printed and circulated in 1471 without the knowledge of either Ficino or Cosimo. It caused a sensation among the intellectuals of Florence from whence *Hermeticism* spread rapidly throughout Europe.[25] Its popularity was undoubtedly largely due to its presentation of taboo topics on alchemy, numerology and magic, in particular the magic of letters of the Hebrew alphabet which were thought to hide coded secrets about the Universe within the Hebrew text of the

Old Testament. The subject was far from being new in 1471, but it was mysterious and exciting.

The works in the *Hermetica* fall into two groups: the "occult" sciences, which includes astral magic, astrology, numerology and so on, and the "philosophical" sciences of which Section XVI has been much discussed in relation to heliocentric cosmology. Nicolaus Copernicus cited Section XVI in relation to the Sun's attractive influence on holding on to the planets in his model of the Solar System. Kepler had extensive discussions with Oxford scholar Robert Fludd who was renowned for his study and support of occult philosophy.[26] Descartes cited it extensively and that in turn directly influenced both Isaac Newton and Robert Boyle. Among the many manuscripts uncovered to the Western World in the 15th century, the *Hermetica* and *De Rerum Natura* would both be among the most influential, but only two among many. In particular, the Hermetic study of Alchemy was of significant interest to both Boyle and Newton.

This late 15th century Renaissance is widely known for the phenomenal rise of Italian painting, sculpture and architecture, but not for science which was still mired in the philosophy of antiquity. It was an Italian Renaissance, led largely from Florence where during the 15th century the city was the fiefdom of the de Medici family, the bankers to the Rome Papacy.[27] In 1469, the 20-year-old Lorenzo de Medici became the ruler of the fiefdom. The following years were the Golden age of Florence, until everything went badly wrong in 1492 when Lorenzo died.[28]

Lorenzo's son Piero took over, who, 2 years later in 1494, ceded to the invasion of Northern Italy by King Charles VIII of France.[29] Piero was exiled from Florence and Florence declared itself a republic. The Dominican friar, Fra Girolamo Savonarola, became the *de facto* ruler of Florence. Savonarola had been invited to Florence by Lorenzo, which proved to have been a huge mistake. The Pope withdrew the papal accounts with the Medici bank.[30] This was the end of the House of Medici and the start of the few years during which major works of art, books, maps, ancient manuscripts, musical instruments and mirrors were publicly burned, simply because Savonarola declared them to be objectionable. This is referred to as the "Bonfire of the Vanities".

Savonarola saw this as a cleansing of a corrupt and ungodly society, famously declaring to the public that "I announce this good news to the city, that Florence will be more glorious, richer, more powerful than she has ever been ...". He claimed that he had made a mystical journey to the Virgin Mary in Heaven and that she had told him of the future in which Florence became the New Rome. The public lapped up his rhetoric.[31] He went as far as to describe the Church as a whore, which annoyed Pope Alexander VI who excommunicated him in 1497. The following year, things got worse when he claimed he could perform miracles and that he could foresee the future. It was suggested that he should show this by walking through fire, an ordeal which had not been seen in Italy for 400 years. After much discussion and pleading, it was decided to set up the ordeal in the central square. A great crowd gathered to see the trial, but after long delays, the show had to be abandoned: a hard rain terminated the proceedings. The crowd became an angry mob. Six weeks later, Savonarola and two of his colleagues were condemned to death by hanging over a burning fire. This was, however, not the end of Savonarola's ideas on faith and politics. His more conservative supporters gathered his writings, sermons and letters, making them widely available. These works were translated into other languages and had some influence, notably a citation from Martin Luther who declared Savonarola to be a martyr.

This was the beginning of the end of the Italian Renaissance. Artists moved to Paris and other great cities and started the trend in Italian Renaissance architecture. Among them was Leonardo da Vinci who moved to Paris in 1516 with some of his most famous paintings. He died in Paris in 1519.[32] The great German painter/engraver from Nuremberg, Albrecht Dürer, spent several of the early years of the 16th century in Venice and Florence learning from and contributing to Italian art. He left Italy for the last time in 1520, returning to Nuremberg to take up the patronage of the new Holy Roman Emperor, Charles V. From there, he made trips to cities in the Netherlands where the Northern Renaissance had been under way since the start of the 15th century.[33] As a consequence, he exerted a considerable influence on the rise of art in the European Renaissance. The role of Italy as a commercial center for the importing of Eastern goods was marginalized by the success of Vasco da Gama's voyage in 1498 from India to Northern Europe via Lisbon. The Italian Wars exacted their toll. In 1527, the Spanish and German troops sacked Rome, ending the papacy and their support for the arts. Later, in 1542, the Sacred Congregation of the Inquisition was formed as part of the Catholic Church's strategy to combat the spread of Protestantism.[34] A few years later, the *Index Librorum Prohibitorum* banned a wide array of Renaissance works of literature. The time was right for the Renaissance to move to Northern Europe.

The biggest driver of the Renaissance in science was probably the invention of the movable metal-type printing press in the 1430s by Johannes Gutenberg. Without that, the many great works of the philosophers and thinkers of the age would have had little visibility. The printing press benefited from another invention: paper. Paper making came from China to the Middle East around the 8th century and then to Europe during the 11th century. By 1398, the paper makers of Paris had formed a Guild of Papermakers, and with the invention of the printing press, the demand for paper soared.[35] By the end of the 16th century, the printing industry was putting out between 250 and 500 million sheets of paper per year.[36]

2.4   The Mechanical Universe

For Epicurus, introducing an element of randomness that scattered atoms ever further from their original paths meant that we had free will and were not victims of the will of the Gods. The whims of the Gods had dominated Greek lives since before the time of Homer.

Aristotle and others have attributed "non-sensible qualities" such as smell, color and sweetness to the perception of those who are looking at, smelling or tasting the object. Observers agree that some such quality is labeled as "sweet" or "red" more or less by mutual agreement. Epicurus' view was that such qualities are not the properties of the atoms themselves or of the observers but qualities of the object they combine to make. In other words, macroscopic bodies behave differently than the atoms they are made of. Their properties and state of motion of bodies are all due to other causes.

Epicurus was arguably the first mechanistic thinker, believing that all things that are non-atomic have an underlying physical reason. One of the issues which was not addressed by these models was about the nature of "heat". We know what it is to be hot or to be cold, but what is the agency that causes this sensation? For the atomists, heat was a manifestation of motion, but they did not have the capability to explain what kind of motion.

Lucretius goes on to talk of the infinitude of the Universe and states that "it is in the highest degree unlikely that this Earth and this sky is the only one to have been created". The infinite nature of the Universe was a central thesis in the early works of Democritus and Epicurus and was in direct conflict with the teachings of Aristotle.

The transition from the 16th to the 17th century in Northern Europe was the start of a veritable revolution of scientific learning. The revolution had already been triggered by Copernicus' *De Revolutionibus Orbium Coelestium,* published the year he died, 1543.[37] The book included an unsigned preface written by Andreas Osiander, suggesting that the new ideas described by Copernicus in the book were not necessarily true, but that they provide a simple mathematical model that explains astronomical observations much better than Ptolemy's system. There is absolutely no doubt that Copernicus believed that his heliocentric model was actually correct.

It exerted a significant influence on philosophers and thinkers such as Giordano Bruno (1548–1600) and Thomas Digges (1546–1595). In 1576, Digges published *A Perfit Description of the Caelestiall Orbes* as an appendix in the reissued version of a perpetual almanac that his father, Leonard Digges, had written years before. This work features a diagram of the heliocentric Copernican system with the sphere of the fixed stars "extending infinitely in altitude" (see Figure 2.2).[38]

According to the British astronomer and cosmologist Edward R. Harrison (1919–2007), this was the first time that a spatially infinite universe had been spoken of.[39]

Earlier, Nicholas de Cusa (1401–1464), known as Cusanus,[40] had declared in the second volume of his *De Docta Ignorantia* (On Learned Ignorance) published in 1440 that the universe must be infinite. De Cusa was a German mathematician and astronomer who rose in the Church to become a Cardinal in 1446 and then promoted to Bishop-Prince in 1448. Given the time when he wrote *De Docta Ignorantia,* it seems likely that he would have known of Poggio's manuscript of the Lucretius poem and probably even seen a copy.

However, Cusa's argument had no scientific basis. It was purely metaphysical, arguing along the lines that we mortals cannot come

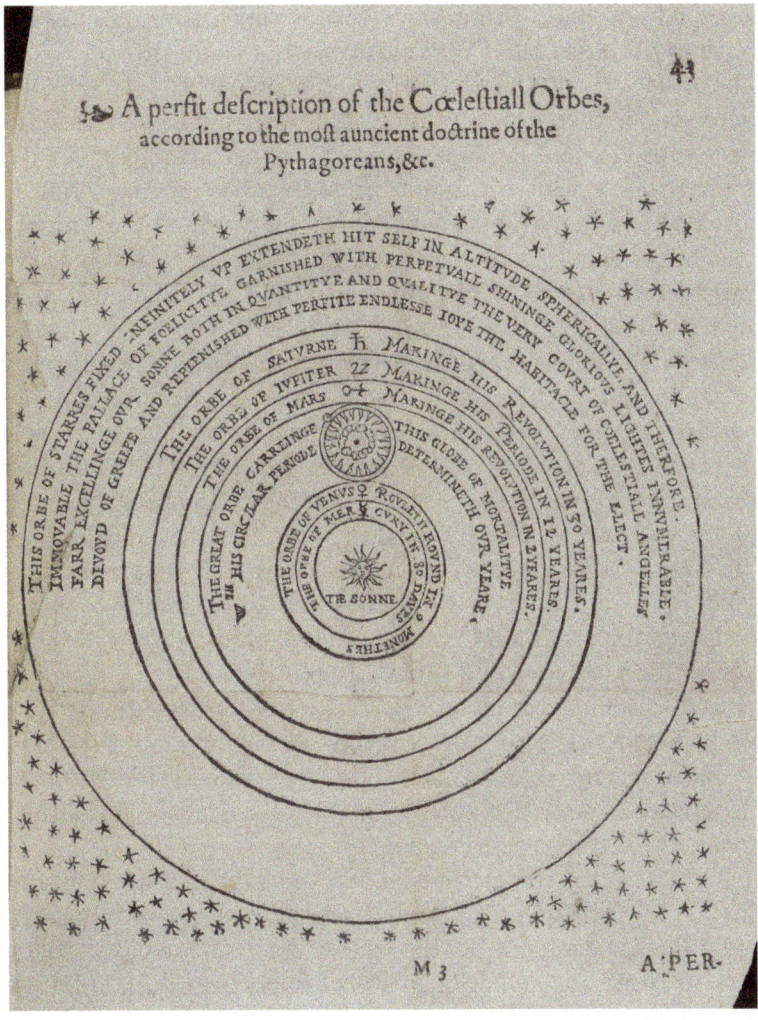

Figure 2.2 The diagram of the Copernican system with the sphere of the fixed stars extending infinitely in altitude. From *A Perfit Description of the Caelestiall Orbes* (1605) by Thomas Digges. Image courtesy History of Science Collections, University of Oklahoma Libraries.

to terms with the notion of anything being infinite,[41] that God, on the other hand, was infinite in all respects and "thus" the universe that He created must itself be infinite. Cusa[42] goes on to say that "... there are stars, of which there is no number". By comparison,

Digges, in 1576, had a framework within which to frame his infinite universe hypothesis: the Copernican view of the universe.

The discovery of the Lucretius' epic poem, *De Rerum Natura*, attracted the immediate more radical forebears of Descartes: Thomas Harriot and Thomas Hobbes. Descartes accepted the atomist doctrine that the world was made of indivisible particles but never accepted the notion that these floated around in the atomist void. The general support from Descartes in the early 17th century was an important factor in Boyle, Huygens and Newton later joining the atomist club.[43]

Without the discovery of the Lucretius manuscript, the revolution in science taking place in the 16th and 17th centuries would undoubtedly still have taken place. Each ingredient to the pot of knowledge has a value, but none is indispensable. What can be changed, and what may have been changed, is the timing of the progress.

There were many editions of Lucretius' poem published during the 16th and 17th centuries, often annotated with biographical notes and different interpretations of the original Latin. Two were particularly notable: the 1511 version of Johannes Pius and the more scholarly, 1563 version of Denys Lambin. These versions certainly influenced the thinking of Galileo, Descartes and Newton on the concept of "inertia": the proposition that atoms and other bodies not acted upon by a force move in straight lines. In addition, the concept of "atomism" flowed through the chain of writings of Montaingne, Descartes, Gassendi and Mersenne.[44]

It was not until the second half of the 17th century with translations of the entire poem appearing in English, French and Italian that the poem became available to a wider audience. The following chapter presents the remarkable story of the English translation by Lucy Hutchinson.

Chapter 3

Bubbles and Influencers

For being subtile, every shape they take;
And as they dance about, they places finde,
Of Formes, that best agree, make every kinde.

Margaret Cavendish, Poems and Fancies, 1653

Scientific advances ride on the back of good communication. The 21st century is the Century of Communication: we recognize the rapid societal changes that are taking place through the action of widespread social networks. This has had a dramatic impact, for better or worse, on politics, science and general social behavior. We see "bubbles" of like-minded people sharing ideas and strengthening their belief sets by virtue of mutual support, often led by a few "influencers". The influencers might belong to several bubbles, so that their influence spreads more widely, from one bubble to another. The process of communicating is referred to as "sharing": you "share" your information on with your bubble "friends" who are encouraged to "like" you or openly dislike you. There is no control over this process so that both information and misinformation can spread freely. There is as much access to important information as there is to conspiracy theories and fake news.

3.1   The Agents of Scientific Advancement

The 17th century opened with the torture and burning of Giordano
Bruno in a square in Rome on 17 February 1600. His crime was
heresy: the denial of orthodox (Catholic) religious views.[1] Bruno was
a Dominican Friar, an influential academic, who had spent time in
Oxford, and was a strong supporter of the Copernican view. His
own cosmology was based on the works of Epicurus and Lucretius:
like Thomas Digges before him, he believed that the universe was
infinite.[2] On a more positive side, William Shakespeare closed the old
century with *Julius Caesar* (1599) and opened the new century with
the publication of *Hamlet* (1599–1601).[3] King James I, in an attempt
to calm the division between the Catholic and Puritan communities,
ordered an English language version of the *Bible* in 1604, published in
1611. This *Authorized Version* of the Bible has been an international
"bestseller" for some 400 years.[4]

The first half of the 17th century was a period of considerable
strife and uncertainty over much of Northern Europe, largely caused
by political and religious repression. The Jews had been expelled from
France in 1615, and the Huguenots were finally suppressed in 1627 at
La Rochelle.[5] The Dutch Republic was still fighting the Dutch Eighty
Years' War (1568–1648) against Philip II of Spain while the Thirty
Years' War was being fought out in central Europe with a death toll
of several million, military and civilians. The Thirty Years' War was
fought between the Hapsburg and Bourbon Kings and was one of
the bloodiest wars in history. It was the war that ended the Holy
Roman Empire at the Peace of Westphalia. Meanwhile, the Franco-
Spanish War (1635–1659) started when France entered the Thirty
Years' War on the side of the Dutch Republic (and Sweden). That
was being fought on several fronts: Catalonia, the Rhineland, the
Spanish Netherlands and Northern Italy. It ended in 1659 because
neither side could afford to continue. In the midst of all that, the
French had their own Civil War ("La Fronde") lasting from 1648 to
1653.[6] Like the English Civil War, this was a war attempting to curb
the power of the monarchy. During this period, the King, Louis XIV,
was living in exile with his Regent mother, Anne of Austria. At the
end of the civil war, Louis returned as King having more power than
he had left with.

In England, the period 1642–1651 was the time of the English Civil Wars, a set of wars between the Parliamentarians and the Royalists that resulted in the death of some 180,000 people, 3.6% of the population. The royal capital moved to Oxford in 1643 but by the following year, the Royalist position was seriously weakened[7] and the Queen, Henrietta Maria, fled with her court to Paris. This was the last time she saw her husband, King Charles I, who was executed in 1649. The Queen's move to France was a significant event in our story,[8] as was the advent of the French Civil War which caused her and her entourage to leave Paris and go to Antwerp.

3.2 Popularizers of Scientific Thought

During the 17th century, there was an accelerating transition from "natural philosophy" to "science" as we know it today. This was the century of the Scientific Revolution in mathematics, physics, chemistry, biology and medicine that had been triggered by the publication of the work of Copernicus in 1543. The question arises as follows: How did science develop and spread in a century that was beset by wars, plagues, religious intolerance and scientific dogmatism?

Within the constraints affected by social privilege, wars, dogma and extremism, it is possible to identify a number of particular individuals who were the drivers in shaping the development of physical science. Their importance is perhaps a little surprising since they were not by any means the greatest scientists of the age. However, they were the key communicators and sat at the centers of the science "bubbles".

The social "bubbles" of the 17th century gathered together in the drawing rooms of the rich. These salons featured music, readings of poetry, discussions of politics or simply talk about science and philosophy. They welcomed notable visitors from other cities who would share news. This had been the social way of life for the rich and educated since the middle ages and continued right into the 20th century. The social circles of relevance to science were those in which the principal subjects of discussion were optics, mathematics, atoms, the void, air, motion of projectiles and perhaps astronomy. The circles tended to have participants who were experts in one or more of these subjects and those might have been participants in other similar circles. It was a complex social system with well-understood rules as to what was expected: joining a circle could only be done with a suitable introduction.

Two of the most powerful "influencers" of science within those 17th century social networks were Pierre Gassendi and Marin Mersenne.[9] Neither came from the wealthy titled classes; both were brought up in religious environments and both were ordained into the church. Both have been described as being from the "peasant" class. But they were brilliant polymaths whose academic promise had been recognized by their schools at an early age. They grew up to dominate the social circles of the most important mathematicians and philosophers of the early 17th century and their influence stretched beyond their own lifetimes. Both had the gift of being able to present new ideas in a manner that was accessible to intelligent people who were not experts. They had a social charisma that encouraged those people to listen and participate. In modern terms, they were the great influencers and popularizers of intellectual thought, individuals with whom the great and wealthy enjoyed conversing.

3.2.1   *The first bubble — Galileo et al.*

It is generally agreed that French polymath Pierre Gassendi provided the trigger for the growth of Science during the 17th century. Astronomers recognize Gassendi for publishing the first data on the transit of Mercury across the face of the Sun on 7 November 1631. Ten years later, he performed the experiment of observing stones thrown in different directions from the top of the mast of a galley sailing in the Bay of Marseilles at a speed of around 25 km/hr. This was the experiment that Galileo is known for: it is said that he dropped cannonballs of different weights from the top of the Tower of Pisa[10] observing that both weights landed on the ground together. Whether or not Galileo did drop cannonballs, he certainly knew about a similar experiment performed by Giovanni Baliani around 1611. Gassendi's experiment went further: he saw that the motion of the ship does not affect the result. In any case, the experiment was not new: Lucretius had written about it in Volume II of the *De Rerum Natura*.[11] Gassendi was the first person to give a full explanation as to why the balls, when shot vertically, struck the deck of the ship at the same time, regardless of the motion of the ship. Aristotle would have said that this was a proof that the Earth was not rotating. The point is that the cannonballs are in fact endowed with what we now call the *momentum* due to the ship's motion. Gassendi published his findings, first with a serious error in 1642 under the title *De Motu* and then in 1646 a corrected version under the title *De Proportione qua Gravia Decidentia Accelerantur.*[12]

There we have our first little social network, our first "bubble": Baliani, Galileo, Torricelli, Pascal, Mersenne and then Gassendi. Galileo was the scientific giant in this group.

3.2.2 *The second bubble — Mersenne et al.*

On a wider front, Gassendi is mainly known for his philosophy of science which was opposed to the views of both Aristotle and Descartes. When Descartes famously said "I think, therefore I am" (*Cogito ergo sum*), Gassendi responded "I walk, therefore I am" (*Ambulo ergo sum*), announcing for the first time the view that the body and mind were inseparable.[13] Gassendi reconciled the atomist philosophy of the Epicureans with the teachings of Christianity and in doing drove the widespread interest in atomism. To achieve that, he had to go beyond the simplistic view of atoms described by Lucretius by refining the concept of "Atomism". In effect, he defined what we might call Neo-Epicureanism, a new version of Epicurus' philosophy. Gassendi was a noted popularizer of science and philosophy and his views appealed to a wider 17th century public without violating their religious sentiments. It was Gassendi's version of atomism, derived from the Lucretius version, that inspired the great scientists of the late 17th and 18th centuries.

Gassendi went to live in Paris in 1640, where he became the leader of a social circle calling itself *La Tétrade*, a group of "free thinkers" including the philosopher Thomas Hobbes and the mathematician Marin Mersenne who brought René Descartes into the group.[14] Mersenne was what we today call an "influencer". He actively corresponded with Descartes, Galileo and Pascal and in 1635 had created the informal *Académie Parisienne* which brought together some 140 astronomers, mathematicians and philosophers[15] from Paris and other places around Europe. When Gassendi published his view of Epicurian philosophy, *De Vita et Moribus Epicuri* (*On the Life and*

Death of Epicurus) in 1647, the members of the *Académie* were quick
to support him and endorse the novel views he was expressing. His
position was reinforced 2 years later with the first release of his work
on the life, morals and opinions of Epicurus, *Animadversiones,* his
own take on the Lucretius manuscript. After his death in 1647, friends
collected and edited his works in a collection known as the *Syntagma
Philosophicum Epicuri*, which divides into three parts: logic, physics
and ethics. It is in the second part that Gassendi reunites God and
science, negating Epicurus' rejection of God. It was the *Syntagma*
that impressed and inspired Boyle, Hooke and Newton.

This is our second bubble: Mersenne, Gassendi, Hobbes and
Descartes, who formed the *Tétrade,* which linked with Mersenne's
other bubble of those who were part of his *Académie.* Through
Mersenne, it also embraced Kepler and Galileo.

3.2.3   *The third bubble — The Newcastle Circle*

In 1644, the situation in the English Civil War forced Charles I's wife,
Henrietta Maria, to exile herself with her court to Paris. At this time,
the King of France, Louis XIV, was a child and his mother Anne
of Austria, the Queen of France, was acting as Regent until Louis
attained his majority in 1651.[16] Among the English Queen's court
was her maid of honor, Margaret Lucas. In 1645, William Cavendish,
the then Marquess of Newcastle, and his brother Sir Charles arrived
in Paris following their defeat at the battle of Marston Moor.[17]
William, recently widowed, met Margaret who was some 30 years
younger than he. He wooed her with poetry, and with the reluctant
permission of the Queen, they were married that same year. The
marriage turned out to be a bringing together of "soul-mates", each
one promoting and supporting the other and only ending with the
death of Margaret in 1673.

Margaret's father, Sir John Lucas, was a landowner and an indus-
trialist and raised his own army to fight for the King in the Civil War.
He was made the 1st Baron Lucas of Shenfield. As typical in that
echelon of 17th century society, Margaret was not formally educated.
On marrying William, his brother Charles, a skilled mathematician,
started to teach her the things she desired to know. She quickly settled into the social and intellectual life of the Cavendish brothers
in Paris, getting to know and listen to some of the great thinkers

of the time, the members of the "Newcastle Circle". She decided to become a writer of poetry, philosophy, nature and science and in 1653 published her *Poems and Fancies*, a collection of diverse essays and poems. These were quite unlike the usual romantic writings of her peers and quite extraordinary for a woman at that time. She discussed philosophy and dressed in an outlandish fashion: she was regarded by some, notably the diarist Samuel Pepys, as an eccentric and some referred to her as "Mad Madge".

In England, the two Cavendish brothers, William and Charles, were renowned patrons of the sciences and in return for that patronage they were at the center of an intellectual bubble. Prior to the Civil War, their group was referred to as the Wellbeck Circle and featured Thomas Hobbes as one of its eminent members. When they exiled themselves to Paris in 1644, their property was sequestered and they lost their wealth but not their social standing. In order to keep up appearances and maintain that social status, they sought to live life as normal, borrowing money and borrowing even more to pay off the interest on the loans.[18] They continued their patronage of Gassendi, Hobbes and Descartes and created the Newcastle Circle, embracing like-minded people from Paris. William's new wife Margaret Cavendish became the hostess for the gatherings in their Salon. They only stayed in Paris three years before moving to Antwerp in 1648, most probably because of the outbreak of the French Civil War, La Fronde. At the same time, the Queen of France and her son, King Louis XIV, fled to Germany with her First Minister, Jules Mazarin. Paris, with the populace barricading the streets, was not a place to be for the nobility.

Yet, those years in the Cavendish household and in the Merssenne's Academie were the melting pot for the consolidation of the atomist point of view as expressed by Gassendi. The "fake news" of the Hermetics was cast aside, though not forgotten, and replaced by a scientific method and a model that provided a framework within which the new science could work.

The Cavendishes only returned to Britain in 1660, following the restoration of the monarchy with King Charles II. William was promoted to the rank of Duke and so he was titled the 1st Duke of Newcastle. However, he was then in his late 60s and sat on the fringes of the Royal court.

Figure 3.1   Portrait said to be of Lucy Hutchinson c. 1643. The artist was likely to be John Souch of Chester. Image courtesy of the National Army Museum, London.

3.3   A Puritan Out of Youthful Curiosity

At the time when Margaret Cavendish was holding her philosophical "circles" and writing her poems, she had a contemporary authoress named Lucy Hutchinson whom she probably never met. Lucy, who's portrait is shown in Figure 3.1, was a Puritain lady who moved in totally different social and religious circles. The Puritans were English Protestants who believed that the Church of England should be purified of all remnants of its Catholic Heritage that were not in the Bible. They rejected what they felt were the excesses of the Catholic Church and advocated extreme personal piety and abstemiousness. Another contemporary of Margaret's and Lucy's was John Milton (1608–1674), also a Puritan, whose poetry Lucy would have read with pleasure.

Lucy's husband, John Hutchinson, was a colonel in Oliver Cromwell's New Model Army that had executed Charles I and established the Commonwealth. He was a signatory to the execution warrant and was later arrested and imprisoned for the crime of Regicide. John Hutchinson was never tried but given a conditional pardon. However, in 1663, he got involved with a failed plot to overturn the monarchy and was returned to prison where he died. Later, Lucy wrote an important biography of her husband's life, *Memoirs of the life of Colonel Hutchinson*, which was only published by her heirs in 1806. She had known the major players in the revolutionary movement and so was an important chronicler of that side of the conflict.[19] In terms of the history of the conflict, the *Memoirs* was more about the people involved than about the military or political strategy of the insurgency.

Unlike Margaret Cavendish, Lucy Hutchinson was exceptionally skilled in Latin.[20] In the period 1649–1660, for an unknown reason, she decided to use her knowledge of Latin to make the first English translation[21] of Lucretius's poem *De Rerum Natura*. It was an excellent translation, but it was never published until long after her death,[22] though unsanctioned copies had been made and were undoubtedly widely read.

De Rerum Natura was an extraordinary choice of manuscript for a person like Lucy to translate, for much of it ran counter to the principles of her life. Indeed, in 1675, she declared that *De Rerum Natura* was not consistent with her Puritan values. Yet, she almost certainly knew before she started what the Lucretius book was about, including its most salacious and ungodly parts. She did in fact excise those parts in her translation, so the translation is not totally complete. She never sought to have it published, but two points are of significance. First, she paid to have copies made. She distributed these but asserts that she lost control of who had copies. Second, she dedicated to and gifted a fair copy of her manuscript to the influential Earl of Anglesey.[23] At that time, she had pride in her achievement and it served to maintain her standing as an intellect in her social circle. She was not prepared to cast her draft into the fire.

In 1853, a copy of her manuscript was sold to the British Library. Her translation was not published until 1996, some 350 years after it was written! The next English language translation was that of Thomas Creech in 1682 which, in the absence of a published version of the Hutchinson translation, became the standard for the 18th century. Nevertheless, there were unauthorized copies of the Hutchinson manuscript in circulation. Not surprisingly, the study and translation of the *De Rerum Natura* was a hot topic in many European countries. In France, Michel de Marolles, the abbot of Villeloin, published a translation into French in 1650 which was revised in 1659, and in Italy, the mathematician Alessandro Marchetti produced the first Italian language translation in 1668.[24]

Until recently, Lucy Hutchinson was known only for the *Memoirs*, which is an important historical document. However, she wrote numerous books and poems most of which were not published until the first decades of the 21st century. 2001 saw the publication of her epic poem *Order and Disorder,* written sometime before 1679, at about the time Milton when wrote *Paradise Lost.* This was the first example of an epic poem written by an English woman. These two epic poems are both retellings of the Creation story as told in the *Book of Genesis,* from very different points of view.[25]

How is her work to be evaluated? There has been a growing interest in 17th century women's writings, and Lucy plays a role in that story.[26] She undertook one of the most difficult tasks in literature: translating an ancient Latin poem into English for the first time. However, from the point of view of her impact on the advancement of science, she had almost no impact at all. This was partly due to her failure to publish her work and partly to the fact that she did not move in a scientifically relevant circle of people. So, the people to whom she showed her work did not pass it on. The key ingredient of good communication was absent.

Lucy Hutchinson is an example of a non-influencer. Like so many others throughout history, she did good work that could potentially have made a contribution to the advancement of science and society but did not do so because it was not adequately communicated. In the words of Thomas Gray,[27] who was a good communicator, *full many a flower is born to blush unseen, and waste its sweetness on the desert air.*

Germaine Greer, in her aforementioned review of the 2001 publication of Lucy's *Order and Disorder*, ends with what might be an appropriate epitaph for Lucy: "Before we can place Hutchinson on equal terms with Milton, Marvell, Waller, Cowley and Butler, we ought to find evidence of how her work was read. What we can be sure of is that she was a serious writer who dared to undertake ambitious works on the largest scale. Regardless of whether she is a great poet or not, she is a hero".

The years of exile in Paris of the young King Charles II and his noble supporters, from 1644–1648, played an important part in the evolution of science at that time. Political wars drew people together who would otherwise never have met. Among these were enthusiastic scientists and philosophers who would exchange ideas that set science in motion for the second half of the 17th century, overcoming the dogma of the previous century and causing a divide between "philosophy" and "science", a gap that would only ever get wider.

Lucy Hutchinson started the translation of the Lucretius manuscript when she was young. She was probably unaware of the danger of this task for her Puritan convictions, but she could be more or less intoxicated for a while by the wonderful ideas she found hidden inside the poem — an example of how the scientific thirst for knowledge has captivated minds through the centuries. She explained her motivation to undertake the translation of *De Rerum Natura* when she wrote[28]: "out of youthful curiosity, to understand things I heard so much discourse of at second hand, but without the least inclination to propagate any of the wicked pernicious doctrines in it". In any case, when she gave the copy of her translation to Anglesey, asking him to hide it, the atomist ideas had already penetrated the European intellectual circles. The idea of the mechanistic universe was gaining adepts not only in England: Galileo in Italy and René Descartes and Pierre Gassendi in France were firm supporters. Finally, a key factor in favor of atomism was the "Great Experiment" of Torricelli in 1643, in which he built a barometer and produced the first vacuum. This was the experiment that marks the second half of what is dubbed "The Scientific Revolution", the period (1550–1750) when the experimental method emerged and when there were a great number of discoveries in many fields of science. The next chapter starts with the engaging story of this experiment.

Chapter 4

The End of Earth, Air, Fire and Water

Nothing exists except atoms and empty space; everything else is opinion.

Diogenes Laertius, Lives of Eminent Philosophers: Democritus, Vol. IX

The most important technological advance in the history of human society was harnessing the use of fire.[1] All early societies recognized the key role of fire, along with the three necessities of life — air, water and food. It is easy to understand why Empedocles asserted that the four basic elements of nature were Earth, Air, Fire and Water, and why that belief was held for nearly 2,000 years. People believed so strongly in the four elements theory that they interpreted all their observations in terms of these entities. Of all the dragons that impede the march of science, prejudice is perhaps the biggest dragon of all.

Change didn't come until people were willing to question accepted dogmas and to look with new eyes at the world around them. The key trigger for this was Martin Luther's religious reformation, after which time people became more willing to challenge the status quo and more receptive to new ideas and approaches.

The "Scientific Revolution" is the term describing the remarkable advances in all sciences during the period 1550–1750. This period can be divided very roughly into two parts. The first 100 years were, in effect, the break from the past, exploiting greater religious freedom and the dramatic rise in printing. Printing provided a wider availability of translations of ancient manuscripts and the wider dissemination of new ideas. This was the time of the giants, Nicolaus Copernicus, Johannes Kepler and Galileo Galilei, all born in the 16th century. It was the time when dogma that had persisted for the previous 2,000 years was questioned.

As we have seen in the previous chapter, the mid-17th century transition to the second 100 years of the revolution was marked by the emergence of social bubbles of scientists. The main drivers were Marin Mersenne and Pierre Gassendi in Paris and the inspiration of their contemporaries René Descartes and Blaise Pascal. This phase saw the emergence of a new form of science driven by quantitative experimentation and mechanistic theories explaining a variety of phenomena. Scientific institutions like the Royal Society of London and the Academie des Sciences of France were founded.[2] It was the birth of science as we know it today, the science that is taught to children in our schools.

4.1   At the Bottom of an Ocean of Air

Galileo died in 1642 following a decade of house arrest. However, his ideas, his know-how and his unfinished works did not die with him. His scientific legacy was passed on to two of his students, Evangelista Torricelli and Vincenzo Viviani. They quickly went to work on one of Galileo's ideas and built the first mercury barometer, a means of measuring atmospheric pressure. This is referred to as Torricelli's "Great Experiment" of 1643 in which he produced[3] and demonstrated for the first time the existence of the vacuum. Not surprisingly, this was highly controversial with the "old guard" strongly defending the Aristotelian adage that nature abhors a vacuum. But it was the beginning of the end of the dominance of ancient Greek scientific thought.

The Torricelli and Viviani experiments depended on the use of advanced glassblowing techniques, which Galileo had been using for some of his own earlier work. As can be seen throughout the history of science, advances in theoretical understanding march side by side with advances in instrumentation and techniques. In fact, later attempts to reproduce the experiment by Blaise Pascal ended in the frustration of breaking glass until he found a specialist glass worker in the town of Rouen.

In the experiment, a number of long tubes of glass were filled with mercury and turned upside down into a bowl of mercury. The tubes were about 80 centimeters in length, with each having a different diameter and being closed at one end. Such tubes would have held at least a kilogram of mercury. This was a perilous process since the tubes could easily break. On being turned upside down, some of the mercury came out of the tube and into the bowl, leaving a vacuum at the top of a column of mercury. The height of the mercury column was 76 centimeters in every tube. The diameter of the tube did not matter nor did the size of the mercury bowl; the height of the mercury column was always the same. This is depicted in Figure 4.1. It was also recognized that the height of the mercury column varied with atmospheric pressure, so the pressure could be measured simply by recording the height of the mercury column in the tube. The barometer had been invented.

So what was the empty space left in the tube above the mercury? Since the end of the tube was sealed, the vacant space at the top of the column could only be a vacuum, now referred to as the "Torricelli vacuum". What was holding up the mercury in the tube? Torricelli argued that the mercury was being held up in the tube by the "weight of the air" above the surface of the mercury in the bowl.[4] The ancient idea that the air is weightless was therefore wrong. As Torricelli put it, we live submerged at the bottom of an ocean of air.[5] Torricelli and Viviani had shown that the air had weight and that a vacuum could be created, observed and discussed, turning the world of classical knowledge upside down.

Torricelli explained the experiment[6] in a letter written in Florence and dated 11 June 1644 to his friend Michelangelo Ricci living in Rome but never published his results. He recognized, in light of what had happened to his mentor, Galileo, that this could cause him a lot of trouble. His team was sworn to secrecy until the experiment was confirmed by others. Nevertheless, the news leaked out and everyone was talking about it.

There was nevertheless still opposition to Torricelli's invoking the weight of the air. Some insisted that air was weightless and that the vacuum itself was somehow responsible for holding up the mercury, perhaps by suction. Other critics suggested that the mercury in the

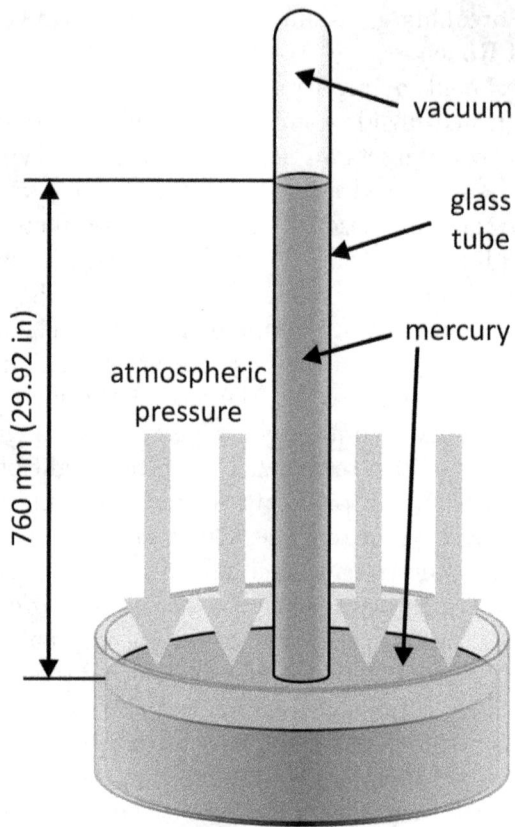

Figure 4.1   The experiment of Torricelli. At the time of the experiment, the atmospheric pressure was attributed to the "weight of air". The mercury tube is in fact some 80 times longer than it is wide, making it rather fragile. The weight of the mercury in such a tube would be around one kilogram (2.26 lbs). Credit: Fernando J. Ballesteros.

tube was being held up by some unknown entity that could somehow leak through the glass into the top part of the tube. The church suggested that this unknown entity was "spirits", while Descartes suggested that the top of the tube was filled with an "ether" that was able to pass through glass. Proof that air had weight was needed. The controversy was resolved by what is now known as the "Torricellian experiment", one of the finest experiments performed up until that time and a wonderful example of what the Scientific Revolution was about.

When the mathematician Marin Mersenne visited him, he and Torricelli repeated the experiment.[7] Mersenne was impressed, but not being well enough to do the experiment himself, told his friend Blaise Pascal in France about it. Pascal decided that the test should be to use two barometers and carry one up a mountain while leaving the other on the ground. There should be a difference in the height of the mercury column since there was less air above the barometer on the mountain. Pascal was himself not fit enough to do such an experiment and so asked his brother-in-law, Florian Périer, to do it with the help of local people, on the Puy de Dôme mountain near the city of Clermont in central France.[8]

Périer obtained the help of the inhabitants of the Minim Friary where the lower experiment was to be performed. While Périer climbed the 3,500-foot mountain with helpers (witnesses), Father Chastin of the monastery stood guard and recorded the height of the mercury barometer at the bottom of the mountain (another witness). Prior to doing the experiment, Périer cleaned the mercury and took great care over other aspects of the experiment, including meticulous documentation. When the experiment was done, a very significant result was obtained: a difference in height of just over three French inches. The experiment was repeated another five times with measurements taken at different altitudes, confirming that the drop in the mercury level was proportional to the altitude. With this, they confirmed the "weight of air" explanation and were able to conclude that the height of the atmosphere was around 50 km.[9]

4.2   A Symbiotic Collaboration

Ten years after Torricelli's discovery of the vacuum, Otto von Guericke, the mayor of the city of Magdeburg in central Germany built himself a vacuum pump and constructed two copper hemispheres that together would make a complete airtight sphere of 50 cm diameter when joined. The pump was the first of its kind. He pumped the air out and showed that two teams of horses could not separate the spheres that were simply held together by "the weight of air" (see Figure 4.2). Although impressive and much talked about, this little experiment had only one significant scientific consequence.

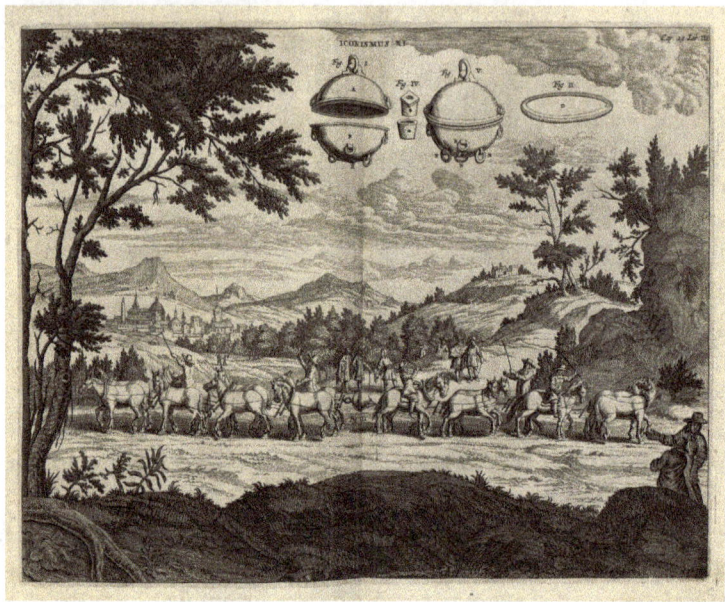

Figure 4.2 Guericke's demonstration was performed on 8 May 1654. Engraving by Gaspar Schott in the book *Experimenta Nova (ut Vocantur) Magdeburgica de Vacuo Spatio* by Guericke, Otto von, Amsterdam, J. Jansson, 1672. Credit: Public domain. Image retrieved via Wikimedia Commons.

In Oxford, Robert Boyle read of Guericke's experiment and then asked his colleague Robert Hooke if he could make a sphere of glass like this and an air pump to evacuate it. Boyle was one of the founding Fellows of the Royal Society of London and is famous for his work on gases. His colleague, Robert Hooke, was one of the greatest experimental geniuses of all time. Together they formed an unbeatable team. This was a defining moment in the study of gases. Between them, they established the culture of the modern experimentalist: ask a question, build some equipment, test it to maximize the control over the setup and get results. Finally, document it so that it can be reproduced lest it enters the realm of irreproducible conclusions.

Their working relationship can only be described as symbiotic, for they came from the opposite extremes of society. Robert Boyle (1627–1691) was a younger son of a very wealthy Irish Earl. He was educated at Eton College and inherited a manor in the county of Dorset in England and land in Ireland. In 1654, at the age of 27, he found himself in Oxford. He did not read of Guericke's experiment until 1657.

Hooke on the other hand came from a much poorer background. He was a sickly child, so much so that it had not been expected that he would live to his teens. He suffered from Scheuermann's kyphosis, a severe and inflexible stooping of the spine, which got progressively worse leaving him deformed. In his later life, he was described as "a sorry sight with his thin and crooked body, his overlarge head, sharp facial features and protruding eyes".[10] At the age of 13, after his father died, he left his home on the Isle of Wight for London, with a sum of £50 that he had inherited from his father and grandmother. By then, he had learned many mechanical and artistic skills from local artisans. He invested his money in his own education at one of the finest schools in London,[11] earning money at the same time from doing diverse skilled jobs. Through the headmaster of the school, who recognized his intellectual capability, he made important contacts, earning a place at Christ Church college in Oxford in 1653, aged 18. A couple of years later he met and was employed by Boyle as an instrument maker. Interestingly, Hooke's education had provided him with skills in languages and mathematics, skills which Boyle lacked. Benefiting from his new position, Hooke was able to perform radical new experiments and collaborate with the most influential English scientists of the day.[12] However, the fact that Hooke had to work for a living meant that he always remained subservient to Boyle and other aristocratic members of the group who would go on later to form the Royal Society of London. Hooke was their academic equal and even an intellectual star among them, but he could never aspire to become their social equal.

Hooke's Oxford life lasted from 1657 to 1662, a short first period during which he established a considerable reputation as an experimentalist through working with Boyle and conducting his own scientific inquiries. A second period followed in London, becoming a Fellow of the newly formed Royal Society in 1663 and having a position at Gresham College where he remained until the end of his life. This phase of his life was highly varied and very productive. He worked within the framework of the Society, first as an unpaid curator of experiments, rising to a paid position, and eventually taking up the important role of Secretary until he was ousted from that in 1682, nevertheless maintaining his position as Curator.

Hooke also benefited from the existence of the newly popular coffee houses, which provided an important meeting place for intellectually interested people of all social and economic classes. The first coffee shops were opened in Oxford in 1651 and London in 1652 by Pasqua Rosée, an immigrant from Smyrna,[13] despite protests from the alehouse owners who saw these as illegal competition. They were highly successful and stimulated the opening of many more coffee shops during the following decades. The Oxford coffee shops were known as "penny universities" and were open to all who enjoyed intelligent debate, were well behaved and could afford the penny entrance fee. The coffee houses became important social centers, social bubbles in modern parlance, for merchants, artisans, journalists, lawyers, academics and traders of shares and commodities.[14] These were radically different from the chocolate houses that opened around the same time[15] and whose clientele were the wealthy, the super-elite and the titled. The chocolate houses have been described as dens of Machiavellian political intrigue, debauchery and gambling at a level that ruined many fortunes of wealthy families. By contrast, the coffee houses were open to all and not restricted to a social elite. King Charles II on returning to take up his throne after the English Civil War, in fear of anti-royalist political plotting, tried and failed to have the chocolate and coffee houses closed.

While in London, Hooke's favorite coffee houses were in Exchange Alley, Garraways (opened 1673) and Jonathan's (opened 1677), the former being the place where the founders and members of the Royal Society would gather. It was through his education and attendances at coffee houses that he was able to meet with influential London people. Hooke was a phenomenal polymath, extending over many fields from the then-new microscopy to the architecture of the new city of London following the Great Fire of 1666. His greatest published work is his *Micrographia*, published in 1665, which combined his scientific, instrument making and artistic skills in a formidable work on his observations with microscopes and telescopes.

4.3 Voids, Pumps and the Law of Gases

In 1659, Hooke had succeeded in making the new pump requested by Boyle, and with this, Boyle started research into physical and biological systems at low pressure. Until that time, Boyle's ambition had been to work on the Torricelli vacuum, but this was his "light-bulb moment". Boyle saw clearly into the possibilities of having a small vacuum environment in which he could perform experiments on gases. Boyle was a supporter of a particular form of atomism: "corpuscularianism" inspired by the earlier atomist views expressed by Descartes, Hobbes and Gassendi.

With his new pump, Boyle was able to show how the boiling point of water dropped as the pressure decreased and to observe how a feather fell in a vacuum. He also showed that a candle or piece of charcoal would not burn without air, which was totally contrary to the Earth, Fire, Air and Water theory of material. He also showed that sound would not propagate in a vacuum. Boyle was doing physics experiments at both high and low pressures.

Boyle recognized, controversially at the time, that air was not a single gas, but a mixture. He went on to study the properties of air under compression resulting in 1662 in what is now known as Boyle's Law of Gases at constant temperature: that the pressure in a gas increased if the gas was compressed while keeping the temperature constant.[16]

This was the start of systematic experimental science, of which Boyle was a leading pioneer. He went on to wonder whether there was such a thing as a minimum temperature. He could see that the volume of a gas decreased as it got colder and obviously wondered at what temperature the volume might shrink to zero. He could not measure temperature: neither the thermometer nor the temperature scale had been invented. The problem of measuring temperature with sufficient accuracy to understand how gases behaved under change in temperature was only solved during the 18th century when the Danish astronomer Ole Rømer (1701), the German physicist Daniel Fahrenheit (1724) and the Swedish scientist Anders Celsius (1742) all developed thermometers and temperature scales.[17]

Hooke's pump was regarded as one of the most sophisticated and costly instruments of the day. A decade after the first pump was built, there were only six or seven of these in Europe, three of which had been made by Hooke. Hooke's pump was the ultimate 1660s hi-tech and would be modified and used in countless experiments for the next 100 years or more.

Boyle's Oxford colleague, chemist and physiologist John Mayow in the late 1670s also experimented with air trapped in a bell-jar inverted over water. Mayow lit candles and other flammable substances in the glass dome using a magnifying glass and noticed that the burning consumed a part of the air in the chamber before combustion stopped. He named that part of the air *spiritus nitroaerus* (nitrous air, later called "oxygen").[18] Mayow went on to examine the result of simply placing a mouse in the jar and noticed that the result was the same as with burning. With this, he established the link between combustion, respiration and the importance of his nitroaerus (oxygen).[19]

More than a 100 years later, in 1783 in Paris, Jacques Charles and the siblings Anne-Jean Robert and Nicolas-Louis Robert launched the first unmanned hydrogen balloon. Later in the same year, Charles and Nicolas-Louis made a two-hour manned flight, carrying on board a barometer and a thermometer. Charles' expertise with balloons allowed him to make several balloons all at the same atmospheric pressure and at different temperatures. The balloon became a state-of-the-art new scientific instrument and enabled Charles to carry out many experiments on gases and other things. Taking advantage of the new temperature calibration techniques, he was able to deduce that the volume occupied by a given quantity of gas was directly proportional to its temperature. This result was presented to the French National Institute in 1802 by Gay-Lussac, who described his own experiments and also paid credit to Charles' unpublished work from the 1780s.

By this time, people had learned about temperature, but not what physical attribute of a gas, liquid or solid the concept of "temperature" referred to, except to say "heat". A better understanding of heat would not come for some time.

4.4   The Problem of Heat

Philo of Byzantium (280–220 BCE)[20] (who lived in Alexandria) wrote extensively about mechanics and discussed an experiment in which a candle is burned in an inverted bell-jar or dome that is submerged under water. As the candle burns, the level of water in the dome rises and the candle goes out: an experiment that is still performed in schools, though frequently explained incorrectly. Philo saw this as the air "perishing", there is a supposed conversion of the air medium into the fire medium which somehow passes through the porous glass: the air in the glass is consumed by fire. As always, observations were interpreted in terms of the accepted dogma.

Philo's experiment was repeated with numerous variations by several researchers, among whom were Robert Boyle himself, Robert Hooke and John Mayow in Oxford. None of these experiments shed any light on the nature of "heat". There were three strands of thought on the matter: that heat was a gaseous substance, that heat was a fluid and that heat was somehow associated with the motion of the corpuscles that were the constituents of all matter.

A German physician and alchemist called Johann Becher published a book in 1667, *Physica Subterranea,* in which he revived the ancient concept of the "fire element", a supposed substance that was emitted during burning. While Becher's work was not particularly distinguished, his student, Georg Stahl, took up the banner of creating a theory based on this substance. It was Stahl who coined the name *Phlogiston* for the fire element. Stahl updated the 1667 edition of Becher's book to include more material on phlogiston. The theory asserted that metals and combustible materials were compounds that contained phlogiston which was released on burning. Air was necessary for combustion because it absorbs phlogiston. Combustion within an enclosed vessel stops when the air becomes saturated with phlogiston. The same is true when a creature confined in a closed container dies. According to Stahl, phlogiston was the substance of heat, manifesting itself as fire.

Phlogiston is obviously a "postulated but not seen" entity, a "dragon" making use of the metaphor of the subtitle of this book. Obviously, there are no pictures to be had because as Forris Jewett Moore[21] has pointed out "It no more occurred to anyone to go out and look for phlogiston than it occurs to us to attempt the isolation of the luminiferous ether".

Belief in phlogiston in one form or another persisted for more than a 100 years and was only eventually dispelled[22] by the work of "the father of modern chemistry", Antoine Lavoisier, in 1789. The phlogiston theory then gave way to the fluid theory of heat.

The origin of the fluid hypothesis for heat goes back to the early years of the 18th century and is attributed to the distinguished Dutch chemist and physician, Hermann Boerhaave. In 1701, he joined the Institute of Medicine at Leiden University and soon developed a Europe-wide reputation as a teacher. He received pupils from all over Europe and even received plaudits from as far as China. He was best known in his lifetime for his work in medicine and chemistry and in 1732 published what is generally regarded as the first textbook to be written on chemistry: the *Elementa Chemiae*. In the book, he gave the opinion that heat should be regarded as a fluid. The heat fluid had the key property that it flowed from hot bodies to cold bodies. The fluid was also thought of as being weightless and being able to pass through solid and liquid materials. These were necessary properties that would conform with our perceptual experience, but there was no underlying mechanistic principle telling us how or why this could work.[23] Boerhaave's influence and reputation were such that this became a widely accepted point of view, alongside the competing phlogiston theory.

The next big advance in the understanding of heat did not come until the advent of kinetic theory in the 19th century. In the meanwhile, attention focused on the discovery of many new elements and their properties. This was the age of chemistry, and its foremost practitioner, Antoine Lavoisier.

4.5   The "Father of Modern Chemistry"

In his repeat of Philo's experiment, Lavoisier used red phosphorus in a bell jar containing air, with its open end immersed in a bath of mercury. He ignited the phosphorous with a shaped red-hot iron stick. The phosphorus burned to phosphorus oxide and he could estimate the amount of what we now know to be oxygen that had been used up by the phosphorus. He improved on this simplistic setup and in 1789 published the details of his experiments and result in his *Traité Élémentaire de Chimie*. Lavoisier had taken a very important step: he devised experiments in closed vessels so that he could weigh the materials that had been burned. In order to measure weights with sufficient accuracy, he invented what we now call "the Lavoisier Balance". The weights of the burned products were found to have increased as a result of burning, while the weight of the vessel remained constant. This was entirely counter to the classical idea that the weight of a burning substance should decrease owing to the loss of phlogiston during the burning process. As Lavoisier famously said, "Nothing is lost, everything transforms: mass is neither created nor destroyed".

It was Lavoisier himself who promoted his own version of the fluid theory when he realized that his experiments on burning were not consistent with phlogiston.[24] He gave the heat fluid a name: *Caloric*. There was a constant amount of Caloric in the universe. Hot bodies had more of it than cold bodies. The fluid was said to be self-repelling and so it flowed from hot to cold, eventually reaching an equilibrium. The idea of the fluid being self-repelling harked back to Newton's view that the particles in a gas repelled one another. Lavoisier designed and built a special piece of equipment to measure the quantity of caloric in objects: the *calorimeter*. His experiments were entirely consistent with the Caloric theory, which was readily adopted by many of the scientific community until the middle of the 19th century.[25]

Lavoisier was born into a wealthy noble family and benefited from his wealth and his connections to practice his science. He added to his status and his wealth by marrying an equally well-connected girl, Marie-Anne Pierrette Paulze. She actively participated in his work in several important ways. She assisted him in his laboratory and made sketches of his equipment. She made translations into French from English letters and documents, including those of Joseph Priestley. Most of all, she hosted parties for eminent scientists and visitors, thereby publicizing his reputation and his results in the manner of the time. After he died, she edited and published his memoirs, *Mémoires de Chimie*, an impressive collection of his works, setting out the principles of the new chemistry.

Although Lavoisier was a liberal and a reformer, a nobleman during the Reign of Terror was never safe. In May 1794, he was arrested, tried and guillotined along with 27 other people. Lavoisier had asked for a stay of execution in order to finish an experiment. The judge simply responded "La Republique has no need for scientists. Justice cannot be delayed".

The Chemistry Revolution was the end of the original idea of Empedocles and the start of a new age in which growing numbers of chemical substances and elements were being identified and studied. It was also the start of a split between the worlds of chemistry and physics. The next century — the 19th century — would see the growing importance of both mathematical tools and sophisticated instrumentation to advance our knowledge across all areas of science.

4.6   The Atomist Point of View

By the start of the 19th century, alchemy had evolved into chemistry, the old view of Earth, air, fire and water had been dispelled and a new approach to science using clever instruments and scientific methods had been established. There was now a calibrated temperature scale, and the theory of phlogiston had been disproved. But there was still no answer to the question "what is heat?"

The corpuscular hypotheses advanced by Boyle, Hooke and Newton in the previous century sat at a completely different level from the Phlogiston and Caloric hypotheses. Their ideas sought to explain the fundamental properties of matter in terms of its atomistic constituents. The views of Boyle and Hooke were phenomenological: they sought to understand the properties of solids, liquids and gases in terms of the atomic structure of matter. Gases expand because the forces between the atoms are repulsive. A piece of metal cannot be compressed because the atoms are packed close together and held there by attractive forces. Atoms of course are themselves hard and indestructible. Thus, *Hooke's law* telling us that the amount by which a wire stretches under tension is proportional to the force applied immediately suggests the existence of attractive inter-atomic forces as the source of elasticity. Such simplistic atomic models are purely phenomenological and do not lead to testable predictions, but they nevertheless offer a way of gaining insight into the processes that may govern the behavior of solids, liquids and gases. It is not surprising that in 1664 in his *Micrographia*, Hooke describes heat as "nothing else but a brisk agitation of the parts of a body".

Newton was one of the founders of the mechanistic movement which held that physical phenomena required a physical explanation. His view of these fundamental corpuscles, or atoms, was far more profound than any of his predecessors[26]: it was based on the scientific principles of his *Principia* and was discussed in his *Opticks*, published in 1704 and revised in 1717.[27] His hypothesis was that a gas can be regarded as an elastic fluid of particles at rest. He endowed his particles with a *repulsive* force that was inversely proportional to their separation and was able to show that this led to Boyle's law.

Such was Newton's stature that the advocates of the phlogiston and fluid hypotheses for heat each tried to incorporate an underlying structure of Newton's atoms into their views.

Newton's view strongly influenced two people who made significant advances on the atomistic view: Daniel Bernoulli (1700–1782), who explained precisely in terms of the motions of atoms what pressure and temperature were, and John Dalton (1766–1844), who explained chemistry in terms of atoms. Dalton's contribution did not bear directly on the nature of heat, but importantly it established the chemistry of atoms by showing how their properties could be determined and how they stuck together to make compounds. Atoms were to move center-stage in the study of various types of matter and of chemical compounds and their interactions.

Daniel Bernoulli was born in the city of Groningen in the north of the Netherlands in 1700. The years 1727–1733 were productively spent working with Leonard Euler in St. Petersburg after which he became the Professor of Experimental and Speculative Philosophy in Basel. In 1738, he published a book that should have made him famous but did not: his *Hydrodynamica*. This was the first mathematical attempt to create a kinetic theory of gases in which thermodynamic phenomena were a consequence of the random motions of atoms.

In contrast with Newton, Bernoulli thought of air as consisting of particles in continual motion. His atoms did not interact with one another but merely bounced off the walls of the containing vessel, exerting a force on the walls: the pressure. The greater the energy of the particles, the more the pressure. From there, he correctly identified the temperature of the gas with the particle energy and then provided the (correct) explanation for Boyle's Law. Bernoulli's version of Boyle's law went one step further: he showed what would happen if the temperature was not kept constant. It was 70 years later, in 1808, that Bernoulli's prediction was proved correct by Joseph Gay-Lussac, by which time Bernoulli's contribution had been forgotten. Today, we would regard Bernoulli as the scientist who predicted what we now refer to as Charles' law, the variant of Boyle's law that takes account of temperature.

Bernoulli's work, brilliant as it was, had relatively little impact or for that matter did Newton's work on Boyle's law. Despite the move towards accepting an atomist point of view of the world, as opposed to the Aristotelian view, there was little general enthusiasm for the central notion of Bernoulli's work that gases were made of randomly moving "atoms". One contributory factor was that, with the work of Joseph Priestley, Antoine Lavoisier and others on identifying gases and elements, this was a golden age for chemistry. Few thought about the mechanics of interactions and even fewer were interested. Moreover, the mathematical arguments in his paper were difficult to read and follow. It was not until the work of James Clerk Maxwell in 1867 that substantial progress on kinetic theory, the study of the collective motions of the atoms in a gas, was made.

Bernoulli's work presaged the birth of kinetic theory and statistical physics, which is today explained in terms of what are called the Boltzmann equations. Although they are named after him, Boltzmann was not the first to discover these equations. The first correct rendition was made by a man named John Waterston, but his work was totally suppressed by contemporary scientists until it was no longer relevant. This was a dragon that held back an important advance in science for nearly 50 years.

4.7   Kinetic Theory and a Great Injustice

John James Waterston was neither an academic nor a person of means, but he was a scientist who had a degree in mathematics and physics from the University of Edinburgh. His interest in kinetic theory started early, around 1830 when he was 19 years of age. After getting a variety of jobs, in 1839, he was sent to India where he was able to develop his ideas, culminating in a self-published book with an incredibly unattractive title *Thoughts on the Mental Functions*. In that book, he develops his kinetic theory on the basis that the average kinetic energy of the particle directly reflects the temperature of the gas taken as a whole. That is the correct assumption and he makes the very important statement that "equilibrium of temperature depends on molecules, however different in size, having the same kinetic energy" which is a clear statement of the equipartition of energy in an equilibrium gas where a temperature is defined.

There is no evidence that anyone ever read the book, which might have been due to the weird title.

Waterston did submit the manuscript for publication to the Royal Society where, although it was not rejected outright, it was sent to the Royal Society archives and a short abstract was published in 1846. This left Waterston in a difficult position since, following the rules of the Royal Society, the manuscript belonged to them and so could not be submitted elsewhere. His predecessor in the quest to formulate a theory of gases, John Herapath, had been advised of this and so chose not to send his paper to the Royal Society. Apparently, Waterston had no copy of the submitted manuscript and so had to start again from scratch to rewrite a short paper for the annual meeting of the British Association for the Advancement of Science in 1851. A short abstract was published and that eventually led to Waterston being formally acknowledged for first stating the equipartition law.

In 1857, Waterston returned to Edinburgh from India due to ill-health but managed to publish 11 papers in the *Philosophical Magazine* between then and 1868. Ten years later, he sent two papers to the Royal astronomical society which were rejected. He must have felt that the world was against him and there is nothing more from him until he disappeared one morning in 1883. It seems probable that this was suicide, but his close friends thought that unlikely.

So what of Waterston's 1851 paper? Waterston was clearly at the forefront of a revolution in statistical physics, he had the right ideas and he clearly understood the importance of the problems that his approach would tackle. One referee of the 1851 paper had said of it that "the paper is nonsense, unfit even for reading before the Society" and another declared that while the paper was not without interest, it was "... by no means a satisfactory basis for a mathematical theory".[28]

Lord Rayleigh in 1891, when he was secretary of the Royal Society, came across the Waterston paper in the Society archive and realized the terrible mistake made by the reviewers while researching the subject. This was half a century after the Waterston paper had been rejected and the only course open to him was to go ahead and publish the paper with an explanation and an apology. The paper duly appeared in 1892. His apology took the form of a severe critique of Waterston's paper, benefiting from the point of view of hindsight, though nonetheless acknowledging that the article should have been published. It was not the sort of apology that anyone would be appreciative of. In any case, Waterston had died 9 years previously and had no living relatives.

This is certainly one of the most appalling cases of abuse of power in science, the ignoring of a good idea by those who have power to determine a person's fate with no redress. We would like to think this would not happen in the 21st century.

Part II

The Dragon's Fire

A New Heaven and a New Earth

Chapter 5

Crystalline Spheres

> *It does not do to leave a live dragon out of your calculations, if you live near him.*

> *J.R.R. Tolkien, The Hobbit*

The first suggestion of the existence of dark matter may possibly have occurred in the fourth century BCE. The idea was put forward to maintain the stability of the cosmos as it was understood at that time. Eudoxus of Cnidus (c. 408–355 BCE), a student of Plato, introduced the concept to justify both the apparent movement of fixed stars in the heavens, which turned in a day, as well as that of the "planets", which, in addition to taking part in this movement, night after night wandered among the fixed stars. The system, as well as the reality of the spheres, was, as it were, crystallized in the writings of Aristotle.[1] Although for Eudoxus, perhaps, the spheres could be only mathematical constructs that had no physical reality, for a naturalist like Aristotle, the spheres would need to have been physical, corporeal entities.[2] In any case, given that they were made of ether or quintessence (a heavenly fifth element), the qualities of terrestrial elements could not be attributed to them, and so, for example, it would not have made sense to state that they were solid or fluid, hard or soft, dense or rarefied and so on. And, of course, they could not change their nature, since changes were not allowed in the celestial region.

The invisible ethereal matter that Aristotle postulated was not dark but transparent: crystalline spheres on which stars were embedded; spheres centered on Earth ("homocentric" where the homo is that of homogenized milk — meaning the same — not the Latin for man). The spheres were thought to be real and to play a role in the explanation of the apparent movements of the Sun, the Moon, the planets and the stars. They were perfect spheres that described uniform and harmonious movements; spheres that allowed astronomical observations to be explained within a geocentric model. Aristotle's description was eventually incorporated into Catholic church doctrine by Thomas Aquinas (c. 1250) and others. In the process, the outer sphere came to be rotated by a divine Prime Mover and the spheres of the planets carried around by angels (also apparently real).

Together with the Aristotelian conception of the heavens, which made them immaculate and immutable, crystalline spheres, a postulated but not visible entity endured for centuries, in part because there was no alternative. Whether crystalline spheres were of a rigid or soft nature was an issue that preoccupied natural philosophers during the Middle Ages. The opinion held by the majority was that they must be solid, this qualifier being understood as synonymous with hard or rigid.[3] Apparently, in the 15th century, Austrian astronomer Georg Peurbach (1423–1461) thought that crystalline spheres were solid and therefore other celestial bodies could not cross them nor was it possible to cast their existence into doubt through the new conception of the world put forward by Nicolaus Copernicus (1473–1543) in 1543, according to which the Sun — and not Earth — is the body that takes the central position.

When did this dragon — the crystalline spheres — disappear from the minds of the natural philosophers?

5.1   Blind Watchers of the Sky

About 30 years after Copernicus' death on 11 November 1572, a young Danish aristocrat who was observing the night sky from Herrevad Abbey, today in Swedish territory, spotted a new star in the constellation of Cassiopeia (see Figure 5.1). He could not believe what his well-trained eyes were seeing. The object shone about as bright as the planet Venus, which, after the Sun and the Moon, is

Figure 5.1 Engraving showing Tycho Brahe observing the new star in Cassiopeia. From *Astronomia Populaire* by Camille Flammarion (Paris, 1894). Credit: Public domain. Wikimedia Commons.

the brightest object in the sky. Tycho Brahe (1546–1601) was not the only one in Europe who was stunned by the star's appearance.

That same night, in the Valencian town of Torrent, shepherds and lime burners, used to working at night and observing the heavens and identifying, with practiced eyes, the Cassiopeia W-shape pattern, were surprised to see this celestial intruder. Convinced that this star had not been in the sky before, they alerted Jerónimo Muñoz, professor of Hebrew, mathematics and astronomy at Valencia's *studium generale*, of its presence.

Jerónimo Muñoz (ca. 1520–1591) had obtained a bachelor of arts degree from the University of Valencia in 1537. He subsequently completed his studies in various European countries and became so well regarded that King Philip II himself, aware of the excitement that the new star had awakened throughout Europe, commissioned Muñoz to brief him on the astronomical phenomenon and offer his observations and interpretation.

But what was this object that attracted so much attention among shepherds, astronomers, philosophers, clergymen, nobles and kings? In 1945, American astronomer Walter Baade (1893–1960), studying the observations made by Tycho Brahe and some of his contemporaries, came to the conclusion that it was a type I supernova. Nowadays, we know that a type I supernova is the explosion of a white dwarf, a star about the size of Earth but with a mass similar to that of the Sun. The density of a white dwarf is enormous: hundreds of tonnes per cubic centimeter. In many cases, white dwarfs are part of binary systems, the companion star of which is usually much larger, but less massive, so the gas from the companion can flow onto the white dwarf and gradually increases its mass. When the white dwarf reaches a mass 1.44 times greater than the mass of the Sun — this point is known as the Chandrasekhar limit — it ignites carbon and oxygen fusion under the force of its own gravity and explodes, no remnant remaining. This explosion is a type I supernova (more specifically type Ia).

To be sure, neither Tycho Brahe nor Jerónimo Muñoz knew what a supernova was. It would take more than 350 years for the stellar evolution theories that explain these fantastic explosions occasionally observed in the sky to be developed. Moreover, these phenomena are rare: the last supernova seen in our galaxy occurred in 1604, and it was observed by Johannes Kepler (1571–1630), who had worked as Tycho Brahe's assistant and by others about the same time. Both Tycho and Kepler referred to these stars as new stars or "stellae novae", using the Latin term.

Jerónimo Muñoz, on the other hand, called his treatise on the new star the "Book of the New Comet", with its full title going on to refer to "the place where they are made; and as will be seen through the Parallaxes, how far they are from earth". It was difficult in academic and scholastic circles to dispute the Aristotelian idea of the immutability of the heavens. The appearance of a new star in the sky and its gradual disappearance contradicted this conception. The supernova's brightness quickly peaked in mid-November 1572, gradually diminishing until it completely disappeared from view in March 1574. Since Aristotle's time, it had been thought that such changes could only occur in the sublunar region. Nothing could change beyond the Moon, for that was the domain of the immutable. The new star was either a phenomenon typical of the sublunar region or it clearly called into question the immutable nature of the heavens that had been accepted for centuries.

It is quite possible that Muñoz shared the idea, held by some astronomers of the time, that comets were an aspect of the heavens and not meteors in the atmosphere. In this context, the natural cosmology that Muñoz supported made it more rational to consider the new apparition a comet rather than a new star, as Tycho did. For Tycho, the reasons for rejecting the idea that the new star was a comet involved both its form — which was typical of a star that was twinkling and not a comet with a luminescent tail — and the fact that it remained still in the sky, without any movement of its own, comets being known at the time to exhibit such movement. Muñoz himself is aware of these characteristics of his "comet" and says in his book, "But it seemed to scintillate like a fixed star"; "In no author do I find a comet similar to this one, which seems to me more like a star than a comet"; "Even now it has inviolably kept to the laws of motion of the Prime Mover, as if it were a fixed star".

To determine its distance, both Muñoz and Brahe attempted to measure the nova's parallax. Geocentric parallax is the apparent displacement of an object relative to the fixed stars when that object is observed from different points on Earth or in the morning and evening from the same place. The absence of parallax confirmed the enormous distance of the new star from Earth. In addition, the fact that it was beyond the lunar orbit meant that the sky (the perfect and immaculate supralunar world) was not as unchanging as Aristotle had believed. Although many astronomers observed the new star, Tycho's work on it brought him fame across Europe as well as earning him the admiration of the Danish king. Nowadays, the new star that appeared in Cassiopeia is deservedly known as Tycho's Supernova.

There was a remarkable exchange of correspondence between European astronomers about the new star. The episode was a prelude to the need for astronomers, like other scientists, to exchange their knowledge, experiences, observations and discoveries. It undoubtedly had the effect of intensifying their relationships and establishing communication networks among them, ones that were extraordinarily efficient for the time. Jerónimo Muñoz's book was translated into

French, and his observations would be collected and commented on, admiringly so, by Tycho Brahe, in a work in which he takes stock of many of the observations of the supernova made by himself and other astronomers. Indeed, *Astronomiae Instauratae Progymnasmata*, a book published in 1602, reviews the works of more than 30 astronomers on the phenomenon. Today, one of the ways to measure the impact of a scientific work is to consider the number of citations it receives from other researchers in the field. Muñoz was also undoubtedly ahead of his time in this regard, as his study was commented on by, in addition to Tycho Brahe, renowned astronomers, such as Cornelius Gemma (1535–1578, in Leuven), Thaddaeus Hagecius (1525–1600, in Prague), William Gilbert (in England) and Galileo Galilei (in Italy).

A new star in the supralunar region was the first major evidence against the then current belief of the unchangeable heavens, but it was not the only one. The story continued only 5 years later when the Great Comet of 1577 made its appearance.

Both Tycho Brahe and Jerónimo Muñoz could observe the comet for weeks. Muñoz, at the age of 11, might even have observed Halley's Comet during its periodic return to the vicinity of the Sun in 1531. Perhaps, the memory of that vision from his childhood led him to think that Cassiopeia's new star was a comet. In any case, though, his observations of the 1577 comet led him to conclude that it was an object far beyond the Moon and, therefore, one that did not fit the Aristotelian explanation of comets as atmospheric phenomena of the sublunar region (see Figure 5.2). Muñoz cited Seneca, Democritus and Anaxagoras as defenders of the celestial nature of comets.

Many European astronomers studied the comet (see Figure 5.3), especially those who lived under the patronage of nobles and monarchs eager to know the omens that the comet foretold. Tycho studied it carefully and came to the conclusion that it was beyond the Moon's orbit, once again contradicting Aristotelian ideas on the unchanging nature of the supralunar world. And there was more: Tycho confirmed that the comet was revolving around the Sun. It is possible that this fact had a marked influence on the production of his own world system, the Tychonic system, in which the Moon and the Sun revolve around a stationary Earth situated at the center of the sphere of fixed stars, while the other planets plus the comets revolve around the Sun. It should be stressed that Tycho's rejection of the

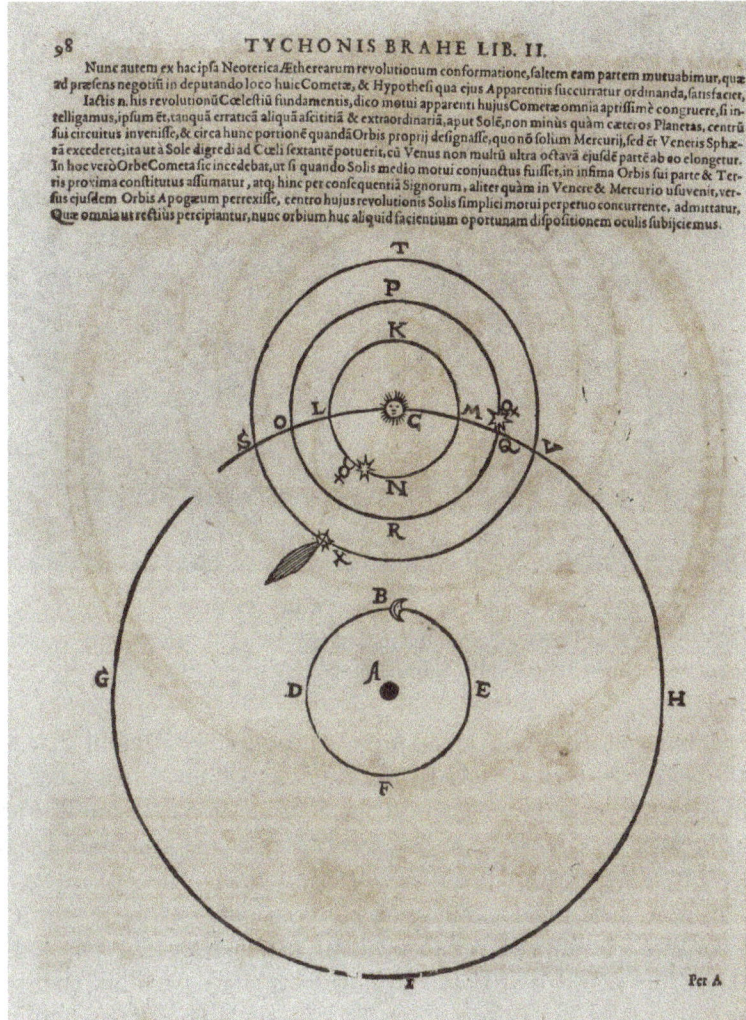

98   **TYCHONIS BRAHE LIB. II.**

Nunc autem ex hac ipsa Neoterica Ætherearum revolutionum conformatione, saltem eam partem mutuabimur, quæ ad præsens negotii in deputando loco huic Cometæ, & Hypothesi qua ejus Apparentiis succurratur ordinanda, satisfaciet. Iactis n. his revolutionū Cœlestiū fundamentis, dico motui apparenti hujus Cometæ omnia aptissimè congruere, si intelligamus, ipsum ēt, tanquā erraticā aliquā ascititiā & extraordinariā, ap ut Solē, non minùs quàm cæteros Planetas, centrū sui circuitus invenisse, & circa hunc portionē quandā Orbis proprij designasse, quo nō solùm Mercurij, sed ēt Veneris Sphæ- rā excederet; ita ut à Sole digredi ad Cœli sextantē potuerit, cū Venus non multū ultra octavā ejusdē partē ab eo elongetur. In hoc verò Orbe Cometa sic incedebat, ut si quando Solis medio motui conjunctus fuisset, in infima Orbis sui parte & Ter- ris proxima constitutus assumatur, atqị hinc per consequentiā Signorum, aliter quàm in Venere & Mercurio usuvenit, ver- sus ejusdem Orbis Apogæum perrexisse, centro hujus revolutionis Solis simplici motui perpetuo concurrente, admittatur, Quæ omnia ut rectiùs percipiantur, nunc orbium huc aliquid facientium oportunam dispositionem oculis subijciemus.

Figure 5.2 Tycho Brahe's system of the world with the path of the 1577 comet published in his two-volume work *Astronomiae Instravratae Progymnas-mata* (*Introduction to the New Astronomy*). The diagram appeared in the second volume entitled *De Mundi Aetherei Recentioribus Phaenomenis Liber Secundus* (Second Book About Recent Phenomena in the Celestial World), originally published in 1588. A is the stationary Earth, BDFE our Moon; C is the Sun, carrying Mercury and Venus (KLNM), Mars (ORQP), and the comet (STVX) with it. The page shown here is from the 1648 edition. Image courtesy of History of Science Collections, University of Oklahoma Libraries.

Figure 5.3 The 1577 comet in a detail from a painting on wood by Jiri Daschitzski, Prague, Petrus Codicilius in Tulechova, 1577. Image courtesy of Zentralbibliotheck Zürik. Wikipedia Commons.

Copernican system is not due to any Aristotelian sentiment. His argument is rigorously constructed from the scientific point of view: if the Earth makes a large circle around the Sun in one year, small changes would be seen in the positions of the stars over the course of the year as a result of annual heliocentric parallax. Tycho did not detect this shift; in fact, more than 260 years were to pass before a star's parallax could be measured reliably (using a telescope). Tycho realized that the stars could be too far away to show any parallax that could be picked up by his observation and measuring instruments, but, as he could not conceive of a universe that was so immense as this theory required, he concluded that the Earth must remain stationary. Although his model of the world was incorrect, Tycho observed the sky and learned to interpret his observations rationally, rejecting preconceived ideas and complaining about his predecessors for not having done the same.

Tycho Brahe verified, with precise and systematic observations, that the comet must be passing through the hypothetical crystalline spheres. He reached the conclusion that these spheres did not exist in the heavens as solid entities. An unobservable entity that had been accepted for centuries was finally put into question by the Great Comet and by Tycho's Brahe scientific insight. The belief in the existence of this kind of invisible matter did not disappear immediately. Changes in science take their time to be accepted, but after a few decades, this dragon was definitively gone.

It should be noted that both Tycho Brahe and Jerónimo Muñoz were honest scientists who did not seek to do face-saving work on behalf of systems and models that did not agree with observations, such as those of Aristotle and Aquinas. This produced tensions with the political and ecclesiastical powers — the guardians of the dragons — since many, due to the credit they gave to established doctrines, were, as Muñoz said, "unable to understand what they could see with their eyes", a comment that calls to mind Tycho Brahe, who in *De Nova Stella's* preface addresses his detractors in a similar way: "O crassa ingenia. O caecos coeli spectatores" or ("Oh thick minds. Oh blind watchers of the sky").[4]

Crystalline spheres stayed with us for centuries because they were necessary to explain the movements of the celestial bodies. Their story is one worth telling.

5.2   Wanderers in the Night

The ancient Greeks attempted to produce rational explanations of the cosmos. They had observed that, when one looked at the night sky, the stars seemed to stick to a large sphere that turned once a day. In fact, it was the Earth completing a rotation on its axis over 24 hours that created the impression that the stars rise from the Eastern horizon and set into the Western horizon. However, the Greeks also observed bodies that moved among the fixed stars. They called them planets, which in Greek means "wanderers". Without the help of telescopes, they could see only five planets: Mercury, Venus, Mars, Jupiter and Saturn. The other two wanderers were the Moon and the Sun. There were therefore seven in total. This is one reason why seven is such a special number. There are seven days of the week, and in many languages, they are named after the planets. Monday, for example, is named after the Moon. There are seven musical notes in an octave scale, and there are seven colors in the rainbow (this was something that Newton decided after shining light through a prism. He counted blue and indigo separately).

Simple crystalline spheres bearing the planets, the Sun and the Moon were not enough to explain the movements of these wanderers against the background of fixed stars. Additional mechanical spheres and systems were postulated to explain the motions of the celestial bodies. The generic term is epicycles. Aristotle imagined a system of concentric spheres with Earth at the center, each of which bore a planet, and the outermost of which had the fixed stars. The movements of the heavenly bodies with the spheres that bore them were thought to be uniform, circular and eternal.

The main goal of this mechanical system was to provide a rational explanation for the observed retrograde motion[5] of the planets. We now know that planets, including the Earth, revolve around the Sun at different speeds. Seeing the planet's motion from the Earth means that eventually the planet changes its direct (or prograde) motion relative to the stars from west-to-east to the opposite direction (retrograde, east-to-west) and keeps this movement for several weeks or months because we pass it or conversely. With the heliocentric view, the retrograde motion of the planets is easily understandable.

Given that Earth revolves around the Sun, we observe the rest of the planets from a planet that is not immobile. So, for example, if we look at Mars, we see that night after night it moves directly from east to west relative to the fixed stars. Occasionally, however, the planet seems to slow down, and as seen from Earth, its movement even becomes retrograde, switching to a west-to-east direction for a few weeks. The reason for this is that Earth, whose orbital speed is greater than that of Mars, "overtakes" it, and during that time we see it move over the celestial sphere in the direction opposite to that which it usually takes.

To explain the retrograde motion in a geocentric system with the immobile Earth in the center, the Greeks found different solutions that accounted for this apparent nonuniformity in the movements of the wandering planet-stars. The challenge set by Plato entailed finding solutions that would explain the observational evidence based on combinations of uniform circular motions.

The first such solution was proposed by Eudoxus of Cnidus himself. He made use of two of his crystalline spheres that we mentioned earlier. Eudoxus imagined that the planet was located at

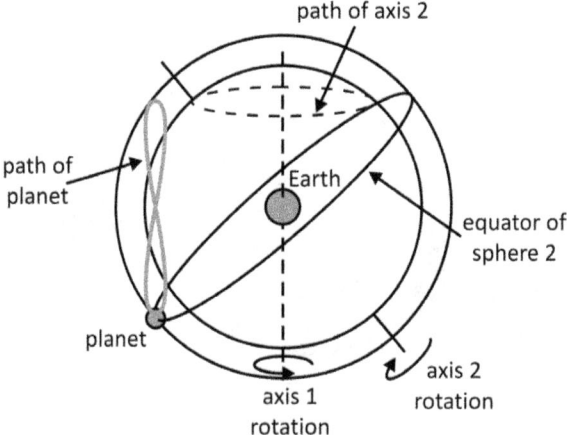

path of axis 2

path of planet

Earth

equator of sphere 2

planet

axis 2 rotation

axis 1 rotation

Figure 5.4 Eudoxus' model to explain the retrograde motion of the planets. The trajectory of the planet follows a eight-shaped curve called hippopede. Credit: Fernando Ballesteros.

the equator of a sphere that rotates with uniform motion. In this model, Eudoxus extended the rotational axis of that sphere, placing its poles on another sphere, this one concentric to the first and rotating at the same speed but in the opposite direction. The outer sphere dragged the axis of the inner one as it followed a smaller circle. The combined planetary movement produced by these two Eudoxan spheres is a figure-eight curve (see Figure 5.4). Eudoxus called it the *hippopede* (horse-fetter). The planet travels half the path along this curve from west to east and the other half from east to west. The British historian of science Michael Hoskin took the view that Eudoxus probably did not believe that the spheres really existed but saw them as an imaginary mechanism to explain the observations of planetary movements. In any case, the spheres that Eudoxus postulated to explain the retrograde motion of the planets were unobservable entities — whether mathematical or real — that lasted for centuries.

Later, we will see how other mechanical spheres and systems simpler than that of Eudoxus were introduced to explain the movement of celestial bodies, but our next stop takes us to the south of Italy, where, according to one tradition, a Pythagorean philosopher named Philolaus (fl. c. 460 BCE) postulated the existence of two unobservable entities: the central fire and counter-Earth. They were two of

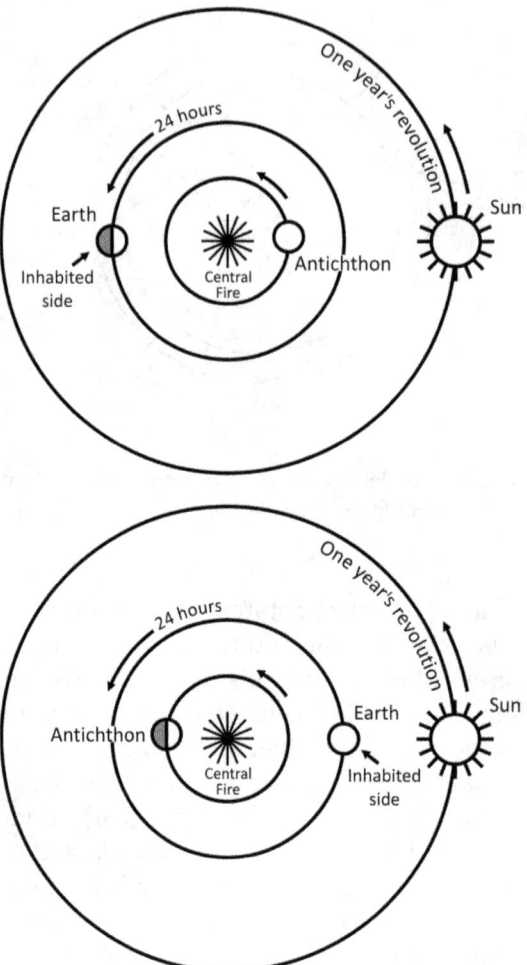

Figure 5.5 Philolaus postulated the existence of a counter-Earth for his non-geocentric cosmology. In this model, all objects, including the Sun, revolve around a "central fire". The diagram shows Earth at night (upper panel) and during day time (bottom panel). Note that probably for Philolaus the Earth and the counter-Earth are flat, and their orbits are very small relative to the orbit of the Sun. Credit: Fernando Ballesteros.

the ingredients of the first nongeocentric world system that we have evidence of (see Figure 5.5).

Philolaus has come down to us only second hand, which is perhaps why seemingly equally authoritative presentations provide two very different versions of his cosmos. Both have a central fire, the throne of Zeus or the heart of the universe and 10 bodies orbiting around it. These are the Sun and Moon, the usual five planets, a sphere of stars, the Earth and the counter-Earth or Antichthon. One possibility (and we think the more persuasive) is that both Earth and counter-Earth are flat, facing outward on opposite sides of the central fire, so that we never see either the fire or our opposite number. In this version, the Sun completes one revolution over the course of a year. This will be more stable than if so ponderous an object as Earth were running around every 24 hours by itself. The Moon, Sun and all planets also orbit the central fire, reflecting its light to us, with periods of a month, a year and the times it takes the planets to move once around their path on the celestial sphere. The alternate interpretation puts Earth and counter-Earth on the same side of the central fire, both with 24-hour orbits, and the counter-Earth completely eclipses the fire, from our point of view, at all times. This allows Earth to be spherical and the heavenly bodies again to reflect light to us, but it seems lopsided!

Philolaus's model is not properly a system centered on the Sun (that is, a heliocentric system), unlike the one that Aristarchus of Samos would introduce almost two centuries later. Rather, it is a system in which 10 spheres (one of them carrying the Sun) revolve around a central fire.

Philolaus, as a Pythagorean, sought beauty and balance as he described the cosmos. Aristotle himself claimed that Philolaus introduced counter-Earth so that his world system was made up of ten spheres — a perfect number — and not just nine.

The way in which scientific research is done has undergone a continual process of change. Aristotle saw the goal of science as being to find out why things happen the way they do. Why are things in a state of motion? Bertrand Russell, in his great *History of Western Philosophy*, argues that during his lifetime, Aristotle's contributions to science and philosophy were so great that his authority could not be questioned. When he died, he left an intellectual vacuum in which his unquestioned authority stood its ground for almost 2,000 years.

That is not to say that there was no scientific progress after the death of Aristotle. Around 280 BCE, Aristarchus of Samos put forward the hypothesis that the Earth and planets all went around the Sun in circles and that the Earth rotated on its axis once every 24 hours.

Most of Aristarchus's works have been lost, and little is known about his life. One astronomical work has survived in which Aristarchus makes use of trigonometric methods (relationships between the sides of a triangle with known angles) to calculate the sizes of the Sun and Moon and the distances of these bodies from Earth. The methods were absolutely ingenious! The size of the Moon and its distance came out about right, although those for the Sun were far from the real values because those methods assumed circular orbits and needed more precise measurements than were then possible.[6]

5.3   A Model that Lasted for Centuries

Almost a century after Aristarchus, around 200 BCE, Apollonius of Perga put forward a geocentric model for the motion of the Moon, Sun and planets; this was an early version of the later "epicyclic theory" for the observed motion of those bodies.

Meanwhile, Greek geometrical cosmology moved on, at least partly with the goal of being able to calculate and predict the motions of the Moon, Sun, and planets against the background of fixed stars.

It was probably the adoption of the epicyclic theory by Hipparchus, arguably the greatest astronomer of antiquity, that signaled the end of Aristarchus' Solar-centric model. In the 2nd century CE, Ptolemy refined the geocentric theory using Babylonian astronomical records.

Ptolemy postulated the existence of geometric entities that were essential for explaining the observed movements of the heavenly bodies, and in particular the retrograde movement of the planets. In Figure 5.6, this motion is naturally illustrated for the heliocentric system. However, to explain the retrograde motion within the geocentric model, two entities have to be postulated. In this model, the planets moved in small circles that were named "epicycles", the centers of which in turn revolved on a larger circle, known as the "deferent" (see Figure 5.7).

In this model, the center of the epicycle moves in a circular and uniform motion around Earth, doing so anticlockwise. The planet moves on the epicycle in the same direction and also displays uniform circular motion, but the sum of the two movements, when they

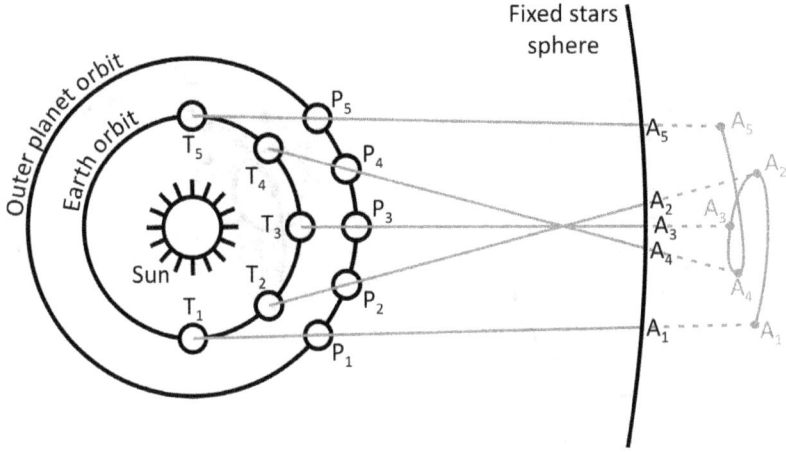

Figure 5.6 Illustration of the retrograde motion of a planet in the heliocentric system. Positions of the Earth are indicated as T_1, T_2, etc., in the left panel, while positions of the planet are indicated by P_1, P_2, etc. These positions are projected onto the celestial sphere at A_1, A_2, etc. The trajectory of the planet as seen from Earth is depicted in the right panel. Note that the orbits are not exactly coplanar. Credit: Fernando Ballesteros.

act simultaneously, makes the planet follow a trajectory known as a cycloid, eventually becoming retrograde relative to an observer located on the central Earth.

To explain observations of the Sun, Moon and planets that exhibit different speeds along their orbits,[7] Ptolemy introduced the concept of the equant. To do this, he put the center of the deferent not at Earth but at an eccentric point, and he placed the equant point precisely in the opposite position from the Earth in the direction of the center of the deferent. The movement of the center of the epicycle had to be uniform (at a constant angular velocity) with respect to the equant so that, ingeniously, the uniformity of the circular motion was preserved; it was uniform relative to the equant and circular relative to the eccentric point.

Ptolemy published his work in the 13 books of his "Almagest", certainly one of the most important scholastic works of the first millennium CE. The original was written in ancient Greek, first translated in the 9th century into Arabic and much later into Latin. His model for planetary motion was so good at predicting where the planets would be in the future that for the next 1,200 or so years,

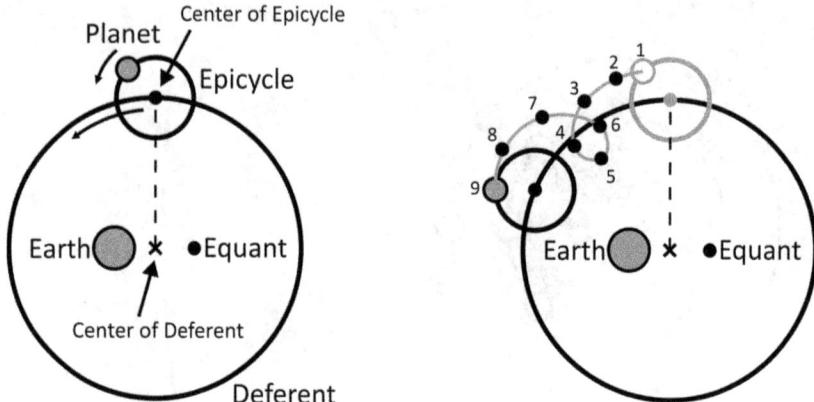

Figure 5.7 The center of the epicycle follows a uniform circular motion (with respect to the equant) on the deferent in the prograde direction (counterclockwise). The planet on the epicycle moves also counterclockwise. The combined motion illustrated in the right panel shows that from points 3 to 5 the planet follows a retrograde path. Credit: Fernando Ballesteros.

it was used for the then-important task of computing accurate horoscopes.

Scientists in Western Europe were ignorant of all of this until the mid-13th century. It was during that century Fibonacci introduced Arabic numerals to Europe and some of the ancient texts were translated from Arabic (notably Ptolemy's *Almagest* was translated in Toledo from Arabic into Latin in the 12th century). Later, in 1252, King Alfonso X of Castille, *el Sabio* (1221–1284), charged a group of scholars with the task of updating and maintaining tables for computing accurate positions for the Sun, Moon and planets using the Ptolemaic model. These were known as the "Alfonsine Tables of Toledo". The work, written in Medieval Spanish, was the continuation of work that had been started earlier by Islamic astronomers and was not published until 1483. In the meantime, another set of Alfonsine tables was produced in Paris in 1320. Importantly, the tables were used by Copernicus to validate his own simpler theory of planetary motions.

At this point, the reader is perhaps expecting to be told that the Ptolemaic system collapsed under the weight of 58 (or some other number of) epicycles. You will find some such tale in many secondary sources, perhaps enlivened by the quote from King Alfonso X when he watched his astronomers compiling the Alfonsine Tables of planetary positions,[8] that if he had been present at the creation, he could have given some advice for its simplification.

5.4 Getting Through the Middle Ages

In order for an idea or other human invention to survive, two things must happen. First, someone has to put it into a durable form — writing of some sort, art, a memorable saga or hummable tune. Second, later, from time to time, someone must think it worth keeping and make a copy or teach it to someone younger, before the papyrus decays or the last minstrel dies. For example, you can read the lines of *The Frogs* by Aristophanes but you will never be able to sing the Croaking Chorus because the tune has not come down to us.

Thus, it is remarkable that the contents of the first encyclopedia of the High Middle Ages, the *Liber Floridus* by Lambert (ca. 1061–1150), Canon of Saint-Omer, completed around 1121 CE are much the same as the contents of the encyclopedia *Etymologiae* (or *Origines*), compiled around 599 by Isidore of Seville (560, Cartagena – 635, Seville) the last scholar of the ancient world and the first Christian to try to compile a summary of universal knowledge. True not much has been added in the intervening 500+ years. But also not much has been lost. (A significant exception to the latter is that Isidore's work was considered so complete that where he had incorporated only part of what some previous scholar had written, the parts left out have now often been lost.) The topics covered by both include the traditional *trivium* and *quadrivium* (as well as shipbuilding, agriculture, medicine and properties of rocks and metals). The ancient coupling of mathematics, music and astronomy in this division of human knowledge brings us what little we know about Greek music, and treating geography as applied astronomy yielded some very curious maps, with east at the top.

The relative absence of both loss and gain was about to change, and change permanently. We think that Isidore of Visigothic Spain and Lambert of Northern France could, after sorting out the changes in church Latin over the centuries, have conversed with a reasonable understanding of each other concerning the everyday applicability of the concepts they discussed and argued about what was important.

Both could presumably also have read the plays of Plautus (c. 254–184 BCE) and Terrence (c. 190–159 BCE), from yet another 500 years earlier. We, on the other hand, need highly trained experts to make sense of any of these.

Yes, much was lost, but also much was gained, and it is those gains that we of the 21st century West think of as having led to all the wonders of modern science and engineering, from atomic bombs to Zoom. For across the Pyrenees began to creep translations of the works of Aristotle and Ptolemy (standing, with your permission, here for all of the Greeks).

These had been preserved — that is, translated into Persian and Arabic — and copied by men (always, it would seem) with names like Avicenna (980–1037, a.k.a. Abu Ali al-Husayn ibn Abdallah ibn Sina or simply Ibn Sina) and Abraham Ibn Ezra (1089–1167) in places that are now the Muslim lands of Uzbekistan, Iran, Egypt and Morocco. The High Middle Ages, (roughly 1000–1300) saw the rise of "Scholasticism".[9] During this period, Arabic translations of the texts of Aristotle became more widely available in Latin. The scholastics set out to reframe the teachings of the Church in a way that was logically consistent with the emerging philosophy of Aristotle and that allowed them to formulate and address previously unanswered questions.

In particular, the translation of a large number of these texts by Adelard of Bath[10] was a key factor driving the intellectual revolution that took place in Europe during the 11th, 12th and 13th centuries. The end of the period is marked by the works of the early 14th century French scholar Jean Buridan (c. 1301–1359) whose introduction of the concept of "impetus" — the forerunner of our concept of momentum — was a blow to the unquestioned supremacy of Aristotlean Cosmology and which is said to have set the Copernican Revolution in motion.

Table 5.1 lists the principal "scholastics" of the period. During this period, Arabic translations of the texts of Aristotle became more widely available in Latin. The scholastics set out to reframe the teachings of the Church in a way that was logically consistent with the rediscovered philosophy of Aristotle and also allowed them to formulate and address new questions. Nearly all the names have variations (e.g. Sacrobosco is also John of Hollywood), and some of the dates are disputed. That you are more likely to recognize some of the names near the bottom than the ones on top is a reflection

Table 5.1 High Middle Ages scholasticism[†]. Scholastics philosophers of the period 1000–1300 CE.

Natural philosopher	Dates	Flourished[*]
Anselm of Canterbury	1034–1109	Founder of scholasticism
Adelard of Bath	c. 1080–1142/1152	*Quaestiones Naturales*, c. 1107–33
Peter Abelard	c. 1079–1142	
Peter Lombard	1096–1160	
Lambert of St. Omer		*Liber Floridus*, c. 1090–1121
Petrus Alfonsi		1106–1120
Raymond of Marseilles		1141
Roger of Hereford		1176–1178
John of Toledo		1181
Robert Grosseteste	c. 1175–1253	
Johannes de Sacrobosco	c. 1195–1256	*De Sphaera Mundi*, c. 1230
Albertus Magnus	c. 1200–1280	
Roger Bacon	c. 1219/20–1292	
Alfonso El Sabio	1221–1284	King of Spain Alfonso X, The Wise
Thomas Aquinas	1225–1274	
Ramon Llull	c. 1232–1315	*Ars Magna*, 1305–1308
John Duns Scotus	1265–1308	Supposed eponym of "dunce"
William of Ockham	1287–1347	Of Ockham's razor
Jean Buridan	c. 1301–1359	Critic of Aristotlean Science, Buridan's ass

Notes: [*]When the birth and death dates are uncertain, we give the approximate year of publication of their best-known work. Those have been intercalated with the others on the assumption that the most important works were written at ages between 40 and 50.
[†]The High Middle Ages is defined loosely as the period from 1000–1300 AD in European history. It was the period during which the universities of Bologna, Oxford, Paris, Salamanca and Cambridge were founded. It was also the time of the seven Crusades. The end of the period is defined by the start of the Mamluk period (1291–1517) and the foundation of the Ottoman Empire by Osman in 1299 and by the extended period of chaos caused by the Black Death pandemic.

of what is included in current educational programs. These versions of the names and dates were drawn from four or five different books and journals and several websites and are to be regarded with due suspicion.

Robert Grosseteste (c. 1175–1253, England) comes somewhere in the middle of this timeline, so, not surprisingly competing experts have described him either as among the last of the Scholastics, with primarily theological interests (he did care deeply about "cure" of souls), or as the first of the moderns, incorporating the idea of "experiments" as a way to choose among ideas. The Latin word that looks like "experiments" certainly appears in his writings, though the meaning could have been closer to what we would call "appearances" or "phenomena". What is clear is that some of what he wrote can be translated into words that allow setting up modern simulations corresponding (perhaps) to his ideas, as expressed in the essays "On light" and "On color". For the historian of science A. C. Crombie,[11] Grosseteste was "the real founder of the tradition of scientific thought in medieval Oxford, and in some ways, of the modern English intellectual tradition". Grosseteste's cosmology is based on light. In his important book *De Luce*, he explains how the diffusion of light carries matter giving rise to the corporeal cosmos.[12]

A group of modern scholastics with a range of expertise, mostly associated with the University of Durham,[13] has attempted some of these simulations. These authors claim "To our knowledge, *De Luce* is the first worked example showing that a single set of physical laws might account for the very different structures of the heavens and Earth, hundreds of years before Newton's 1687 appeal to gravity to unite the falling of objects on Earth with the orbiting of the Moon".

Grosseteste's "color" is three-dimensional, with the dimensions closer to "hue, brightness and saturation" of modern color theorists than to the red-green-blue of additive television systems or the magenta-yellow-cyan subtractive systems of color motion pictures.[14] The Durham group has managed some two-dimensional projections into a "hue — linear combination of saturation and brightness" plane that perhaps illustrate part of what Grosseteste had in mind. We can, anyhow, be reasonably sure that, unlike Dalton, he was not color blind.

The other set of ideas describes a beginning to the universe (required by church doctrine but absent from Aristotle and all) in which light, originating from a point, spreads out, taking matter with it, to establish various shells of different densities. These become the crystalline spheres that carry the seven planets (Sun and Moon yes, Earth no), the stars and the prime mover around the Earth. The parameters of the simulation have to be quite finely tuned, as is, the authors note,[15] the case with our current lambda-cold-dark-matter universe if it is to evolve in 13.8 Gyr to look like what we see around us.

Some have described Grosseteste as the first to propose the modern experimental method of science. Surely, however, 10,000 apprentices for 10,000 years have said to their master craftsmen, "suppose we make the arrows, or the pots, or the mortar for laying bricks, or the spears this way instead of that?" and been told "well, go ahead and try it, and we'll see what happens". Thus, our modern idea of experimenting would seem to have come from craftsmanship and engineering rather than from natural philosophy or science, while astronomy is the purest example of observations. The new idea, which has had to be rediscovered or reinvented many times, is that the consequences of your model or hypothesis or theory or scenario should agree with the results of those observations and experiments.

Roger Bacon (c. 1219–1292, England), who followed Grosseteste at Oxford (though they probably did not overlap and indeed may never have actually met), went further, declaring that proper astronomy and astrology should yield predictions that actually came true, whether of planetary positions, eclipses or the deaths of kings. Did he also test recipes for gunpowder? Perhaps, although the recipe he published would not have exploded under available conditions.

In any case, the dictum that your theories and observations or experiments should agree gradually caught on and pervades most of modern science under the label of testability or falsifiability.

If there is a dragon here, it is the residual notion that the elegance of your theory matters more than its ability to yield testable predictions. String theorists or multiverse supporters (see Chapter 18) have been accused of that notion.

5.5 Muslim Astronomy

While ancient Greek astronomy had reached Western Europe by the middle of the 13th century, the path had been a wandering one, through several countries and languages, and with old dragons — postulated unobserved entities — revived and new ones added. In particular, as Prof. Noel M. Swerdlow[16] has pointed out, "From the 8th to the 15th centuries, the most important and original astronomy — observational, mathematical, and physical — was from the Islamic world".

The drivers for Muslim involvement in astronomy appear to have been (a) many of the 8th and 9th century converts to Islam had Greek among their languages and so could read Ptolemy and his predecessors and (b) astronomical methods were of use in determining the correct orientations of Mosques (so as to point towards Mecca), the times for the five daily prayers and the start and end of the fast month of Ramadan. Some early Christian astronomy was similarly driven by the need to establish the correct date for Easter ("the first Sunday after the first full moon after the vernal equinox" as one of our mothers taught us very long ago).

The contributions included better astronomical instrumentation,[17] better observational values for the inclination of the ecliptic[18] and precession of the equinoxes,[19] and better mathematical methods for predicting planetary positions from geometrical models of the cosmos. But the key point in our inventory of essential, undetectable structures was a gradual trend back from Ptolemaic cosmology to homocentric, uniform circular motion often expressed in epicycles. Of the important people, we mention only three. Their three locations were Cairo (the capital of Fatimid Caliphate), the observatory at Maragha (in present-day northeastern Iran), established by a grandson of Genghis Khan after the Abbasid Caliphate was conquered by the Mongols in 1256–1258, and Muslim Spain, though Muslim astronomers also worked in Baghdad, Damascus and many other places.

First Ibn al-Haytham (965–1040), known as Alhazen, spent many of his most productive years in Cairo. He objected specifically to the nonuniform motion required by equants. For each planet, in its zone between Earth-centered shells, he suggested adding a second epicycle with motion equal to the first but in the opposite direction. Both spheres are concentric and that is the reason this model is also known as homocentric so that the path is a sort of figure eight moving around the sky (called a hippopede, as was mentioned at the beginning of this chapter) accounting for the variations in planetary latitudes above and below the ecliptic in much the same way as was done by Eudoxus more than a millennium before.

Alhazen is renowned for his seminal text on optics, the seven-volume *Book of Optics* and his *Treatise on Light*, works that stood the test of time until Newton and Leibniz wrote on the same subject. The book on optics was translated into Latin in the late 12th century under the title *De Aspectibus*. The work was quickly taken up by Robert Grosseteste, Roger Bacon and a Polish philosopher, the Silesian Friar, Erazmus Witelo (1220–1278) (alternatively known as Vitello), who published a seminal text on optics, the 500-page *Perspectiva*. Kepler owned a copy of this book, writing his own addendum to it in 1604, and Rheticus (1514–1574) presented a copy to Copernicus. Rheticus' version had been published by Johannes Petreius (c. 1497–1550) in 1535. As a result of this, Petreius was selected as the publisher of *De Revolutionibus*. Other readers of Alhazen's *De Aspectibus* and Vitello's *Perspectiva* were Tycho Brahe, Galileo and Descartes. By 1572, a volume containing both works was published by the mathematician Friedrich Risner (c. 1533–1580) under the title *Opticae Thesaurus*.

Second, Abu Jafar Muhammad ibn Muhammad ibn al-Hasan Nasir al-Din al-Tusi (b. Tusi 1201, died in Baghdad in 1274) was the "founding director" of the Maragha Observatory and the only one of this group whose name we give in full. He is generally called al-Tusi for his hometown. Ibn means son of, and some of the more spectacular names include four of these, typically more than a century of genealogy. He showed that you can combine two uniform circular motions to get a straight line, by following a point on the circumference of a circle rolling inside a circle of twice its diameter. This "Tusi-couple" appears in the more mathematical, Part II, of Copernicus's *De Revolutionibus*, though not under that name. With both circular arcs and straight-line portions, you can approximate most shapes.[20] Incidentally, some authors consider that al-Tusi was a precursor of Charles Darwin's ideas on evolution (almost 600 years before the birth of the British naturalist).[21]

Third, meanwhile, back in Toledo and Cordoba, Zarqali (c. 1029–1100, with only a single Ibn to his name, but a student of Sa'id al-Andalusi, who had four) devised structures to separate four motions of the Sun: trepidation (variation of the precession of the equinoxes, a motion that does not exist),[22] variation in the obliquity of the ecliptic (which really does oscillate), motion of the solar apogee at about 1° every 279 years, and variable eccentricity of the apparent solar orbit around the Earth. That his system was complicated is perfectly true, but most of these motions actually occur, though we now describe them as changes in the eccentricity, orientation and inclination of the Earth's orbit relative to its rotation axis. Thus, a sun-centered model cannot be much simpler, nor were these Copernican models free of divine intervention: the planetary orbits were powered by connection to the outermost one, which was powered by God.

Astronomical knowledge and practice never quite died out in medieval trans-Pyraneean Europe, but the practitioners were certainly aided and abetted by leakage of people, ideas and writing from Spain as the Christian monarchs gradually retook the Iberian peninsula from the 11th century down to 1492, with additional inputs directly from Byzantium and through Sicily. The distinguished 14th century French Philosopher/Mathematician Nicole Oresme (c. 1320/25–1382), the first person we know of to use coordinates in geometry, wrote about the Tusi Couple in his *Quaestiones de Spera* some time before 1362.[23] His version can best be described as confusing, but it is clear that he had access to a — possibly equally poor — Latin translation of the original Arabic.

The main point to be made here is that virtually all the structures needed in Ptolemaic or Tusic models of the cosmos are also needed in a heliocentric model. Nonetheless, elementary texts and authors, who should know better, often depict only the Copernican universe of "Part I" of *De Revolutionibus* — seven perfect circles around the Sun plus one for the Moon (see Figure 5.8). This is enough to produce retrograde motion but nothing of the other variations of longitudinal and latitudinal motions of the Moon and planets.

An important question in this chapter is how much, if any, of the methods of Muslim astronomy did Copernicus know about. Some historians of science are sure that the Muslim work in Latin translations was known to Copernicus (who was in Italy from 1496–1503) and was essential to his (almost!) heliocentric model. Other authors think that Copernicus's discovery was independent. It is clear however that Copernicus used the Tusi couple in both *De Revolutionibus*

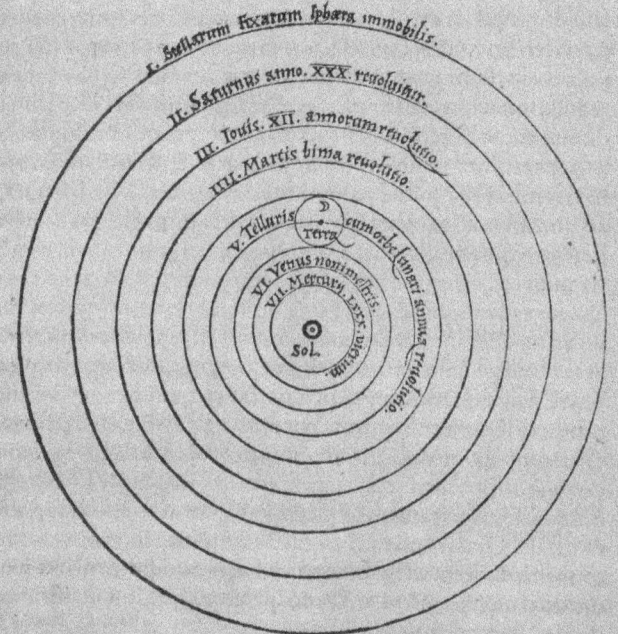

NICOLAI COPERNICI

net,in quo terram cum orbe lunari tanquam epicyclo contineri diximus. Quinto loco Venus nono menſe reducitur.; Sextum deniꝗ locum Mercurius tenet,octuaginta dierum ſpacio circũ currens,In medio uero omnium reſidet Sol. Quis enim in hoc

pulcherrimo templolampadem hanc in alio uel meliori loco po neret,quàm unde totum ſimul poſsit illuminare? Siquidem non inepte quidam lucernam mundi,alij mentem, alij rectorem uo= cant. Trimegiſtus uiſibilem Deum,Sophoclis Electra intuentẽ omnia.Ita profecto tanquam in folio re gali Sol reſidens circum agentem gubernat Aſtrorum familiam. Tellus quoꝗ minime fraudatur lunari miniſterio, ſed ut Ariſtoteles de animalibus ait,maximã Luna cũ terra cognationẽ habet.Concipit interea à Sole terra, & impregnatur annuo partu. Inuenimus igitur ſub hac

Figure 5.8 A page from *De Revolutionibus Orbium Coelestium* Part I (1543), with a diagram of the Copernican heliocentric system. Image courtesy of History of Science Collections, University of Oklahoma Libraries.

and in the less well known but earlier treatise *Commentariolus* to deal with variable precession and obliquity, as well as with the motion of the Moon and of the inner planets.[24]

Nor, of course, was Copernicus the first to put some things in orbits not around the Earth. The original manuscript of *De Revolutionibus* mentioned both Aristarchus (who put the Sun at the center) and Philolaus (with his central fire), though the published version omits these. The *Commentariolus* version of the motion of the Moon is that of Ibn al-Shatir (1305–1375) who used secondary epicycles rather than eccentrics and deferents to achieve uniform circular motion. The combination of structures he used to describe and predict the motion of Mercury in the sky is the same as that employed by Copernicus about 150 years later.

As the historian of science Stephen Blake[25] states, "It was the astronomers and mathematicians of the Islamic world who provided the theories and concepts that paved the way from the geocentric theories of Claudius Ptolemy in the 2nd century CE to the heliocentric breakthroughs of Nicolaus Copernicus and Johannes Kepler in the 16th and 17th centuries".

Martianus Capella (c. 365–440), who, like Tycho Brahe 1,200 years later, put Mercury and Venus in orbit around the Sun, but the Sun and the rest of the planets around the Earth, was also mentioned by Copernicus.

The transition is nicely bracketed by Georg Peurbach and Erasmus Reinhold (1511–1553) who, on the one hand, wrote a commentary on the work of Peurbach and, on the other, used the Copernican model to calculate the Prutenic Tables of planetary motion without actually endorsing[26] a heliocentric cosmos but rather emphasizing the importance of uniform circular motion and the improved ease of calculating things if you pretend the Sun is at the center.

Chapter 6

The End of the Heaven's Immutability

> *Do not look at stars as bright spots only. Try to take in the vastness of the universe.*
>
> *Maria Mitchell*

By the end of the 16th century, much scientific research was funded by private benefactors, as was the case for Kepler and Galileo. The drive for better planetary predictions and an inherent interest in astronomy was what drove King Frederick II of Denmark to fund Tycho Brahe's "Uraniborg" observatory during the period 1580–1597. During that period, Tycho designed unique equipment with which to measure accurate positions of stars and planets. It is said that the sum of money expended was a substantial fraction of the Danish GDP at the time, but this was surely one of the greatest scientific experiments of history. The most important outcome of the work was the production, in Prague, of the "Rudolphine Tables" which Kepler later used to establish the laws of planetary motion. Using Tycho's accurate measurements, Kepler was able to discover a 1/6th-degree discrepancy between the predictions of the Ptolemaic model and the observed positions of the planets, especially Mars. This marked the beginning of the end of some 1,400 years of use of the Ptolemaic system.

6.1　On the Shoulders of Giants

When the Great Comet of 1577 was still crossing the sky, on the 27 December, Johannes Kepler was celebrating his sixth birthday. His mother took him to the top of a hill to see the arched shape of the comet's impressive tail. People in Europe probably were talking about the bad omens linked to celestial body. After several years working as a teacher of mathematics and astronomy at the Protestant school in Graz (Austria), in 1600, the young Kepler that became as an assistant to Tycho, who was by then living in Prague as an imperial astronomer to King Rudolph II. As Kepler wrote, "Just one of his astronomy instruments alone is worth more than all my family's possessions". On Tycho's death in 1601, Kepler inherited his fabulous astronomical data, enabling him to determine the three laws describing planetary motions. The observations on the positions of the planets that Tycho Brahe had painstakingly recorded for 38 years were a unique treasure that fell into exactly the right hands, those of an astronomer like Kepler, who was a highly talented mathematician. Kepler began poring almost reverently over Tycho's manuscripts, the testimony of many nights of observation.

In many aspects of astronomy, it is crucial to build up observational records over long periods of time to be able to interpret the data correctly and obtain convincing results. Tycho hoped his observations would serve to confirm his own world system, and this was his plea to his assistant as he lay on his deathbed, repeating over and over again "Let me not have lived in vain". His request was honored but not quite as he had hoped. His data led Kepler, in the words of the historian of science Alexander Koyré (1892–1964), "to break the spell of circularity", as a discrepancy of only eight arc minutes in the position of Mars meant that Kepler opted for elliptical instead of circular orbits, in Copernicus's heliocentric framework. This was a fantastic testimony to his tenacity and honesty in scientific practice. It would have been much more convenient to think that the small discrepancy was due to Tycho's observational errors and be content with making a merely adequate fit with the Copernican model. But Kepler was certain of the extraordinary quality of the astronomical data in his possession, and his mind was sufficiently open to enable him, following laborious work over 8 years, to publish in *Astronomia Nova* that planets really had elliptical rather than circular orbits around the Sun and that the Sun is in one of the foci of the ellipse. This is Kepler's first law. The same book contains the second law in which Kepler states that the speed of the planets throughout their trajectories is greater when they are closer to the Sun (perihelion) than when they are further away (aphelion), while the areal velocity remains constant: the line that joins the Sun to the planet sweeps out equal areas in equal intervals of time.

Science historian Owen Gingerich states that Galileo Galilei,[1] "after being only a timid Copernican, was converted by his own astronomical findings, using the telescope, into a flaming Copernican". Galileo did not invent the telescope nor was he the first to build one. It appears to have been a Dutch invention. When Galileo learned about its lens-based construction in 1609, he was able to build one in a few months.

In 1610, Kepler via the Tuscan ambassador in Prague, received the book[2] in which Galileo Galilei described the discoveries he had made using his telescopes. The story was as follows. On 7 January 1610, Galileo, who had considerably improved his telescope (now he had one with 20x magnification and a tripod that held it up), observed the planet Jupiter and was surprised to see three stars next to it. Small but luminous, they were, curiously, aligned, two on the eastern side and one on the western side, along a straight line parallel to the ecliptic (the name given to the apparent trajectory of the Sun on the celestial sphere throughout the year). The next day, the three stars were found to be on the western side of the planet. Galileo believed that they were fixed stars and that perhaps Jupiter was experiencing retrograde motion. On 9 January, Galileo waited patiently for nightfall to make his observation. He was intrigued and eager for an answer. But that night was cloudy. The next day, 10 January, only two of his "stars" could be seen on the eastern side, and Galileo, "his perplexity turning into admiration", realized that the third was hidden by the planet. It was not a movement of Jupiter that produced different arrangements. The stars, rather than being fixed, were satellites of Jupiter. And there were not three but four of them, something he observed on the nights of 13 and 15 January (the night of 14 January was also cloudy).

We might not be able to pinpoint the moment when Galileo became aware that they were satellites orbiting Jupiter, but we can be sure that by 15 January he was totally convinced of the importance of his discovery. From that day on, the entries in his diary go from being written in Italian to being written in Latin, the language that was then used for scientific communication. Galileo collectively named these satellites "Medicean planets" in honor of the Grand Duke of Tuscany, Cosimo de Medici II. After the publication of *Sidereus Nuncius*, Galileo painstakingly observed the satellite system, trying to determine the periods of revolution around the planet and the orbital radii. His results were published in 1612 in his *Discourse on Floating Bodies*. The values he stated are very close to being correct.

These four moons of Jupiter, Ganymede, Io, Callisto and Europa, which Galileo was the first to recognize, enabled both Kepler and Galileo to reiterate their certainty about the accuracy of the Copernican system, which said that the moons revolve around Jupiter the way the planets revolve around the Sun. Galileo had also used the telescope to observe the phases of the planet Venus, similar to those of the Moon, and he concluded — rightly — that these phases can be observed in Copernicus's heliocentric model but not in Ptolemy's geocentric system. Kepler pointed out that the phases of Venus could also be observed with the planetary arrangement put forward in the Tychonic system, as they are the result of the relative movements of the Sun, Venus and Earth (which are similar in the Tychonic and Copernican systems). Emmanuel Kant was justified in referring to Kepler as "the most acute thinker ever born". This acuteness culminated in 1619 with the publication of *Harmonices Mundi* (*The Harmony of the World*) in which Kepler discusses the laws that govern the movement of the planets. The third law relates the orbits of the planets to each other, by stating that for each, the square of the orbital period of the planet is proportional to the cube of the semi-major axis of its orbit. Armed with this law, we can calculate the mass of Jupiter by simply observing the orbital period of its moons and the distance that separates them from the planet, or, in the same way, measure the mass of a spiral galaxy from the speed of a star near the edge of its disk to complete a full circuit around the center of the galaxy. Kepler finished the Rudolphine Tables, the astronomy tables that Tycho Brahe had started years earlier. He presented them at the Frankfurt book fair in 1627. More than half the volume is devoted to teaching the reader how to use a set of tables that could predict astronomical events better than any other before its time. For example, even today applied to the planet Mars, they would enable the correct current position of the planet to be predicted with an error margin of only three arc minutes.

Galileo was the first to observe across the sky with a telescope, discovering sunspots and that the Sun spins on its axis. He was the first to spot something around Saturn, although it was a Dutch optical expert Christian Huygens who, years later, would correctly interpret the features as rings. Galileo observed the lunar landscape and concluded that the Moon's surface was not an immaculate sphere but that instead it was pitted and pockmarked. When he pointed his spyglass at the Milky Way, he found that the instrument allowed him to see an enormous number of individual stars that could not be seen with the naked eye. Galileo was also a great publicist for science. His book published in 1632 *Dialogo Sopra i Due Massimi Sistemi del Mondo Tolemaico, e Coperniciano* (*Dialogue Concerning the Two Chief World Systems Ptolemaican and Copernican*), written in vernacular Italian, was the first work on cosmology that could be described as a quality scientific publication for a general reader. Galileo's original writing style used approachable, captivating language to present his ideas on the systems of the world as a dialog between Salviati, Sagredo and Simplicio.[3] Sagredo is an educated man keen to grasp reality and understand the mysteries of the cosmos; Simplicio is a follower of Aristotle who bases his answers only on the authority of the eminent Greek philosopher, without discussing either reasons or evidence; and Salviati represents the ingenuity of the scientist, Galileo himself. Salviati's wit is so sharp that he manages to find more faults with the Copernican ideas than Simplicio, subsequently rebutting them in a masterly fashion by employing even more convincing arguments.

When Newton stated in 1687 that he had "stood on the shoulders of giants", he probably meant the ancients, but also he knew that the publication of *Philosophiae Naturalis Principia Mathematica* (*Mathematical Principles of Natural Philosophy*) was the culmination of a scientific revolution started by Copernicus 144 years earlier and that, in the intervening century and a half, leading figures such as Tycho Brahe, Kepler, Galileo and Robert Hooke (1635–1703) made decisive contributions to the birth of modern astronomy and physics.

When he formulated the law of universal gravitation, Newton put a definitive end to the Aristotelian distinction between the sublunar and supralunar regions, stating that things on Earth are governed by the same physical laws that describe the movements of celestial bodies. Using this law and Newton's second law of motion, it is possible to derive the three phenomenological laws that Kepler, after many years of hard work, deduced from Tycho Brahe's observations.

Kepler had a telescope for only a few days on loan from the Duke of Bavaria. But despite this, he wrote a treatise entitled *Dioptrice*, in which he described the characteristics of the refracting telescope, which uses a lens as its objective (the optical element responsible for capturing light). In his book *Opticks*, Newton presented a new type of telescope, the reflecting telescope, whose objective was a concave primary mirror that was hit by light from the stars; the light converged on the focal plane after the reflection by a secondary mirror whose job was to redirect the light so that the focal plane fell outside the telescope tube and stayed outside the path of the incident light.

In subsequent years, telescopes became increasingly widely used and some 18th century astronomers like William Herschel (1738–1822) started developing an interest in astronomy by building their own. Many European observatories were built during this period, and some are still in operation today. Soon, scientific societies devoted to astronomy started to appear, along with enthusiasts indulging their passion for the night sky, all wanting to acquire telescopes for their own use; as we will see in this chapter, in some cases, these people would find themselves at the center of major astronomical discoveries.

6.2   Sunspots

In the early 17th century, sunspots constituted a primary challenge to the Church and Aristotelian doctrine of the incorruptibility of the heavens. The largest groups are naked-eye objects under suitable (foggy, dusty, etc.) conditions and were recorded by Chinese, Arab and medieval European astronomers and quite often attributed to transits of Mercury, Venus or previously unknown planets.[4] Ascribing to them shapes of astrological significance that cannot actually be seen without magnification was also fairly common.

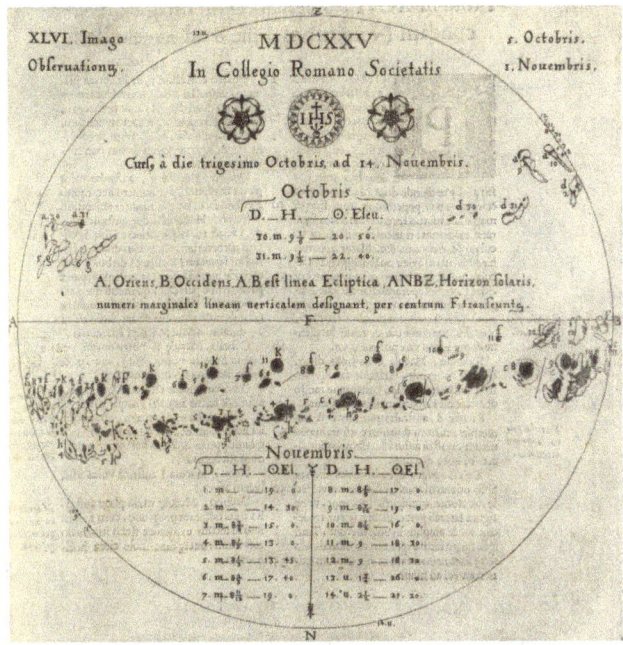

Figure 6.1 Sunspots depicted by Christoph Scheiner in *Rosa Ursina*. Credit: Public domain. Wikipedia Commons.

Then came telescopes, and four European astronomers independently began observing sunspots from 1610 onward. The English astronomer Thomas Harriott (c. 1560–1621), in December 1610, and the Frisian/German astronomer Johannes Fabricius (1587–1616), in October 1611, were the first to see and to publish respectively, but it was the Jesuit Christoph Scheiner (1573–1650) whose January 1612 publication,[5] under the pseudonym Apelles to avoid problems with the Church, firmly advocated transits by previously unknown planets (see Figure 6.1). Finally, the fourth Galileo, handed this opportunity to promote anti-Aristotelianism (and so pro-Copernicanism), settled down to systematic observations and interpretation, thereby showing, in a 1613 publication,[6] that the spots were part of the solar surface and they therefore demonstrated the rotation of the Sun. Critical items were the slower rate of progress across the solar disk and distorted spot shapes close to the limbs of the Sun in comparison with spot groups as they crossed the center of the disk. He suggested very tentatively that the spots might have the nature of clouds and pointed out that the circles of spot motion, and so presumably the solar equator, were roughly parallel to the ecliptic.

Modest spotting and staining of the Sun having been accepted, later conceptions of the cosmos required attributing the spots to some mechanism consistent with the later models. René Descartes, for instance, required exactly three kinds of matter: one to make up the Sun and stars, a second to make up the heavens and a third for the Earth and planets, known by his time (c. 1644) not to be self-luminous. The spots were then bits of type three, like scum on boiling liquid, atop the mostly type-one sun. And he supposed that transient very large spot areas might account for stellar variability and even, when large areas cleared after a long time, for novae like the Tycho and Kepler events. We have not come across any peculiarly Newtonian view of sunspots; he seems to have concentrated on how one might measure the Sun's mass and density.

In contrast, William Herschel wrote on many occasions about the nature of the Sun and sunspots. He felt firmly that all the "planets" must be inhabited. This same view surely contributed to his frequent reports of active volcanoes, land and water, and artificial structures on the Moon. Kepler had also generally favored other celestial objects being as much like Earth as possible. Herschel thought the Sun must be mostly solid in order to be able to exert gravitational forces. Sunspots, therefore, were temporary holes in the shining atmosphere, enabling us to see down to the solid, dark, cooler, inhabited surface. The "lucid fluid" incidentally was the product of chemical reactions of a sort that reveal Herschel to have been a late supporter of phlogiston (see Chapter 4). Openings down to a solid surface were still the "best bet" model in 1859 when the English amateur astronomer Richard Carrington (1826–1875) recognized differential rotation[7] and variation of spot latitudes through the 11 years cycles. Curiously, Herschel's interpretation of sunspots led him to the conclusion that the Sun should be brightest when most spotty. This is, of course, true (because extra emission from plages trumps light missing from the spots), though not enough to affect the price of wheat.

Other natural philosophers, less well known than Descartes and Herschel, also put forward interpretations of sunspots that postulated opaque bits in a fluid mass that rises and falls or holes in a luminous layer through which one sees down.

6.3   Variable Stars

Faced with a new astronomical observation that in principle may seem paradoxical, it is reasonable to posit the existence of entities that, although not initially detectable, could provide a convincing explanation for the observations.

The "novae stellae" introduced in the previous chapter and associated with the names of Tycho (1572) and Kepler (1604) had, in principle, established the nonimmutability of the translunar heavens. But the idea of "no more variability than absolutely necessary" and the invention of some unobservable/unobserved items — dragons — to defend it seem to have persisted for many years.

In their wake, a good many novae seem to have been announced and, sometimes, retracted. But even very late, popular mechanisms appear to have been designed to minimize the amount of real change involved. In particular, the German astronomer Hugo von Seeliger (1849–1924) in 1892 proposed that novae were white dwarfs wandering into previously existing nebulae. Even in the 20th century, the astronomer Harlow Shapley was still endorsing the idea at the time of the "Great Debate" with Heber D. Curtis (1920), a story that will be told in Chapter 15. It is perhaps also significant that, in the intervening centuries, very little attention was paid to quantifying or explaining variability.

The French astronomer and mathematician Ismaël Boulliau (1605–1694) established the approximate 11-month period of the star Mira (Omicron Ceti) which had been previously reported as two different novae by the Dutch astronomers Johannes Fabricius (1587–1615) in 1596 and Phocylides Holwarda (1618–1651) in 1638. Boulliau suggested that, like the Sun, it rotated (but slower) and had spots (but many more), so that no real change was required. But the first periodic variables really belong to two English amateur astronomers John Goodricke (1764–1786) and Edward Pigott (1735–1825). They systematically studied the stars that nowadays we describe as "variable", the brightness of which periodically changes with the passage of time; they initially suggested that the variations in brightness were a consequence of the passage of dark stars (or unenlightened stars) in front of them.

The second brightest star in the constellation of Perseus is Algol. The name derives from the Arab term *ra's al-ghūl*: "head of the ghoul". It probably got its name from its strange behavior, as it doesn't always shine with the same brightness. In an era when the heavens were thought to be unchanging, no one could understand why a star's brightness would change like this. In fact, it turned out to be a variable star. Every three days, Algol would suddenly become dim by a factor 3.3 and it would stay dimmed for 10 hours before recovering its original brightness. Nobody had been able to come up with a satisfactory explanation for the way this demon star behaved.

In 1782, Goodricke, who had been deaf from the age of 5 due to scarlet fever, studied the star Algol for many nights and found that the variations in its brightness were extremely regular, occurring every 68 hours and 49 minutes. To explain this phenomenon, he put forward a hypothesis that was revolutionary for its time, suggesting that Algol had a dark companion orbiting around it (see Figure 6.2). He postulated the existence of an unobservable item to explain the observations. Algol's brightness suddenly dimmed when this other celestial body passed between it and the Earth, eclipsing most of its light. He had just discovered the first eclipsing binary star.

It is interesting to note that this argument was advanced when pairs of stars that are gravitationally bound to form binary systems were not yet known. Goodricke and Pigott had the good luck to start with Algol (Beta Persei) and catch it in multiple fadings, from 2nd magnitude[8] to 4th. Their first suggestion was that it was being partially eclipsed by an "unenlightened star", put forward in papers by them separately in 1784 and 1785 with Goodricke first.

The second star Goodricke followed was Beta Lyrae, for whose 12-day period occultation was also a plausible fit. But the third was Delta Cephei, whose asymmetric light curve[9] cannot be matched by a uniform-brightness sphere being eclipsed by a dark sphere. The variability of Delta Cephei, the fourth brightest star in the constellation of Cepheus, is therefore more complicated. In fact, it is another variable star, but in this case, its regular changes in brightness are due to pulsations inside the star that make its outer layers expand and contract rhythmically, like a heartbeat, but at intervals of 5 days, 8 hours and 47 minutes. When the light curve of Delta Cephei defied explanation by eclipses, Goodricke and Pigott retreated to the rotating, spotted model of Boulliau.

The young Goodricke was made a member of the Royal Society for his discoveries. Sadly, only 14 days later (and 5 months short

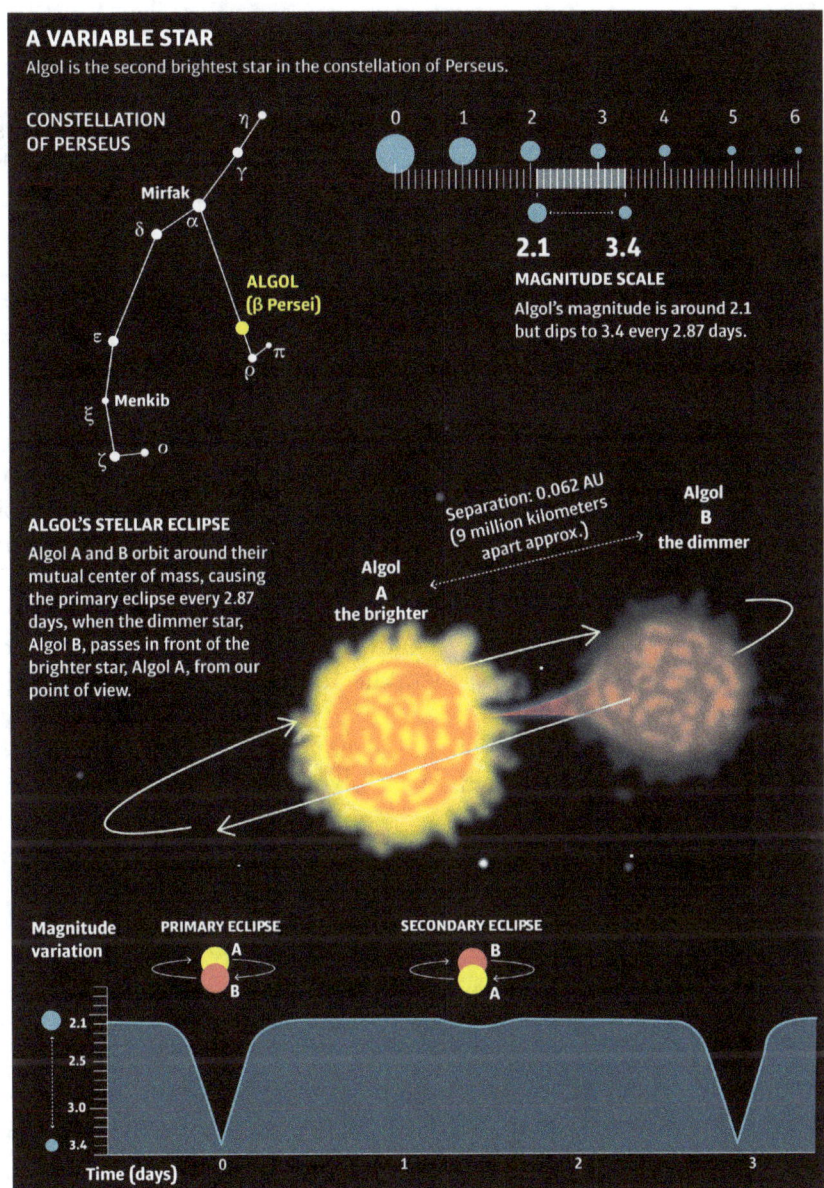

Figure 6.2 The variability of Algol, the Demon Star. Credit: Infographic sketched by the authors and elaborated by Javier Pérez Belmonte.

of his 22nd birthday), on 20 April 1786, John Goodricke died from pneumonia.

In 1805, Pigott went on to publish a catalog of 50 variable stars and also to suggest that unenlightened objects in large numbers might account for phenomena like the Coal Sack (connecting up with the idea of interstellar absorption). He was under French detention at the time, which is the sort of thing likely to make people think of darkness as palpable!

Although the variability of many of these stars is certainly due to internal pulsations, as Sir Arthur S. Eddington (1882–1944) would demonstrate in 1926, the unenlightened stars that overshadow others stopped being mere hypotheses when, by making use of spectroscopy, precise measurements of the radial velocities of stars were carried out and it became possible to verify that there were eclipsing binary stars. The Sun, of course, also flares, and in an 1885 discussion of variable stars, the Canadian–American astronomer Simon Newcomb (1835–1909) suggested that this is another "like the sun only more so" mechanism for nonperiodic changes. Goodricke's mechanism (put forward when no pairs of stars were known to be gravitationally bound binaries) resurfaced in 1889 when the director of the Astrophysical Observatory in Potsdam, Hermann Vogel, determined the first radial velocity curve for Algol. Large numbers of these soon came from observers at Harvard and Lick observatories. Newcomb then added eclipses to his inventory of variability mechanisms by 1902. If you think of the Sun being slightly dimmed for a distant observer during a transit of Jupiter, this is yet another "like the sun, only more so" mechanism.[10]

Notice that none of these — eclipses, spotted rotation or flares — will do for Delta Cephei. The pulsational mechanism was put forward first by the English astronomer Henry Crozier Plummer (1875–1946) in 1912 and Harlow Shapley in 1914 and further developed by Eddington in 1926. All four processes, of course, occur, sometimes for the same star, including (if you will allow us helioseismology) the Sun. But pulsation is surely the furthest away from any immutability of the heavens, at least until we get to the nuclear explosions and gravitational collapses of supernovae well into the 20th century.

To summarize from the point of view of undetected/undetectable entities, "unenlightened stars" were at the beginning purely hypothetical and counted as observed with the development of radial velocity spectroscopy and spectroscopic binary orbits. Spots on the stars other than the Sun have become observable only quite recently, with the discovery of radial velocity tomography, wherein small changes in line profiles are seen in spectra of stars passing one behind the other in eclipsing binaries. Flares on stars other than the Sun are not known to arise from magnetically active regions. But it would be remarkable if they did not. The pulsation theory was born with the radial velocity variations in place that we would regard as confirmation of it if they had not already been known. A few defenders of binary orbit velocities for the Cepheids held their ground into the 1920s, though Eddington, for instance, showed that one star would have to be inside the other.

Also in the 1920s, the English physicist and astronomer James Jeans (1877–1946) attempted to make some variable stars (including the Cepheids and RV Tauri stars[11]) part of a sequence driven by rotational spin-up, until the shapes became very distorted and the stars eventually broke into close binaries. The rotation speeds needed probably exceed those allowed by spectral line widths in most of the stars. Rapidly rotating stars do, however, exist.

6.4   Defending Newton

It is essential that scientific ideas be tested and questioned: most often that is done through hard work and criticism by little known scientists, and through them we gain deeper understanding. The revolutionary ideas must be consistent, not only with what was known prior to their publication but also with new data gathered subsequently. This is an important principle for all of modern science: there is nothing that cannot be criticized, and everything must be tested against independent data sets.

Without a doubt, the largest set of astronomical entities posited to support a particular conception of the cosmos came about as a result of the discovery of the law of gravity by Newton.

That set includes Neptune, dark companions of the stars Sirius and Procyon, Vulcan and other items associated with the advance of the perihelion of Mercury, Pluto, Nemesis, Planet X and shepherd satellites in the rings of Saturn. Notice that some, not all, of these turned out to exist, more or less as postulated. The first "Newtonian defense" actually preceded Newton. This was Ole Römer's 1675 demonstration of the finite speed of light, if the moons of Jupiter were to follow Kepler's laws (see Chapter 7). Observed immersions could come almost half an hour early or late relative to the orbital elements published by the eldest (Gian) Cassini in 1668, with a specific prediction of 10 minutes late for the 9 November 1675 event calculated by Römer.

It was indeed late. Cassini was apparently not pleased and did not update his orbital elements until 1698. Newton's Principia (1687) came in between, and Cassini remained a firm opponent of Newtonian gravitation. Indeed, he was not even entirely a Copernican. Precise predictions and observations of the Jovian satellite eclipses were taken seriously in their day because of (successful) applications to determining precise longitude on land and unsuccessful ones at sea. Does 10 clock minutes matter? Oh yes. It is 2.5° of longitude or 150 miles at the equator, quite enough for you to run aground overnight.

The most spectacular example was the discovery of Neptune in 1846. Years earlier, in 1781, William Herschel and his sister Caroline (1750–1848) had discovered, somewhat by chance, the planet Uranus. The deviations from Keplerian motion observed in Uranus's orbit led to the existence of an eighth planet being posited. Urbain Le Verrier (1811–1877) of the Paris Observatory and, independently, the Englishman John Couch Adams (1819–1892) would conclude, using Newtonian laws, that this anomaly was due to the gravitational pull produced by another, more distant planet. They even calculated where and when it should be observable, based on Le Verrier's calculation, an astronomer at the Berlin Observatory, Johann Galle (1812–1910), who had the right telescope and star charts, discovered Neptune in 1846. Many factors contributed to the discovery: the observations of the anomalies of Uranus, the sharp minds of Le Verrier and Adams, their imagination, the mathematics necessary to perform the calculations and technology. It is also true that chance lent a helping hand. Just a few years before, Uranus had overtaken Neptune as the two turned around the Sun. This proximity was what made the anomalies in Uranus's orbit appreciable. If the two planets had been on opposite sides of the Sun, no one would have noticed anything, and this discovery would have had to wait.

Similarly, the German astronomer and mathematician Friedrich W. Bessel (1784–1846) stated that the movements of two of the brightest stars, Sirius and Procyon, implied that each should have a dark companion of a mass comparable to that of those stars. As Bessel died in 1846, he did not have the chance to say, "Just like I told you!" when the American astronomer and telescope maker Alvan Clark (1804–1887) discovered Sirius B in 1862 and the German–American astronomer John M. Schaeberle (1853–1924) identified Procyon B in 1895.

The successes had been followed by some failures to detect what had initially been predicted. For instance, in 1859, Urbain Le Verrier recognized an anomaly in the orbit of Mercury. Its perihelion (the point in an orbit closest to the Sun) undergoes a secular advance (precession), mostly due to the gravitational force of Venus and an unfortunate choice of coordinates.

But the measured value was greater than expected with Newtonian gravity by about 40 arc seconds per century. Today, we know from Einstein's theory of gravity (general relativity) that there is a natural explanation for the small difference, but to justify the excess, in the 19th century, different entities — ones that were obviously never detected — were proposed: for example, a planet inside Mercury's orbit that was named Vulcan, a ring of asteroids that were also intramercurial, a sufficiently flattened sun, an ethereal impulse and deviations from Newton's law of gravity in the vicinity of the Sun as Newcomb proposed in 1906. The American astronomer Asaph Hall (1829–1927) in 1894, for instance, suggested $1/r^x$ law with x = 2.00000016. All these postulated "dragons" were slayed for those who accepted Einstein's general relativity in 1917.

You have perhaps been used to thinking of Clyde Tombaugh's (1906–1997) 1930 discovery of Pluto as part of a series of successes, but (a) he found the new wanderer well away from the part of the sky the Bostonian businessman and amateur astronomer Percival Lowell (1855–1916) had predicted, (b) it would not have been massive enough, even if in the right place, to account for the deviations in motion of Uranus and Neptune that Lowell and others were trying to account for, and (c) those deviations turned out to have a large contribution from observational error so that there is no longer thought be the need for some more massive, more distant planet.

Incidentally, W. Pickering at various times thought he had seen evidence for no fewer than seven trans-Neptunian planets. None was confirmed, but Tombaugh was perhaps encouraged in his search. You have perhaps never noticed that discovering planets seems to be good for people. William Herschel died at 83, Tombaugh at 90 and Galle at 98!

Nemesis was another hypothetical, distant planet, with an orbit period near 30 million years (at about 10^5 times the distance from the Earth to the Sun) and large eccentricity, whose job was supposed to be to pass through the Oort cloud[12] and send many new comets inward to account for periodicities in the cratering record of Earth and in major extinction episodes (see Chapter 10). Again, in the end, there has been no observational confirmation, though a few are still looking. As University of Chicago dynamicist Peter Vandervoort said on a visit to the University of Maryland many years ago, it is an interesting orbit, there just does not happen to be anything in it (nor does purely Newtonian gravity require there to be).

Shepherd satellites clearing out narrow gaps in the rings of Saturn have been seen often enough that observers will postulate them for newly discovered gaps. If the gaps are purely gravitational products, then such satellites have the same status as Neptune and Sirius B before they were imaged.

A final item seems to belong here. It is also definitely Newtonian, but in addition, the people concerned were committed to pictures of stellar dynamics of the Milky Way that turned out not to correspond to reality. First was a very massive central object, which would support the rotation of the galactic disk analogous to planets orbiting the Sun. The Swiss–German mathematician, astronomer and physicist Johann Lambert (1728–1777) thought of this in 1749 and finally published it in 1761. The German astronomer Johann von Maedler (1794–1874) put forward the same idea, using proper motions from a combination of star positions measured by the third Astronomer Royal James Bradley and his own measured at Dorpat (Tartu Observatory in Estonia). He concluded that the center of the stellar system was in the Pleiades cluster.

Using the same database, Marian Kovalsky (1821–1884) refuted Maedler's theory of the central super-sun and, in turn, proposed solid body rotation (for which velocities would increase linearly from the center).

James Bradley also played an important role in the measurement of the speed of light. The story of this important measurement is told at the beginning of the next chapter.

Chapter 7

Tripping the Light Fantastic

Com, and trip it as ye go,
On the light fantastic toe.

John Milton, L'Allegro, 1645

From an everyday point of view, the speed of light seems infinite. So, it is not surprising to hear Aristotle arguing that the speed of light must be infinite. However, Heron of Alexandria (ca. 10–70 CE) argued that nothing could be in two places at the same time, from which he concluded that the speed of light must be finite. That point of view went unheeded for a long time, and we find, as late as the early 17th century, both Kepler and Descartes arguing for an infinite speed of light.

The astronomer Giovanni Cassini (1625–1712) spent much of his life researching the motions of the four Galilean satellites of Jupiter and was the first to report anomalies in the timing of their occultations by the main planet. In particular, he found that the anomaly of the timing of the occultations for the satellite Io, as we now call it, depended systematically on the distance of Jupiter from the Earth. At first, he suggested that this might be explained if the velocity of light was finite but soon after retracted his opinion. No alternative solution to the phenomenon was offered at the time.

7.1   The First Measurements of the Speed of Light

The Danish astronomer Ole Rømer who had been at the Paris Observatory since 1671 took on this problem. Analysis of his own data and that of Cassini led him to predict, in September 1676, that an eclipse of Io would take place on 9 November of that year and that the difference in times between the predictions assuming an infinite speed of light versus a finite speed of light would be around 10 minutes. This was roughly what was observed, but calculating what value of the speed of light this implied required a good knowledge of the then poorly known distance of the Earth from the Sun. That distance was to be measured in any of a variety of units: leagues were commonly used at that time. There are diverse opinions about what value for the Earth–Sun distance Rømer had used, but the oft-quoted number he derived for the speed of light was an amazing 220,000 kilometers per second. But it is not the actual value that is important so much as the fact that the value was finite.

Nonetheless, although Rømer published his work in the widely read *Philosophical Transactions of the Royal Society of London*, he was not widely believed. Even Cassini did not change his mind back to his original suggestion. That's how strongly the Aristotelian belief in an infinite speed of light was held, even by the experts. The one expert who supported Rømer at the time was the great Dutch scientist Christiaan Huygens. The finiteness of the speed of light was only finally accepted when, on 1 January 1729, James Bradley determined that speed from his work on the apparent motions of the stars throughout the year.

Bradley had been attempting measurements of parallax on the positions of the stars that were due to the Earth's motion around the Sun. The parallax would tell him the distance of a star, but there was an anomaly in the pattern of displacement of the star he was studying, Eltanin, the brightest star in the constellation Draco, which Bradley attributed to an important phenomenon that became known as "aberration".[1] From that, he derived a value of 301,000 kilometers per second for the speed of light, which is very close to the currently adopted value of 299,792.458 kilometers per second. Not only did Bradley's work solve a problem that had been around since ancient times but also proved that the Earth is not stationary but moves around the Sun.

The first terrestrial measurement of the speed of light was made by Armand Fizeau (1819–1896) in 1849. His experiment was remarkably simple: he used two mirrors separated by a distance of 8 kilometers between which passed a beam of light. A wheel with teeth was imposed between the mirrors so that the teeth would interrupt the light path when the wheel was rotated into one position and pass the light when the tooth was rotated out of the way. If the wheel was rotated fast enough, the next tooth would get in the way of the return beam, and if it was rotated even faster, the light could be blocked by alternate teeth. So, at some appropriate speeds of rotation, no returning light would be seen. With this simple device, he obtained a value of 313,000 kilometers per second.

Later refinements, and both competition and collaboration with Leon Foucault (1819–1868), improved Fizeau's experimental accuracy, and in 1851, Fizeau was able to pass the light through two channels of moving water one approaching the source and the other receding at the same speed. With this, he showed that the speed of light was unaffected by the motion of the water. The motion of the water did not hinder nor help the passage of light: the speed of light was therefore a constant, independent of the motion of the medium in which it was moving but not independent of the nature of the medium (light is slower in glass or in water than in a vacuum or in thin air).

This had wider consequences for the luminiferous ether, the medium that was thought to pervade the vacuum and allow light waves to propagate. It was difficult for scientists of the time to conceive how light waves could propagate across empty space without there being some medium to support the wave motion. Sound is a wave of varying preasure in the air in which it propagates, so by analogy, it was thought that light had to be a movement of some invisible all-pervading medium.

Many people thereafter worked with experimental setups like Fizeau's, notably Albert Michelson and Edward Morley. As explained in Chapter 1, they used two Fizeau-style setups with a single light source to measure the difference in the speed of light in two perpendicular directions. Since the Earth was moving around the Sun, it should have been possible to detect this motion if the speed of the Earth boosted the velocity of light. Even more, the same effect would be detected if the Solar System were moving through the hypothetical luminiferous ether. This was, in effect, a search for an "ether wind". The search failed. The speed of light in perpendicular directions was independent of direction. These puzzles would not be solved until 1905 when Albert Einstein put forward his Theory of Special Relativity.

7.2   A Very Special Property of Light

Einstein's inspiration did not come out of nowhere. The result of the Michelson-Morley experiment, explained in Chapter 1, had stimulated many physicists to rethink the question of the propagation of light: what was so special that the speed of light was independent of the speed of the source? The Irish physicists Joseph Larmor (1857–1942) in 1897 and George Fitzgerald (1851–1901) in 1899 published papers putting forward a new radical explanation for the null result of the Michelson–Morley experiment. The Dutch physicist Hendrik Lorentz had proposed a similar idea in 1892. In essence, they produced a law for the addition of velocities that would be consistent with the velocity of light in a vacuum being the maximum attainable velocity.

They addressed the issue of the mathematical relationship between two observers in relative motion observing (with light, of course) the same event. At least one of the observers has to be moving relative to the event. Prior to the result of Michelson and Morley, we might have thought that the moving observer saw the event happening with a larger or smaller speed of light than the other observer would. If the additive law of velocities had worked for light, the moving observer would see the event happening earlier if moving towards the event than if moving away.

What these three scientists arrived at is commonly referred to as the Lorentz–Fitzgerald contraction. Consider two observers, one at rest relative to an event and the other moving towards the event. Simply put, they asserted that measured lengths for the moving observer would be shortened relative to what an observer at rest relative to the object being measured would see. So, even if the observers were at the same place, one moving relative to the event and the other not, they would find different numbers for size and duration of what they were measuring.

Perhaps even more extraordinary, not only would distances be measured differently but also intervals of time between two events occurring at the same place would look different to the two observers. This was the phenomenon known as time dilation. Here was the break with the Newtonian world: time as well as space was no longer absolute.

The relationship between what observers in relative motion would measure, when expressed mathematically, became known as the Lorentz Transformation.

If the speed of light is the same for all those who measure it, are there any differences in the light as seen by different observers? If a person moving along a railway track at half the speed of light shone a yellow light beam in the direction of motion, both that person and another person standing on the ground would measure the same speed for the beam: the "speed of light". The only observed difference between the two people would be that the person on the ground would see the light as blue, not yellow.

The apparent change in the frequency of the light when the separation between you and the source is changing was known as the "Doppler Effect", named after the Austrian mathematician and physicist Christian Doppler (1803–1853) who in 1842 provided an explanation for the change in the pitch of the sound heard by people standing on a railway platform as a train sounding its whistle passed through the station at speed. As the train passed, the pitch would suddenly drop. This analogy between light and sound inevitably raised the question that, if this happened because sound was traveling through the air, then what was the medium through which light traveled?

What Einstein did in 1905 was to bring all these effects into a single coherent framework embracing the whole of physics, including Newton's laws of motion. Inevitably, these ideas were not easily accepted, and the nonabsolute nature of time was particularly irksome. However, despite the early uncertainty about Einstein's 1905 ideas, some well-established physicists rapidly jumped on board to explore the new ideas, notably Lorentz and Larmor, who had first proposed the laws relating to what different observers would measure, and the German mathematician Hermann Minkowski (1864–1909), who in 1909 put special relativity into the mathematical form that we use now.

There is a beautiful visualization of the extraordinary consequences of such a theory in the little book *Mr Tompkins in Wonderland* written by the Russian-born American nuclear physicist and cosmologist George Gamow. This is a story of everyday life in a world where the speed of light is mere 4.5 meters per second (16 kilometers per hour or 10 miles per hour).

7.3   The *Annus Mirabilis*

In 1905, Einstein published five papers each of which would contribute to a significant change in physics as then understood. One of those was the Nobel-cited paper on the photoelectric effect: the theoretical idea that luminous energy consists of discrete packets called "quanta". This was a radical idea that was completely at variance with the long-accepted notion that light was a wave phenomenon. Accepting Einstein's idea that a light beam consisted of quanta might have made the constancy of the velocity of light and the Doppler Effect easier to come to terms with. Full acceptance of quanta of light only came with the experiments performed by the American physicist Arthur Holly Compton (1892–1962) in 1923, for which Compton received the 1927 Nobel Prize. So, yes, Einstein was given his Nobel Prize for an idea that had not at the time been experimentally verified and not for relativity!

The other four 1905 papers were on Brownian motion,[2] special relativity, the determination of molecular dimensions (his doctoral thesis) and the equivalence of mass and energy: the famous $E = mc^2$ equation. The article on special relativity formalized the then-controversial idea that no velocities could exceed the speed of light.

7.4   The Dark Night Sky

It was often the case that when encountering situations that challenged the *status quo*, scientists turned to some mysterious hitherto unseen agent as a possible explanation. Invoking a mysterious ether to explain the propagation of light is an example. Another is the so-called "Olbers' paradox" which can be simply rephrased as a question: "Why is the sky dark at night?"

In an infinite, homogeneous, static, mostly transparent, universe whose contents are infinitely old, and where there is no new physics, the night sky should be as bright as the surface of the Sun, because any direction you chose to look, your sight line should hit the surface of a star (see Figure 7.1). This scenario is called Olbers' Paradox (though it is really a riddle or a puzzle), and the complete story is complicated enough to fill a whole book. Luckily there is such a book: Edward "Ted" Harrison's *Darkness at Night: A Riddle of the Universe* (Harrison 1987).

Also, of course, Olbers was not the first person to worry about the observable consequences of such a universe. Predecessors include the early advocates of an infinite, Copernican cosmos, Thomas Digges (1546–1595) and Giordano Bruno (1548–1600), whose non-infinite life ended with his being the last person burned at the stake in Rome, at least partly for that particular heresy. Kepler,[3] in his 1610 *Conversation with the Starry Messenger*, simply turned the paradox around and concluded that the universe could not be infinite[4]: otherwise the sky would be bright. A century later, Isaac Newton was willing to support the infinite universe view, as was his friend Edmund Halley (1656–1742).

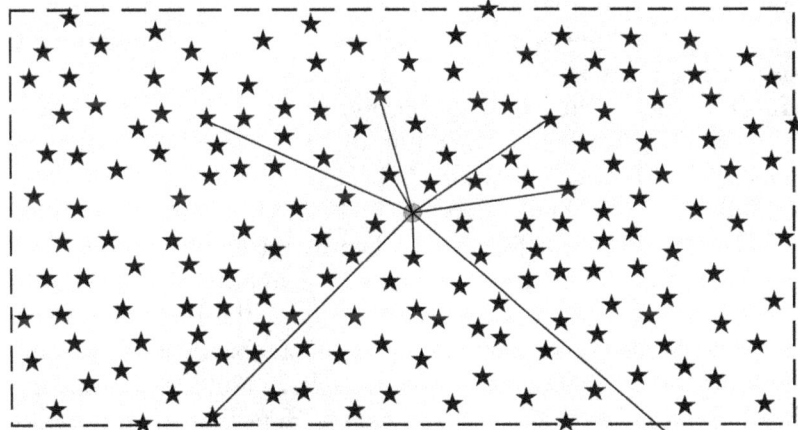

Figure 7.1 Illustration of the Olbers' paradox: in an infinite, homogeneous, static and infinitely old universe, in all lines of sight, an observer sees a star. Kepler used this as an argument for a finite universe, while Digges and Halley, who thought of the universe as infinite in both time and space, argued that the very distant stars provided too little light to be able to "move our senses".

Halley tried to estimate the night sky's brightness starting from the knowledge that the brightness of a single star would fall off as the square of the distance. In 1720, he published two back-to-back articles confirming that, just as Kepler had said, the sky darkness was indeed a serious issue if the universe was infinite.[5] Halley wrote a second article on the subject proposing a way out of the difficulty, but his explanation was rather confused and seems somewhat like an attempt to justify the unjustifiable. A short time later, in 1744, Swiss astronomer Jean-Philip Loys de Cheseaux (1718–1751) proposed a resolution of the paradox or puzzle.

De Cheseaux is best known for his work on the Great Comet of 1744, a comet that is noteworthy for displaying a fan of six tails after it had passed around the Sun. He was one of the three independent discovers of the comet and wrote an extensive text of over 200 pages about it. His text contains an Appendix which has nothing to do with comets entitled *Sur la Force de la Lumière et sa Propagation dans l'Ether*[6] (*On the Power of Light and Its Propagation in the Ether*). That Appendix rephrased the paradox and presented the view that the problem could be solved if there were holes in the ether through which light would not pass: then not all of the light from great distances would reach us. Later suggestions invoked the idea of an interstellar medium that could play the role of the absorber, also not a solution to Olbers' paradox but essential for understanding the apparent brightnesses of stars in our Galaxy.

In 1823 absorption was also proposed by Olbers,[7] who owned a copy of de Cheseaux's book but did not refer to it, and independently by the then director of the Dorpat (Tartu) Observatory, Friedrich Georg Wilhelm Struve (1793–1864). Olbers was inclined to invoke absorption of light by a then-hypothetical physical interstellar medium, rather than by holes in a hypothetical ether. Just as de Cheseaux had done, he had to guess the distance between stars and to assign an arbitrary degree of absorption.[8] However, as first remarked in 1831 by John Herschel (1792–1871, son of Sir William Herschel and nephew of Caroline Herschel), this hypothesis is unsustainable for thermodynamic reasons: the absorbing material, dust, would simply be heated until it glowed as brightly as the stars illuminating it.[9]

John Herschel went on to propose a different solution to the puzzle based on abandoning the idea of uniform distribution on a large scale.

By 1833, he had cataloged the nebulae of the northern sky, going on in 1847 to compile an equivalent catalog for the southern sky. This all-sky sample of 3812 nebulae formed the basis for his study. Herschel was struck by the lack of uniformity in the distribution of nebulae: the distribution was far from homogeneous. He could see that a substantial fraction of nebulae were concentrated in the huge aggregates that we now know as the Virgo Cluster and the Perseus–Pisces supercluster.[10]

Herschel's map suggested a hierarchy of groupings of cosmic matter: stars come together to form galaxies, these in turn form clusters of galaxies and so forth. This was an idea that had already been set out by Immanuel Kant (1724–1804) in 1755, and in modern terminology, it is known, following Benoît Mandelbrot (1924–2010), as "fractal structure". In a fractal, the average density of matter decreases as we average over ever-greater volumes. Even if there were no end to this hierarchy, there would always be starless gaps in the sky, and the night would continue to be dark. However, in the current context, this solution is not valid since from the study of large-scale cosmic structure, we have learned that the fractal system is observed in the distribution of galaxies only at particular scales, and it disappears at the largest scales. Today, there is therefore no basis for considering a fractal distribution to keep the sky dark.

The idea of extinguishing light on its way towards our telescopes from distant sources was revived in the 20th century. In 1929, the Swiss-American (though born in Bulgaria) astronomer Fritz Zwicky (1898–1974) postulated a mechanism that today is referred as "tired light" to explain the redshift observed in remote galaxies — photons from distant galaxies lose energy and therefore increase their wavelength on the long journey. This explained the observation of the American astronomers Vesto Slipher (1875–1969) and Edwin Hubble (1889–1953) that the light from distant galaxies was shifted to lower frequencies without invoking some "hypothetical" cosmic expansion. It would also provide an alternative cosmological view of Olbers' discussion, albeit subject to the same criticism. The Tired Light, non-expanding universe, cosmology was pursued into the 1950s, notably by the distinguished French astronomer Jean-Claude Pecker (1923–2020) who argued that not enough attention was being devoted to the idea that the universe was not in fact expanding.[11] We now know that this hypothesis of tired light is also not true.

Interestingly enough, the first person who sensed the real reason for the darkness of the night was not a scientist but a poet. The American writer Edgar Alan Poe (1809–1849), in his essay *Eureka: A Prose Poem*, published in 1848, writes the following:

> The only mode, therefore, in which, under such a state of affairs, we could comprehend the voids which our telescopes find in innumerable directions, would be supposing the distance of the invisible background so immense that no ray from it has yet been able to reach us at all.

Indeed, there is a simple explanation for the darkness of the night, and in a way, it is contained in Poe's text. Since the speed of light is finite, and the lifetime of stars is, too, the light of many of the stars needed to cover the sky has not had time to reach Earth. Not enough time has passed since those stars started to shine, and up until now, for their radiation to cover the immense distance that separates them from us. A similar idea is contained in a general readership book on astronomy published by the German astronomer Johann von Mädler (1794–1874) in 1861. Strangely enough, this reasoning went unnoticed, as did the article that Lord Kelvin (1824–1907) published in 1901 where he demonstrated in quantitative terms that the travel time of the light that comes from the most distant stars in an infinite universe is greater than their lifetime as luminous heavenly bodies (although for Lord Kelvin, stars lived only about 10 million years!).[12]

7.5   The Olbers' Riddle in Modern Times

Since those early times when we knew little about our universe, a new cosmology paradigm has emerged: the Hot Big Bang theory in which the universe emerged from a hot state of unimaginably high density at a finite time in our past. The experimental evidence gathered since 1965 supporting this view seems almost incontrovertible. So, what do we think about Olbers' paradox now, if we even think about it at all?

We can view the sky using light having wavelengths ranging from high-energy gamma-ray photons to low-frequency radio waves. Much of that comes from discrete sources like stars, a variety of types of galaxies including quasars.[13] We can remove all of these sources to reveal a extragalactic background light which consists of a true background that is not due to individual sources and to unseen sources such as hugely distant galaxies which ever more powerful telescopes will find. By far, the dominant contribution to this "true" background appears at microwave frequencies, the frequencies at which our mobile telephones and old-fashioned analog broadcast television worked. This radiation is so strong that its signal was seen as "snow", "noise", on the screens of home TV sets when they were not tuned to a broadcast channel. Nobody realized or thought to ask where this was coming from. The discovery of what we now call "The cosmic microwave background radiation" (usually referred to simply as the "CMB" radiation) was made by Penzias and Wilson and interpreted by Dicke, Peebles, Roll and Wilkinson in 1965 and is discussed in greater detail in Chapters 17 and 18. Modern Physical Cosmology was born in 1965.[14]

The correct modern cosmological solution to Olbers' paradox was put forward by Edward Harrison in 1965: *Olbers' Paradox and the Background Radiation Density in an Isotropic Homogeneous Universe*. Although published after the announcement of the discovery of the Cosmic Background Radiation by Penzias and Wilson in May 1965, Harrison's article had been submitted in September 1964. Harrison described the general problem of radiation in the universe within the framework of Einstein's theory of General Relativity and so his work was equally applicable to the old problem of Olbers' paradox and to the new cosmological paradigm. The basic idea is easy to grasp: the stars that populate the expanding universe do not live forever and their light travels at a finite speed. This means that the sum of their light cannot, in any moment of cosmic history, light up the night sky. Harrison calculates that "in a static universe, the stars must radiate for 10^{23} years in order that the radiation level is raised to the level of the surface of the stars." He goes on to say that if the stars radiated for only 10^{10} years, then the sky's brightness would be consistent with what is currently observed.

We are seeing the afterglow of the flash of light that came with the start of the cosmic expansion. The expansion of the universe has cooled down that radiation considerably. When a gas compresses or decompresses, it gets hotter or cooler: most of us have experienced how a bicycle pump gets hotter to hold as the pressure of the tire increases. Expansion brings about the opposite effect. The hot gas of photons from the primordial universe cooled as it expanded. Indeed, the primordial photons started out on their long cosmic journey when the first stable atoms formed and have been traveling for some 13,000 million years. During the course of that journey, the wavelength of this radiation has increased owing to the expansion of space. The temperature associated with this radiation is today roughly 1,100 times smaller than it was at the time of the emission and so is now a little under 3 K. It is still the dominant contribution to the brightness of the night sky.

The sky is glowing like a cosmological fossil that shows us, in Georges Lemaître's (1894–1966) words, the "evanescent brilliance of the origin of the worlds". Cosmic background radiation (see Chapter 17), invisible to our eyes but detectable in the millimetric region of the electromagnetic spectrum, makes our sky shine with a faint brightness not because of the stars but because of the Big Bang itself.

7.6   Cosmological Fossils

In 1917, Einstein applied his equations of general relativity to describing the cosmos. This is undoubtedly the theoretical starting point of modern cosmology. Princeton University Professor Emeritus and Nobel Prize winner in Physics Jim Peebles explains this eloquently in his book[15] *Cosmology's Century*:

> Modern cosmology grew out of Albert Einstein's search for how the general theory of relativity might apply to the large-scale nature of the universe. Einstein's (1917) thought was that a philosophically reasonable universe is the same everywhere and in all directions, apart from minor irregularities, such as the observed concentrations of matter in planets and stars.

This hypothesis asserts that, on a large scale, the universe is homogeneous and isotropic. Later, in 1933, the British astrophysicist Edward Arthur Milne (1896–1950) suggested referring to this assumption as the "cosmological principle" that the universe is homogeneous implies that there are no special or preferential locations — that is, observers located in different places in the universe would see the same large-scale properties. That the universe is isotropic implies that there is no special or preferential direction — that is, regardless of the direction of observation, the universe will appear very similar to us.

It is important to note that the assumptions of homogeneity and isotropy apply to large scales. For example, at the scale of the Solar System (light-minutes), there is clearly a preferential direction: towards the Sun, a star that accounts for 99.8% of the mass of the entire system. On much larger scales, for example, those of the Local Group,[16] which is about 10 million light-years in diameter, the universe is still anisotropic. More than 80 percent of the luminous matter in this environment is concentrated in two galaxies: the Milky Way and Andromeda. The direction that connects these two majestic spiral galaxies would be a special direction. At scales 10 times larger — 100 million light-years — the direction exhibited by the Virgo cluster, the highest concentration of galaxies in the local supercluster, seems to be a preferential direction. We have to go up to scales in the order of 500 million light-years to begin to detect isotropy in the distribution of galaxies.

Similar reasoning could be applied to homogeneity. If we were to place spheres measuring 500 million light-years in diameter in different locations in the universe, they would contain approximately the same mass, and their density would approach the average density of the universe. Of course, they would host clusters and superclusters of galaxies, surrounding large spaces nearly empty of luminous matter (voids), but in such a way that the pattern formed by these large-scale structures would be similar in all these gigantic spheres. The same thing would not happen in the case of smaller spheres. Consider spheres 5 times smaller, 100 million light-years in diameter: some of them might contain only the densest regions of a supercluster, while others could be centered in one of the large voids and contain practically no galaxies, meaning that the properties of the various spheres, particularly their density, would be very different. At these scales, therefore, the universe would still be nonhomogeneous.

However, in 1917, for Einstein, the universe was the Milky Way galaxy. Therefore, when he postulated the cosmological principle, he did so based on the distribution of stars in the sky and the apparent absence of systematic motions. The postulate was necessary to

develop his theory, but it was not supported by astronomical observations when he put it forward. In this case, the hypothesis is not
about the existence of a physical entity (such as ether, crystalline
spheres or the dark companions of the eclipsing binary stars that we
have talked about in previous chapters). It is more of a philosophical "dragon" — an idea of what the nature of the large-scale cosmic
structure ought to be. Einstein's intuition would be confirmed little
by little as cosmological observations were carried out during what
Peebles, in the title of a book, calls "Cosmology's Century".[17]

The first strong proof of the universe's extreme isotropy was
obtained by a team led by George Smoot, who analyzed data from
the COBE satellite in the early 1990s.[18] The relative fluctuations
in the temperature of this radiation coming from different directions
(anisotropies) are very small — on the order of one part in a hundred
thousand — but they too are essential for our cosmological model
because they reveal the existence of the small fluctuations in density that were the seeds of the cosmic structures that we observe
today. Indeed, gravity made regions that are slightly denser than
their surroundings attract the surrounding matter to gain more mass,
so that, in this process of gravitational instability, they transformed
into the galaxies and clusters that constitute the building blocks of
the universe today.

The cosmic background radiation, to which we will return in
Chapter 17, is a cosmological "fossil". The helium abundance is the
other. Analyzing them allows us to understand the state of the primitive universe and is key to understanding how the universe evolved,
somewhat like the way fossils that palaeontologists analyze allow
them to probe the history of life on Earth and its evolution. In the following chapters, we will talk about how the discovery of these fossils
has revolutionized our understanding of the history of life on Earth.
We will begin with the fantastic story of England's Mary Anning,
who, with strikes of a hammer and chisel, extracted fossils on the
beaches of the county of Dorset (UK) in the early 19th century.

Part III
The Dragon's Flight
How Discoveries Occur

Chapter 8

The Girl Who Found Primeval Dragons

"You're not allowed to call them dinosaurs any more", said Yo-less. "It's speciesist. You have to call them pre-petroleum persons."

Terry Pratchett, *Johnny and the Bomb*

Dinosaurs, the terrible lizards, are hugely admired and amusingly feared by millions of children the world over. The images of these ancient and fearsome monsters make the study of dinosaurs the first encounter with science experienced by children and young adults, whether it be to spell their strange long names or simply to play with plastic models. The name "Tyrannosaurus Rex" has instant recognition and the ultimate challenge is to correctly spell "Epidexipteryx" or "Ichthyosaurus".

Interest in fossils goes back over 2,500 years to Aristotle and Xenophon to whom they were mere curiosities, products of the "sports of nature". This view only changed with the writings of Leonardo da Vinci in his *Codex Leicester* journal spanning the years 1504 to 1510. The journal was never published. With typical pre-science, Leonardo argued that the fossils were related to animal life that existed even before biblical times. The fact that the fossils were sometimes found high up in mountains he attributed to changes in the Earth itself, happening over long periods of time. The mountain top where fish fossils were found must have once been under an ancient sea. In so saying, he was denying the biblical story of creation.

The ideas of Leonardo were not taken up until the late 17th century, by Robert Hooke in England and particularly by Nicolas Steno (1638–1686) in Denmark who is regarded as the father of the branch of science known as "Stratigraphy".[1] Steno's great contribution is his "law of superposition": meaning that the layers of rock were laid down one upon the other in a time-ordered sequence. The oldest is therefore at the bottom. This is the key to ordering fossils by age.

By the start of the 19th century, fossil collecting had become the latest craze. That was driven by a number of factors, not the least of which was the discovery in 1812 of an almost complete Ichthyosaur skeleton by two children. The skeleton was uncovered, buried in a chalk cliff at the seaside town of Lyme Regis in southwest England by a young girl, Mary, 12 years of age, and her older brother, Joseph. Both children had known of the existence of the skull buried in the chalk for at least a year, and it was Mary who went on to expose the rest of the skeleton in early 1812 (see Figure 8.1).

This was a windfall for the impoverished family whose father had died 2 years earlier and left them heavily in debt. However, the discovery in the first instance only brought notoriety rather than enough cash to clear the debts. Clearing the debts would happen only many years later with Mary's ever-increasing fame as a fossilist.

8.1   The Lyme Regis Ichthyosaurus

Mary Anning was born in 1799. With her brother Joseph, she went foraging in the chalky rocks, armed with a small rock-hammer, to find "curiosities" that were locally known as "snake stones", "devil's fingers" and "verteberries". These "curiosities" were sold to tourists and fossil collectors visiting the Lyme area by Mary's parents from a small stall outside their house. The crumbling cliffs were somewhat dangerous,[2] but the more intrepid of those tourists would have purchased a small rock-hammer to go finding fossils: they were known as "amateur fossilists".

In the decades marking the end of the 18th century and the start of the 19th, Lyme Regis was a popular sea-bathing resort equipped with small bathing huts along the beach. The period was dominated

A SCHOOL-GIRL MEETS ICHTHYOSAURUS

AT LYME REGIS TOOK PLACE ONE OF THE MOST ASTOUNDING MEETINGS IN THE HISTORY OF GEOLOGY.
MARY ANNING, AGED 12, COMING UPON THE FIRST ICHTHYOSAURUS FOUND IN ENGLAND

Figure 8.1 Depiction of young Mary Anning working on her Ichthyosaurus fossil in 1811, by Charles Edmund Brock. Credit: Arthur Mee's *The Children's Encyclopedia*, London, 1925.

by the aftermath of the French Revolution and the Napoleonic Wars: holidaymakers were encouraged to spend their vacations in England rather than in Europe. Fossil collecting was then a popular pastime and so Lyme Regis with its sea air and bathing facilities saw a good flow of wealthy tourists come to the area, which is now referred to as the "Jurassic Coast".[3]

The public interest in fossils surged during the decades prior to 1800 and persisted throughout the early 19th century. This was largely due to the first discoveries of the skeletons of dinosaurs and their graphical depiction as monsters or "terrible lizards". So, Mary was born into a place where fossil hunting was at its height and she was born into the place which has been one of the richest sources of fossils in England. She was also mentored by an educated lady having strong connections with the London intellectual elite who were to form the London Geological Society.

Mary, who was largely uneducated, met her mentor-to-be Elizabeth Philpot (1780–1857) in 1805. Elizabeth Philpot's brother, a wealthy London lawyer, purchased a large house in Lyme Regis for Elizabeth, then in her early 20s, and her two sisters, Louise and Margaret. The Philpot sisters were members of the upper-middle-class gentry; they had money and did not need to work. But notwithstanding the vast gulf in social standing, Elizabeth worked closely with Mary on fossil hunts and encouraged Mary to learn to read and write and, in particular, to read about geology. That, together with the unearthing of the entire 5.2-meter Ichthyosaur skeleton in 1812, was Mary's first step towards acquiring a worldwide reputation as a leading but academically unacknowledged paleontologist.

The Philpot family worked with some of the museums emerging in London, Cambridge and Oxford to establish important fossil collections. The Philpot Museum in Lyme Regis has a Mary Anning Wing, opened in July 2017, recognizing the work that Mary undertook. A beautiful fictional account of the life and work of Mary Anning and Elisabeth Philpot and their relationship can be found in the delightful novel *Remarkable Creatures* by Tracy Chevalier.

8.2 Mary Finds a Fish Lizard and More

A few Ichthyosaur skulls had been unearthed in other parts of England prior to 1812, and indeed it seems likely that Mary and her brother Joseph knew about the skull of the Lyme Regis specimen before then. But during 1812, she managed to uncover the rest of the skeleton, thereby making it the first almost complete specimen.[4] The specimen was described as having the "the muzzle of a porpoise, the head of a lizard, the jaws and teeth of a crocodile, the vertebrae of a fish, the breastbone of an ornithorhynchus (the name given to the then newly discovered platypus which was believed to be the last extant member of that family), the paddles of a whale and the trunk and tail of an ordinary quadruped". The fossil was sold soon after and shortly found its way to the then-new London Museum of Natural History, where in 1816 it was classified and given the informal name "Ichthyosaurus" ("fish-saurian").[5] Mary found a second Ichthyosaur skeleton in 1821; this one measuring 6.1 meters. Following the academic discussion of this latter specimen, the name *Ichthyosaurus* officially became the name of the species.

The catalog descriptions of these specimens did not state that the Annings had found them, nor were their names cited in subsequent publications discussing them. Only once was she ever publicly acknowledged. Indeed, Mary Anning's name was never cited in a publication in relation to any part of her work as a fossilist until recent times. The Annings were poor and uneducated; this was social exclusion in which it was inconceivable that such a person, let alone a woman, could understand such technical matters. The situation has not improved much 200 years later in the 21st century. As late as 2020, the *Wikipedia* article on *Plesiosauria* dismissed the Annings as the "commercial fossil collector[s] Mary Anning and her family", which is remarkable considering that Mary Anning discovered the superb complete skeleton which is showcased today in the Natural History Museum in London (with due acknowledgment to Mary). The *Wikipedia* page for Mary Anning page has since been totally revised. There are now signs of greater recognition for Mary following the opening of the Mary Anning Museum in Lyme Regis, her work being acknowledged on a commemorative set of English 50p coins in 2021, and with exhibitions of many of her fossils at the London Natural History Museum.

Yet, she was certainly well versed as regards both the geological and biological aspects, almost entirely due to her life-long mentoring by Elizabeth Philpot. She had developed the skill of reconstructing skeletons from separate bone fragments and was able to recognize different species. She was widely reported as being able to engage in detailed conversations about fossils with the experts who were her clients.

The height of Mary's achievements came in 1823 with the discovery of an almost complete skeleton of a Plesiosaurus,[6] see Figure 8.2. Part skeletons of plesiosaurs had previously been found and described as far back as 1605 by Richard Verstegen (c. 1550–1640) from Antwerp and in 1719 by William Stukely (1687–1765). The latter's

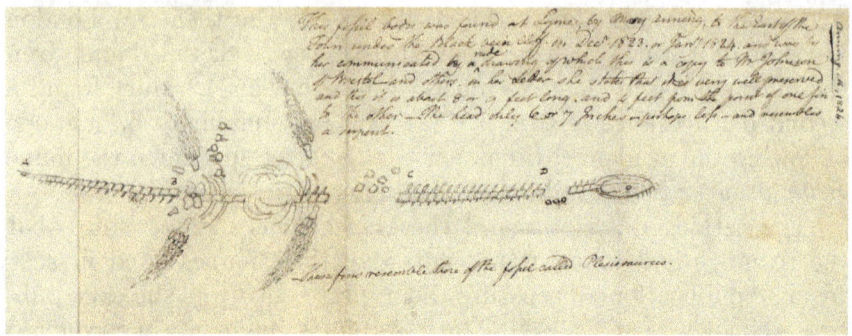

Figure 8.2 Mary Anning's sketch of her first Plesiosaur. The fossil was at first declared to be a fake on the grounds that no creature could have so many bones in its neck. Credit: Trustees of the Natural History Museum, London.

specimen was an almost complete skeleton found in a vicarage in Nottinghamshire where it was described as the remains of a sinner drowned in the Great Biblical Flood.

Mary's was the first plesiosaur discovered at Lyme Regis. The total length of the skeleton was just under 3 meters; it had a small head and an extraordinarily long neck consisting of 35 vertebrae. With its four flippers and small head, it looked like a long-necked turtle. The fossil was formally discussed at a meeting of the London Geographical Society before a packed and excited audience in February 1824. There it was described and ratified as a genuine specimen and was soon sold on. That established Mary's reputation once and for all as a reliable purveyor of important fossils: she became a celebrity, receiving visitors from all over the world. The Anning Plesiosaur is now a featured display at the London Natural History Museum.

Importantly, she recognized as early as 1824 that some of the stone-like samples that she had found within the abdominal region of the Ichthyosaur skeletons were in fact fossilized feces (formally referred to as "coprolites"): she broke them open to reveal fossilized fish bones and scales. This discovery was presented to the Geological Society in 1829, where her contribution was duly acknowledged. That started the science that formally became known as "Coprology".[7] Ever since that time, fossilized droppings have been used to discover what animals ate and the status of their health.

In 1828, she also found her first fossilized Pterodactyl, the first known type of flying reptile, and the first such fossil found outside of Germany. In 1830, she unearthed another almost complete plesiosaur. Throughout the rest of her life, until 1847, she made more discoveries.

8.3   Enter Georges Cuvier

Another important player in this story was the eminent French paleontologist, Georges Cuvier. Cuvier was born in 1769 and lived through the French Revolution and Napoleonic Wars until 1832. Despite the acclaim and excitement that surrounded Mary Anning's 1824 plesiosaur, there was an important controversy arising from the fact that the specimen had 35 neck vertebrae (see Figure 8.2). Cuvier had been sent an accurate drawing of the fossil and declared it to be a fraud, on the grounds that no modern beast had more than 25 vertebrae. Indeed, the giraffe has only seven neck vertebrae and the sloths may have up to 10, while ducks may have up to 16, and swans around 24. Several years later, he admitted that he had "acted hastily". His haste could easily have ruined Mary Anning's standing as a seller of important fossils from Lyme Regis.

"Eminent" is perhaps an understatement about Cuvier's worldwide reputation: he was, and still is, widely regarded, even beyond France, as the founder of the science of vertebrate paleontology. In 1812, he was elected as a foreign member of scientific societies in several countries, including Britain, Sweden and the Netherlands and in 1822, the United States. He had a huge list of publications and was many honors.

He was the first person to believe that the reason why we saw no descendants of the fossils was because the animals had become extinct. Specifically, he imagined "the existence of a world previous to ours, destroyed by some kind of catastrophe". His theory was thus described as "Catastrophism".

Some "catastrophists" thought in terms of a single catastrophic event that created the world as it was known, but Cuvier's belief evolved to thinking that there had been many catastrophes in the past, each responsible for creating or annihilating species of plants and animals. There were many ideas about what caused either the creation or extinction of species, most of them being based on a notion of world flooding by rising sea levels or wipe-out due to tidal waves.

Cuvier linked the species he found to their host geological layer, the start of biostratigraphy. He was not in favor of the theory of evolution as it stood in the 1820s. In that pre-Darwinian time, the prevalent thinking was expressed in the ideas of Jean-Baptiste Lamarck (1744–1829), that creatures changed over time as a consequence of physical laws. Lamarck had, in 1809, postulated two drivers of change: first, an unknown force that drove organisms to evolve towards higher levels of complexity ("orthogenesis") and second, an environmental force that caused adaptation to changes in the environment through the use or disuse of structures that were either useful in adapting to the changing environment or not.

As evidence for his opposition to Lamarck, Cuvier cited the manifest cycles of creation and extinction observed in the geological strata: those cycles appeared to violate Lamarck's continuity of physical law. A quarter of a century after Cuvier's death, Charles Darwin (1809–1882) presented in his *On the Origin of Species* the idea of the evolution of the multitude of species from a few common ancestors. In contrast with Darwin, who saw the evolution of species as a branching process, both Cuvier and Lamarck supported the view, described today as "independent lineages", in which each species evolves in its own way, independent of other species.

In 1811, Cuvier collaborated with geologist Alexandre Brongniart (1770–1847) on one of the most important projects either of them ever carried out: they cataloged and studied the fossils in the geological structures in the region around Paris. They found they could identify nine separate layers that were common to all the samples. The commonality was established by noting which layers were alternately occupied by sea water and by freshwater molluscs. This very important discovery showed that the strata were not simply laid down by lowering of the sea level. That was not all: by matching layers at different places around the city containing similar fossils, they were able to show that the strata covered large areas, which is what William Smith (1769–1839) had done in Britain a decade earlier.

8.4 So How Did It All End for Mary?

For one raised in the poorest of circumstances who lost a father while still young and had no formal education, Mary was extraordinarily productive in later life (see Figure 8.3). People from that background do not normally get to influence the course of science, which is what she achieved. In modern times, she would be regarded as a highly effective data scientist, acquiring and curating data for analysis by others — her collaborators. She would be a co-author of the important papers that exploited her data. That of course did not happen. It is also likely that had her brother Joseph done what Mary had done, he would not have been any more successful: the Annings were of the wrong social class. She was a contemporary of Michael Faraday (1791–1867) who came from a poor London background and had no formal education, but Faraday had opportunities that simply were not available to Mary.

Compare Mary with her wealthy friend Elizabeth Philpot. Both were providers of fossils to the scientists but neither got any public acknowledgment in the papers that resulted from their specimens, simply because they were women. However, owing to her wealth and access to society, Elizabeth Philpot did get acknowledgment through the subsequent construction of museums that bear her name and museums that have benefited from donations of duly acknowledged items from her personal collection of fossils. Social exclusion of Mary is quite evident. More recently, since July 2017, the Philpot Museum in Lyme Regis now has an Anning Wing.

8.5 Time Told by the Rock Strata

The Danish scientist Nicolas Steno, born in 1638, made contributions to several branches of the science of the times. His contribution to geology described in his *Dissertationis Prodromus* of 1669 was to argue for the notion that the surface of the Earth was built in successive layers called "strata". The oldest strata were therefore at the bottom of the pile and the most recent were at the top: they were laid down in strict chronological order. At the time when a particular

Figure 8.3 Portrait of Mary Anning with her dog Tray and the Golden Cap outcrop in the background. Credited to *Mr. Grey* in Crispin Tickell's book *Mary Anning of Lyme Regis* (1996). Credit: Public domain. Wikipedia Commons (Natural History Museum, London).

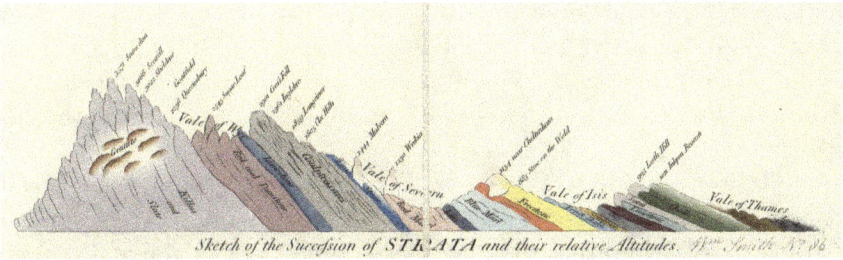

Sketch of the Succession of STRATA and their relative Altitudes. Wᵐ Smith Nᵒ 86

Figure 8.4 A geological section from London to Mount Snowdon in Wales measured and drawn by William Smith identifying the strata of a cross-section of England and depicting their inclination. The vertical scale reflects altitude. This is from the geological cross-section of his map of 1815. Credit: Trustees of the Natural History Museum, London. See also William Smith's Maps: http://www.strata-smith.com/.

stratum was being formed, none of the upper strata existed. A similar idea was put forward by Robert Hooke.

The Steno–Hooke picture gave strong support to the work of William Smith, another self-educated person who worked on surveying old coal pits during the period 1790–1799. He noticed that the pits were arranged in a series of layers of coal and rock in which there were fossils which he collected and documented. Importantly, he found that the layering and the fossil contents of the strata were identical in pits located in many parts of England. He also noticed a systematic easterly dip in the layers, smallest at the top and increasing towards the bottom. When he finished his job as a pit inspector, he moved on to the canals that were being dug all over Britain and found the same phenomenon. By 1801, he was able to publish the first geological map of Britain and had amassed a huge database of the fossils he had found. He published another map of England and Wales and part of Scotland, measuring 2.5 m by 1.8 m, in 1815. The map was acquired by the Sedgwick Museum that same year and is still exhibited today (see Figure 8.4).

It would be easy to think that such an achievement would have brought Smith fame, fortune and recognition, but that was not the case. Up until 1819, he had been paying for a mortgage and for the publication costs of his work. His work was plagiarized, cheaper copies of his maps were sold by copyists, and the establishment was either unaware of him or simply ignored him. In 1819, unemployed and heavily in debt, he ended up in a debtor's prison. Even the sale of his fossil collection to the British Museum could not pay for those debts. He did not gain recognition until 1831 when the Geological Society awarded him a medal and described him as "the Father of English Geology".

1831 was the final year of the Reverend Adam Sedgwick's presidency of the Royal Geological Society. Sedgwick had bought several Ichthyosaur skeletons for the museum from Mary Anning. In 1865, he famously opposed the admission of women to Cambridge University, describing aspiring female students as "nasty forward minxes".[8] Charles Darwin had been one of his students and great friends, but Sedgwick refused to come out in favor of Darwin's theory of evolution describing Darwin's book as "From first to last, it is a dish of rank materialism cleverly cooked and served up". Apparently, they remained friends. The Sedgwick Museum of Earth Sciences in Cambridge is named after him.

What brought about the eventual recognition of Smith's work could have been the highly acclaimed work of Cuvier and Alexander Brongniart (1770–1847), first published in its final version in 1811. Both became well-known members of the scientific establishment. William Smith could never have entered the doors of that society, let alone the Geological Society of London, no matter what he had achieved.

Of course, the same phenomenon of similar rock layers, "strata", was being found in many other places and was not unique to England. In the United States, it was discovered that rock strata found in New York State were duplicated in Wisconsin. Over such vast distances, it was possible to notice that the strata thickened in an easterly direction. It was also noted that the layers in the Appalachian mountains were often "folded" and twisted. What could have caused that?

Chapter 9

The Backbone of the Ocean

In astronomy everything has essentially been done. . . . and besides, astronomy offers no opportunity for physical activity.

Alfred Wegener, Frankel: The Continental Drift Controversy

The land masses of the Earth are divided into seven vast continents.[1] To suggest that these vast landmasses are in motion may seem preposterous, yet this is what had been proposed over a century ago and what we have since been able to measure. Speeds are in the range of 1–10 cm per year.

The continents (with surrounding ocean crust) make up "tectonic plates". These cover the surface of the Earth, like cracked egg shell. Underneath crust of 50–100 miles thickness is the mantle, solid but slightly soft rock capable of convective motion (like slightly softened cold butter or the white of a partly boiled egg). Rising portions of mantle sometimes melt when exposed to smaller pressures because fluids are generally less dense than solids. The molten material, sometimes including melted crustal rocks, is called magma, and lava when it flows across the Earth's surface.[2]

When plates move apart, magma wells up to fill the cracks at mid-ocean ridges and in rift valleys. When plate collide, ocean crust can slide under continental crust, melting and upwelling to

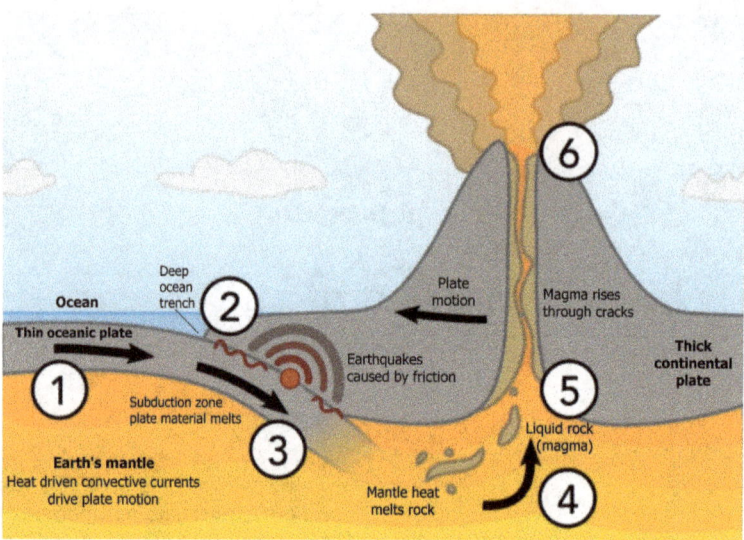

Figure 9.1 Modern Plate Tectonics. At the intersection of an Oceanic Plate (1) and a Continental Plate (2), the denser Oceanic plate dives beneath the Continental Plate — subduction (3). The plate melts (4) and magma is driven upwards into cracks in the crust (5), sometimes creating a volcanic event (6). Credit: Adapted and annotated version of a public domain image retrieved via Wikimedia Commons.

form mountains like the Andes, with associated earthquakes and volcanoes. The collisions of continental crust on both sides of a plate boundary lifts up taller mountains with little melting but vigorous quakes, like the Himalayas. Magma also wells up from mid-plate hot spots, where the magma columns are tied to the mantle, so that we get a chain of volcanoes from old to young, like the Hawaiian islands, marking the directions of plate motion over tens of millions of years (see Figure 9.1).[3]

9.1   Wegener's Drifting Continents

The idea that the continents are in motion is widely attributed to the German scientist Alfred Wegener (1880–1930),[4] who wrote a book on the subject over a century ago when the First World War was just starting. He got little positive recognition during his lifetime. Indeed, he appears on many "they laughed at..." lists, along with Copernicus, Galileo and Darwin. It was an extreme case of "NIH"[5] coupled with a strong anti-German feeling among those countries that had joined the First World War against Germany.[6] Wegener viewed himself as a Polar explorer and meteorologist; he was not really a part of the geology community.[7]

His 1905 PhD was in positional astronomy,[8] and the first of his papers easily accessed was a quarter-page in *Astronomische Nachrichten*, reporting the ephemerides (positions) of minor planet 511 Davida as seen from Berlin. For much of his career, he focused on meteorology (with a special interest in tornados) and polar research. The latter led to his death, sometime in November 1930, in a failed attempt to return from his fourth expedition across Greenland. Part of the goal had been to set up an over-wintering station to record temperature, precipitation and wind conditions. His work in these territories was perceived as mainstream and respected.

But, from about 1910, Wegener started devoting serious attention to the corresponding Atlantic coastlines of the Americas and Africa and Europe. He was far from the first to do so, anticipators dating back to those who first mapped the western hemisphere, followed by Alexander von Humboldt (1769–1859) of the Humboldt current.[9] Wegener also noted correspondences in shapes of parts of the coasts of Australia, Antarctica and India. What is more, he started collecting data on rock formations and fossils, especially the fern Glossopteris, and found matching forms across those mapped boundaries. He was by no means the first to recognize these coincidences (see Section 9.2), but he was the first to discuss the idea that the continents were in motion.

Beginning in January 1912, he began lecturing and writing about these data and suggesting that the explanation was that the current continents had once been parts of a single *urkontinent*, or *super-continent* as we would now describe it, which he named *Pangea*.

Pangea was formed some 335 Mya (millions of years ago) from the merger of two continents: Gondwana in the South and Laurasia in the North. This super-continent started to break up during the Mesozoic Era,[10] first into a northern part and a southern part (called Gond-wanaland) and then along an axis that became the Atlantic Ocean. India broke loose from the southern continent, traveled into the northern hemisphere and crashed into Asia to produce the Tibetan plateau and Himalayas.

Wegener's travels and data collecting were interrupted by the First World War (two wounds in quick succession took him out of direct combat), though the first edition of his magnum opus, *The Origin of Continents and Oceans* (in German, of course[11]) appeared in 1915.

Later editions included reconstructions of ancient climates on the various land masses, indicating, he felt sure, that some areas now ice-bound had once been tropical, and conversely. The final, definitive edition of his book comes from 1929. Meanwhile, in 1926, he gave a talk before the American Association of Petroleum Geologists in New York City. With one or two exceptions, the participants rejected, ridiculed and laughed at him. One participant reportedly stated that "If we are to believe Wegener's hypothesis, we must forget everything that has been learned in the past 70 years and start all over again".[12] The basic data could not be doubted, for many mainstream geologists and paleontologists had contributed to it.

But the accepted explanations included land bridges, rising and falling sea levels, and the occasional catastrophic flood, separating continents that had always been in their present positions but sporadically connected.

One of his ideas, that the mid-Atlantic ridge was spewing out fresh lava and forcing apart the continents on either side of the Atlantic, is now part of our world picture. Unfortunately, he estimated the rate of separation at something like 250 cm/year (the modern value is more like 2.5 cm/year, measured by radio very long-baseline interferometry with stations on opposite sides of the ocean). Curiously, his ideas were taken seriously enough in the mid-1920s for an attempt to be made to measure the changing longitudes of astrometric stations on either side of the mid-Atlantic ridge, in Denmark, Greenland and North America. Just possibly the proposed large rate would have been detectable. But in an era when longitude generally came from "transportation of the timepiece", the correct one was not.

The "what if" department invites us to consider a longer lifetime for Wegener. This might not have been possible, since his 1930 death (explained after the body was found) was probably due to a heart attack rather than freezing or starvation. But, reaching his 80s, he could easily have been hailed as a living hero rather than as one long gone, when work on magnetic stripes in the ocean floor and all the rest made continental drift, plate tectonics and mantle convection a "true" theory (meaning the best understanding we have of something) rather than a mere hypothesis. The demi-heroes who contributed to the revolution are so numerous we will not try to mention them, one, perhaps, of the reasons that geophysics has never received a Nobel Prize, while astrophysics crept in long ago.

The new data and concepts answered the major objection to Wegener's ideas: "How can solid continents plow through solid oceanic crust?" The answer is that they don't. It is crustal plates of both continental and oceanic crust that move around. The two most prominent 20th century pundits of geophysics took opposite sides through it all. Arthur Holmes (1890–1965) already in 1944 had ended his long-lived text, *The Principles of Physical Geology*, with a chapter on continental drift, and was himself instrumental in contributing the idea of mantle convection to carry the crustal plates around on the Earth's surface. In contrast, Sir Harold Jeffreys (1891–1989) went to his grave firmly denying that anything of the sort could be going on.

But Alfred Wegener was right: the Himalayas are the highest mountains on Earth because India is ploughing into the main body of Asia.

9.2   Shrinking Earths and Drifting Continents

If Wegener's hypothesis was so firmly rejected, what were the alternatives on offer? Was his rejection merely "NIH" xenophobia and lack of a mechanism for pushing plates, or was there a better established idea with stronger scientific support in place?

The idea that the continents had once been connected was far from being new and can be traced back as far as the early Flemish mapmaker, Abraham Ortelius (1527–1598). In 1596, Ortelius publish a map of the world with a commentary that the East coast of South America looked as though it would fit snugly in the West coast of Africa like pieces of a jigsaw puzzle. Ortelius suggested that the two continents might have been torn apart by earthquakes and floods. The fitting together of the continents was impossible to ignore.

The following 400 years produced a multitude of ideas about the separation of the continents but none gained much traction. Each had its own defining concept, its own particular twist, of what happened and why. Each was heavily criticized by others defending their own points of view. No global consensus could emerge and most suggestions are now relegated to the position of "other ideas were ...". However, from time to time, a few new ideas did rise to the top to become supported by relatively strong group-think bubbles.[13]

As early as 1801, Alexander von Humboldt[14] provided his own comments on this pointing out the similarities in rock formations at corresponding points on the two sides of the Atlantic. Thomas Dick (1774–1857), a prolific amateur scientist and Presbyterian minister, wrote in 1838 that "There is a striking correspondence between two sides of the two continents ... it [is] not altogether improbable that these continents were originally conjoined, and that, at some former physical revolution or catastrophe, they may have been rent asunder by some tremendous power, when the waters of the ocean rushed in between them, and left them separated as we now behold them." This catastrophe refers to the "universal, deluge", i.e. the Biblical Flood.[15] At a more scientific level, the French naturalist Antonio Snider-Pellegrini (1802–1885) remarked on the similarity of plant fossils and coal deposits at corresponding points on the two continents and was the first to publish "before and after" maps of the world (Figure 9.2). He also explained that the two continents had been ripped apart at the time of the Biblical Flood because, he argued, the Earth was at the time lop-sided. The flood was the ocean that filled the gap left behind and, as a consequence, the Earth was no longer lop-sided.[16]

The Austrian geologist Eduard Suess (1831–1914)[17] wrote in 1861 about a time past when all of today's continents were part of two large masses that he named "Gondwana" and "Angarland".[18] They were located in a primeval sea that he later named the "Tethys Ocean". Wegener later argued that these two land masses once formed a single mass that he named "Pangaea".

Suess' claim was that the Earth had shrunk during its cooling, causing some of the primeval Gondwana landscape to be submerged by the rising sea level. The rising sea level, and the associated

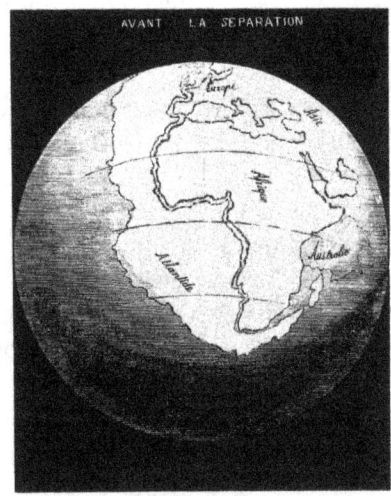 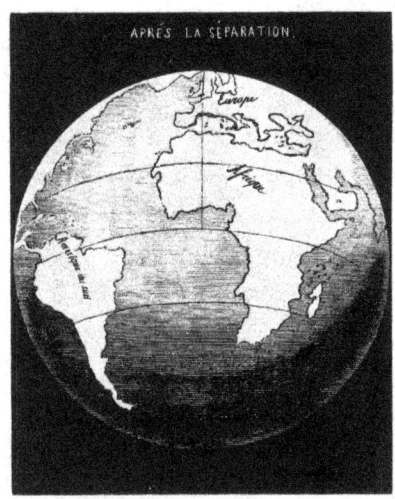

Figure 9.2 The first known illustration of the Opening of the Atlantic Ocean, by Antonio Snider-Pellegrini, 1858. Credit: Public domain. Wikipedia Commons (https://en.wikipedia.org/wiki/Antonio_Snider-Pellegrini).

subsidence of land, caused the land bridges between the higher land masses to sink, cutting off communication between the continents we see today, Suess also thought that more local geological processes could yield rise and fall of sea levels. The early land bridges would account for the similarities that had been observed between the flora and fauna of the now-isolated continents. He resisted invoking any lateral movement of the continental masses, arguing that the Earth had changed through the rising and falling of the sea level. His textbook, *Das Antlitz der Erde* (*The Face of the Earth*), published in four editions between 1883 and 1909, became standard reading in the early part of the 20th century and had a strong impact on the later development of the continental drift theory.

Suess had not invoked the motion of the continents in his writings, yet in 1873[19] he gave the opinion, based on his own explorations, that the portion of the Earth's crust that was Europe was acted on by a force and therefore moving. Subsequent to that, the first scientific paper asserting that the continents were in motion was by the American geologist Frank Bursley Taylor (1860–1938) who, in 1910, published a 47-page paper, "Bearing of the tertiary mountain belt on the origin of the Earth's plan".[20] Taylor's paper had been submitted 18 months earlier, at the end of 1908, which would indicate that the paper was peer-reviewed prior to publication, which in turn was 2 years prior to Wegener.[21] Taylor was an experienced geologist and provided the expected arguments for why one might think the continents had moved, arguments that were similar to those put forward later by Wegener. Most perceptively, he noted that "It is probably much nearer the truth to suppose the mid-Atlantic ridge has remained unmoved, while the two continents on opposite sides of it have crept away in nearly parallel and opposite directions."[22] Fifty years later, as we shall see in the following, this would be one of the key points supporting the tectonic view of the movement of the continents.

If Taylor's geological arguments were impeccable, his explanation for the cause of the movements was speculative. He invoked tidal and centrifugal forces to move the continents away from the poles and towards the equator. According to Taylor, the continents would plow their way through the crust towards their present positions. In so moving, the front of the moving continent would be pushed up on the equator-facing side by the resistance of the crust. This was the first mechanical theory for mountain building, albeit not the correct explanation. Wegener himself had not offered any mechanism for the movement of the continents.

Taylor's drift hypothesis got a slightly better treatment than Wegener got for his offering: Taylor was simply ignored. Wegener did not refer to Taylor's work except to cite it in 1911 and to update the 1920 edition of his book to include Taylor's explanation of the movement. Taylor thought he deserved some of the credit for presenting the idea of continental drift. Wegener did, however, refer to the earlier works of Roberto Mantovani (1854–1933)[23] who had suggested that Earth's expansion would tear apart a large continent leaving rifts between the smaller fragments. Even before Mantovani, Alfred Wilks Drayson (1827–1901) had, in 1859, written a book devoted to the subject of the expanding Earth and together with his colleague, William Thorp (1804–1860), published a paper in a local *Geology* journal.[24] Their work was promptly forgotten until it was resurrected by the mid-20th century advocate for Earth expansion, S.W. Carey (1911–2002). The early 20th century saw a number of expanding Earth supporters and even some who thought the Earth's diameter would oscillate.

9.3   Understanding the Formation of Mountains

The renowned Scottish geologist James Hutton (1726–1797) was one of the founders, along with David Hume (1711–1776) and Adam Smith (1723–1790),[25] of what has been called the "Scottish Enlightenment". This was a period of over 100 years spanning roughly the last half of the 18th and the first half of the 19th centuries and beyond. By the 1750s, the Scottish university towns had developed a thriving intellectual café society in which books and libraries played a large part. While Denis Diderot (1713–1784) in France started compiling his great *Encyclopédie* in the 1750s, the first editions of the *Encyclopedia Britannica* were being produced in Scotland for publication between 1768 and 1771.

Looking at the geology of the land around him in the 1780s, Hutton argued that there had to have been an uplifting of the land to expose those lower levels that were manifestly visible. He had no specific mechanism for this process of uplifting, but felt on the basis of what he observed, that it had to be cyclic. Areas of land would rise and then sink while their top layers were eroded by wind and rain. Finally, a given land mass would sink below the water to begin the cycle again. Hutton proposed that as the land sank, it would be subject to heating and molten rock would be injected into the gaps in dislocated strata to rise to the top again.[26]

Inevitably, there were serious questions raised by Hutton's ideas. Neither he nor anyone else at that time had any idea about a possible cause of mountain uplift. Moreover, Hutton's cycles would take enormous periods of time which would imply that the Earth could not possibly be a mere few thousands of years old as suggested by the 1650 Bible-based calculation of James Ussher (1581–1656), Archbishop of Armagh.[27]

Nevertheless, Hutton's ideas were not an intellectual fantasy. They were based on data derived from rocks that he himself had explored: those observations were reliable scientific data derived by an expert who knew what he was looking at.[28]

The astronomer John Herschel[29] knew of and respected Hutton's work. John was a noted scientific polymath[30] and took an active interest in geology, corresponding with the Oxford geologist Charles Lyell (1797–1875) about the first edition of Lyell's book *Principles of*

Geology.[31] In the book, Lyell presented his view that climate, geology and even life were connected through evolution. Lyell took the remarks made by Herschel seriously, modifying subsequent editions accordingly. There was extensive correspondence between the two.

Herschel left England with his family to establish an observatory in the Cape Colony (South Africa) in January 1834. During the journey, Herschel had studied the third volume of Lyall's book. He had with him an 18″ reflecting telescope, which at that time was one of the largest reflecting telescopes in the world. There was enough time to set up the telescope and prepare to observe the reappearance of Halley's Comet the following year. He could also begin work on the southern extension of the catalogs of stars and nebulae created by his father William and his aunt Caroline. Significantly, he also spent much of his time exploring the surrounding region, drawing maps and writing extensively on the Cape geology.

On 20 February 1836, he wrote a remarkable 11-page letter to Charles Lyell, in response to receiving a copy of a new edition of the *Principles of Geology.*[32] Herschel's letter is a dense, in-depth discussion connecting his discoveries about the Cape geology with Lyell's point of view on geological evolution. There is even a paragraph that begins "Now for a bit of theory." after which Herschel uses his knowledge of mathematics and physics to go deeper into some of Lyell's concepts. The letter is famous for Herschel's statement that Lyell's view provides a framework within which to understand "... that mystery of mysteries the replacement of extinct species by others." Darwin's use of the phrase "mystery of mysteries" in his *Origin of the Species* is, in effect, an acknowledgment of Herschel's role in establishing a basis for the Darwinian theory of evolution.[33]

Lyell had presented Darwin with a copy of his book to read during his voyage on the *Beagle* and suggested that on his way he visit Herschel in South Africa for discussions on geology. During the second voyage of the *Beagle*, in May 1836, Charles Darwin and the ship's captain Robert FitzRoy (1819–1851) stopped off at the Cape and spent time with John Herschel to learn about the geology of the area.[34] Significantly, Herschel had seen Lyell's approach to geology as possibly providing a driving mechanism for the origin of new species.

Herschel's reputation as a scientist was impeccable, yet he published relatively little and was therefore acknowledged far less than his contributions deserved. However, his legacy lay in his astute descriptions of the geological phenomena he studied. He certainly influenced the American paleontologist James Hall (1811–1898)[35] who, in his Presidential Address to the American Association for the Advancement of Science in 1857,[36] presented a paper on the formation of mountains by a mechanism he referred to as "uplift". The scientific paper describing the Uplift theory was published in its first form in 1859. However, Hall's paper was poorly written and also highly controversial and it generated little attention until the work of James Dwight Dana (1813–1895).[37]

Dana created a concept which in 1866 he referred to as the *"geosyncline theory"*. It was not so much a "theory" of mountain formation as the recognition of the importance of a particular kind of geological structure found at the margins of continents. He proposed that the long valleys surrounded on either side by ridges would become the coastal mountain ranges. He gave the structure the name "geosyncline". The name was in common use for a century in the context of mountain building. A host of different types of geosynclines were proposed by others.[38] Dana's underlying idea was that the Earth had shrunk due to the early cooling of the planet. During the shrinking process, the initially smooth crust became wrinkled, the way the surface of an inflated balloon becomes wrinkled as it loses its internal pressure.

Evidence for the shrinking of the Earth's crust came from several sources. Around the mid-1880s, Richard George McConnell (1857–1942) mapped the cross-sectional structure of the Canadian Rockies to show, among many other things, that the mountain range had been squeezed horizontally. By unfolding maps, he could show that points 25 miles apart were originally 50 miles apart. A solid Earth could not account for such huge horizontal motions. Moreover, this was consistent with Lord Kelvin's model of an ever-cooling and shrinking Earth.

There were a large number of technical difficulties with this view, not the least of which was that the Earth has an internal heat source due to radioactivity and has not cooled recently. Theories were subsequently developed by many others, each presenting a variation on the same theme. There was no consensus, and we get the impression of conflicting views held by small and assertive factions, each of which was arguing that the views held by other groups were simply wrong. The factions spent much of their time arguing. The simple fact was that there was no credible mechanism: they were all wrong.

Nevertheless, by the end of the 19th century, the shrinking Earth hypothesis was supported by a majority of scientists, with many authoritative figures like Lord Kelvin among them. Support for the shrinking Earth idea continued into the 1980s with books and articles by Raymond Lyttleton (1911–1995) and Sir Harold Jeffreys (1891–1989) who both supported the idea that the Earth shrank because of physical changes taking place in the core. They had throughout their lives vociferously rejected Wegner's ideas.[39] Max Planck (1858–1947) was correct when he asserted that "A new scientific truth does not triumph by convincing its opponents and making them see the light but rather because its opponents eventually die, and a new generation grows up that is familiar with it", often paraphrased as "Science advances one funeral at a time".

9.4   Plate Tectonics

Eventually, during the 1950s and 1960s, another more profound idea arose to replace the simplistic vision of Taylor and Wegener: plate tectonics. There was no immediate leap into this new idea; many theories were proposed and forgotten on the way, while the quantity of relevant geophysical data grew. The first problem, of course, was the question of the mechanism for the migration of the continents. Only Taylor had made any serious suggestion and he had been ignored.

Arthur Holmes, who was already famous for his 1912 work determining the ages of rocks using radioactive dating, published a paper in 1931 describing how the continents might be driven by deep underground currents of radioactively heated magma below the surface crust.[40] That region is referred to as the "mantle". It extends from the base of the Earth's crust right down to the top of the Earth's core[2] and accounts for 68% of the Earth's mass. The paradigm shift proposed by Holmes was that radioactive heating in the Earth's core would drive convection in the mantle, which would in turn drive the continents (see Figure 9.3). Harold Jeffries, a key supporter of the shrinking Earth theory, famously conceded that Holmes' proposal, while highly improbable, was not impossible. Holmes had presented both a model and a plausible mechanism for the movement of the

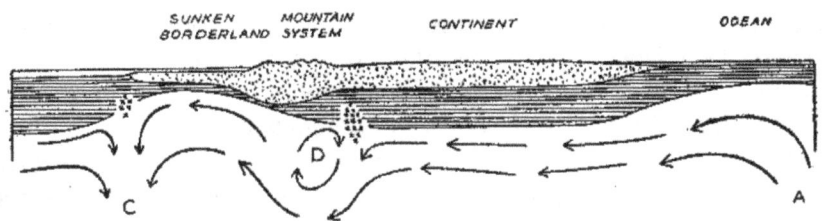

Figure 9.3   Arthur Holmes' 1929 figure showing how mantle convection would drive the motion of tectonic plates. Credit: Holmes' lecture notes as published in 1931 as Figure 4 in the *Transactions of the Geological Society of Glasgow.*

continents. This was, in essence, the inspiration behind the theory of "Plate Tectonics" as developed by Harry H. Hess (1906–1969) and others in the late 1950s and the early 1960s.

Harry Hess was a World War II Naval Submarine commander and a geologist. In 1962, he put forward a mechanism, very similar to that of Holmes, explaining the phenomenon of seafloor spreading: the missing part of Wegener's continental drift. It provided the driving mechanism which caused the continents to move apart.[41] This was the first great step towards plate tectonics. With further developments during the following decades, the theory has become a central part of modern geology to an extent that its originators would hardly recognize.[42]

Hess argued that the ridges discovered in the oceans were indeed the result of the rising part of convective flows in the mantle and that the deep ocean troughs were where there were descending mantle currents. We would therefore see what is called "seafloor spreading" at the bottom of the Atlantic Ocean, marking the place where new oceanic crust material was being produced to fill the gaps left by the continental movements. Hess's model explained why rocks older than 200 million years were never found in oceans, while rocks on the continents can be as old as 4 billion years. The evidence for this was found deep in the North Atlantic Ocean by Marie Tharp (1920–2006) in 1952, using measurement based on magnetic field reversals in basaltic lava flows.

In 1905, Bernard Bruhnes (1867–1910) discovered that the polarity of the Earth's magnetic field switches, the North Magnetic Pole becoming the South Magnetic Pole and vice versa. This phenomenon was discovered when he measured the magnetic fields in samples of basaltic lava flows in the mountains of central France.[43] The pole flip happens relatively quickly and happens up to several times every million years, but the rate seems to be highly variable. As lava flow cools, the magnetic field at that time gets frozen in, and so the solid rock will show a banded pattern of field directions — a bit like the familiar barcode. In the mid-1950s, Keith Runcorn (1922–1995)[44] an expert on terrestrial magnetism recognized that it was possible to use the Earth's magnetic field within the oldest rocks to find the location on Earth where particular rocks were formed. In effect, rocks had a built-in magnetic compass after they solidified, which continued to point to what had been north and south then, no matter where they traveled or how they were turned around. Using sensitive magnetometers that he built in his laboratory, he was able to show that the rocks on the East coast of North America had once been very close to those on the west coast of Europe and Africa: the rocks were formed in almost the same place (see Figure 9.4).

The seafloor shows such a pattern of magnetic field reversals on both sides of the rift from which the lava has been exuding, with the same pattern on each side. The pattern furthest from the rift was printed long before the pattern that is closest, so we can see and time the moving apart on either side of the rift. This is key evidence for continental drift.

Marie Tharp measured this magnetic pattern, produced maps of the North Atlantic seafloor, and discovered deep sea mountain ranges between which there were magma flows that had driven the neighboring continental shelves apart.

9.5   Marie Tharp and Seafloor Spreading

The name of Marie Tharp, born in 1920, is one that became known only in the later years of her life, a considerable time after she made her great discoveries about the ocean floors. In the post-World War II years, she was in her early thirties. She graduated in 1943 from Ohio University, receiving a BA in English. At Ohio, she had taken a course in geology. As a consequence of that, she was recruited into a wartime program aimed at bringing women into the petroleum industry. The program took her to Michigan where, at the University of

CONTINENTAL DRIFT

Oceanic ridges create new crust that moves away the continents.

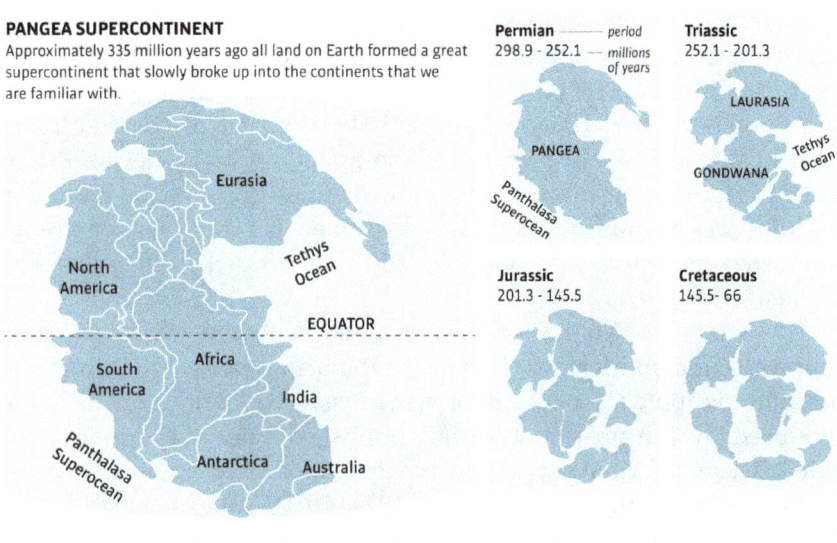

PANGEA SUPERCONTINENT

Approximately 335 million years ago all land on Earth formed a great supercontinent that slowly broke up into the continents that we are familiar with.

Eurasia

North America

Tethys Ocean

EQUATOR

South America

Africa

India

Panthalasa Superocean

Antarctica

Australia

Permian ——— period
298.9 - 252.1 — millions of years

PANGEA

Panthalasa Superocean

Triassic
252.1 - 201.3

LAURASIA

GONDWANA

Tethys Ocean

Jurassic
201.3 - 145.5

Cretaceous
145.5- 66

CONTINENTAL DRIFT AND TECTONIC PLATES

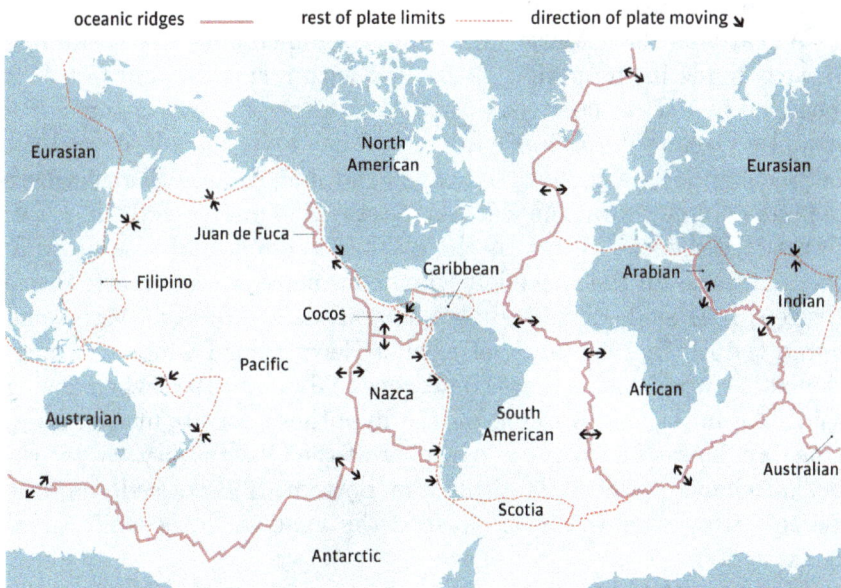

oceanic ridges ——— rest of plate limits ········ direction of plate moving ↘

Eurasian

North American

Eurasian

Juan de Fuca

Caribbean

Arabian

Indian

Filipino

Cocos

Pacific

Nazca

African

Australian

South American

Australian

Scotia

Antarctic

Figure 9.4 The tectonic plates, showing where land masses were at various times in the past (top) and where they are now (bottom) and how they are moving. Credit: Infographic sketched by the authors and elaborated by Javier Pérez Belmonte.

Michigan Ann Arbor, she completed a Master's degree in petroleum geology, a field previously barred to women.[45] In 1949, she moved to a job in the Lamont–Doherty Earth Observatory project of the Earth Institute at Columbia University in New York. The director there was Maurice Ewing (1906–1974), a distinguished geophysicist and oceanographer. Her position in 1949 was that of "drafter": she would draw technical diagrams of the results of mapping data. She gained a number of promotions from that position, but in 1968, her position was terminated and she continued her work under grant-supported contracts at Columbia until she retired from university life in 1983.[46]

By 1949, the "Cold War" between the USA and USSR was getting started, a political war in which submarines played an important role and mapping the ocean floor was considered a priority. The maps produced by the project were not published since they would provide critical information to the Soviet Union. For about 20 years, Marie worked with Bruce Heezen (1924–1977), converting the data he obtained from surveys made by the research oceanographic vessel *Venna* into maps and tracings.[47]

As early as 1952, Marie Tharp had combined data from sounding surveys made in the period 1948–1952 with a small amount of data gathered in 1921 to create six East-West scans of a part of the North Atlantic Ocean (shown in Figure 9.5). They revealed a V-shaped rift between under-sea mountain ranges, the mid-Atlantic Ridge where magma came up from the Earth's mantle. She interpreted the rift in terms of Wegener's theory as the place where material rose up from the mantle to fill the space left when continents moved apart. Bruce Heezen, it is said, dismissed this as "girl talk". Indeed, Marie had found the rift in 1952, and yet neither Heezen nor Ewing even mentioned her work until a press conference in January 1957 when Ewing said "We have been working on the hypothesis for about 5 years ... One day, Marie Tharp, a cartographer at the Observatory, was working on charts of the Atlantic Ocean bottom. She noticed that the deepest rifts in mid-Atlantic formed the locus of an oceanic earthquake belt."

Heezen had been interpreting the maps in terms of the variant of the Expanding Earth hypothesis for the motion of the continents that had been put forward by the Australian geologist Samuel Carey (1911–2002).[48] Like many other scientists of the time, he believed

Figure 9.5 Marie Tharp with her map of the North Atlantic seafloor. Illustration by Isabel Gálvez (www.isabelgalvez.eu) inspired by a picture from the Lamont–Doherty Earth Observatory and tracings of ocean depth at various points across the Atlantic.

that Wegener's continental drift was a fantasy. However, when a map of earthquake epicenters became available, it was clear that in the North Atlantic they coincided with Marie's rift. Marie was then convinced that she had been right and managed to persuade Heezen to abandon the expanding Earth picture on the basis of the rift-quake connection.

In 1956, Heezen had been credited with the discovery of seafloor spreading with no mention of Marie Tharp. The first glimpse of Marie Tharp's insight came only at a press conference.[49] Between 1959 and 1963, there were several important papers published, again based on her work on under-ocean mapping data, none of which included her as an author. She was obviously viewed as a research assistant, doing the bidding of her superiors without adding anything of importance to the science.[50]

Henry Frankel (1944–2019), in his seminal 2012 history of the subject, gives the opinion that the group of Ewing, Heezen and Marie Tharp, "did more than any other group in tracing the continuity of oceanic ridges and their rift valleys throughout ocean basins." Sir Edward Bullard (1907–1980), the director of the Cambridge University Department of Geodesy and Geophysics, declared that "Ewing and his associates" at Lamont–Doherty "discovered more things about the seafloor than any other group has ever done before."[51]

9.6   So How Old Is the Earth?

There were two fundamental questions that needed to be answered: How old is the Earth and how can we date the rocks in the strata where the fossils were found? The answer to the first question tells us how much time is available for a scenario like that of James Hutton (1726–1797) to work, but it tells us nothing about the rocks themselves. The answer to the second provides a timetable for events in the history of the Earth and its inhabitants, providing dates for the various fossils.

Neither question is easy to answer, and in the 19th century, scientists did not have the necessary knowledge to answer the questions. The second question could be addressed and partially answered using Steno's 1669 Law of Superposition[52] that the oldest rock layers lie below the younger ones, and the layering is time-ordered. Combining this with the observation that fossils of a given type are only found in some rock layers, we have a way of matching rock layers of the same age. This was the procedure used by William Smith in his coal pits and canals. It allowed his and allowed us to say which species co-existed and to put them into a time sequence. Thus Cuvier came to his theory of extinctions: a given species appears, lives for some time, and then simply disappears. and we can find the order in which the species corresponding to the fossils evolved. However, ordering by itself does not address the problem of knowing how many years in the past the various layers were created.

Lord Kelvin (Sir William Thomson at the time), in 1862, made a famous attempt to estimate the ages of the Sun and the Earth using models for their formation. He estimated the age of the Sun on the basis of how much energy could be extracted from its gravitational contraction as it radiates and cools. He knew of no other source of energy for the Sun and got an answer of 20 million years. A similar calculation for the Earth, assuming that it started in a molten state and cooled to its present temperature, gave him a range of 20–400 million years. That was not really consistent with the age he assigned to the Sun. Other astronomers, notably Henry Norris Russell (1877–1957) and Simon Newcomb (1835–1909), also got a solar life span close to 20 million years, not surprising since they assumed the same gravitational energy source that Kelvin used.

Shortly after the First World War, Arthur Stanley Eddington realized that gravity could not be the only source of energy for stars, in light of data on their masses, temperatures, and brightnesses. He added radiation pressure to models of stars and concluded that they must have an unknown source of energy which he speculated might be the hitherto unthought of nuclear fusion of hydrogen. When the old guard complained that the centers of stars could not be hot enough, he simply told them to go and find a hotter place! With this addition to the theory of stellar structure, the stars would fit the data and have lifetimes that in the light of the radioactive dating of Earth rocks seemed reasonable: billions of years.[53]

What had been missing from Kelvin's work was the knowledge of radioactive atomic processes. It was Rutherford who first suggested in 1904 that radioactivity could be used to date rocks and the Earth. That started a flurry of activity that was not, at first, very successful, and the early enthusiasm declined. However, in 1911, the year after he graduated, Arthur Holmes used the decay of radioactive uranium to lead to give an age of 350 million years for rocks from Norway classified as coming from the Devonian period. Holmes speculated further that rocks from the Archaean Eon[54] could be at least 1,600 million years old. Such timescales had never before been contemplated, but by 1927, he presented ages for rocks that he determined to be in the range of 1.6–3 billion years. More recent dating of Earth and moonrocks and meteorites that have landed on Earth now gives an estimate of 4.54 billion years, with a possible error of a few tens of million years either way.[55]

Perhaps, it should not surprise us that during the first quarter of the 20th century, radioactive dating aroused a considerable amount of skepticism in the more conservative corners of the geological community. The reported numbers were beyond the scope of their comprehension. They preferred to stick with Kelvin's compromise value of 100 million years.

Since Holmes had used the radioactive decay of uranium to lead for his measurements, only rocks containing uranium could be dated. In the years that followed, radioactive decays of other elements could be used to measure different ranges of times with higher precision. Other examples of radioactive dating use the decay of potassium to argon and of unstable isotopes of lead to stable ones. This allows different rocks to be dated thereby providing additional estimates, confirming the dates of geological formations. Nevertheless, the original uranium to lead decay remains the gold standard.

Chapter 10

Great Extinctions

Extinction is the rule.
Survival is the exception.

Carl Sagan, The Varieties of Scientific
Experience

We live in a constantly changing world. The history of the Earth is punctuated by the ebb and flow of everything we see about us. It is not only the mountains and oceans that change with time. All of the lifeforms on Earth are constantly changing too and sometimes in calamitous waves of change. With his theory of evolution, Darwin achieved lasting worldwide fame, but it was his friend Alfred Wallace who recognized that the rise and fall of species is a continual and rapid process.

10.1 Cold Case Forensics

Dinosaurs are children's favorite, and most people today have heard the claim that these great beasts that lived millions of years ago were wiped out when an asteroid struck the Earth. This was what is called the "K-Pg extinction event": it took place 66 million years ago.[1] What is less well known is that it was just the most recent of five major extinction events which have occurred during the past 500 million years and that there have also been many lesser extinction events. Were all these extinctions due to impacts? If not, then what? Is there a viable alternative theory? What is the proof? None of these questions has a simple answer.

There is also an apparent paradox: we are told that 99% of species that ever lived no longer exist, yet we now live in a world where we have greater species diversity than ever.

In order to understand what happened at the K-Pg event, we have to look hundreds of millions of years into the past where there are no testimonies of witnesses and the deceased are just piles of bones and bone fragments. Moreover, all that remains of the "crime scene" is a stack of rocks containing fossils and minerals: the "fossil record". The study of the extinctions is a cold case exercise in forensic science.

Fortunately, advances in science have provided tools with which to reconstruct much about these extinction events. We can set reasonably accurate dates on the rocks and even infer what the makeup of the atmosphere was by using our understanding of chemical processes. We can look to see what plants existed and so infer the availability of food for the animals. More astonishingly, we can reconstruct a map of the world at these times and run our climate models.

We begin by looking back at what we know about the past 541 million years. That sounds like a long time, but it is about 12% of the time since the Earth formed 4.54 billion years ago. The other 88% is not a total blank, but fortunately, we can address the question of those five great extinctions without going back beyond that 541 million year mark. What is so special about the 541 Myr?[2] It marks the beginning of the Cambrian geological period. This is the time when the first fossils appear which have hard body parts (and hence become more easily seen): there was an explosion of life forms, referred to as the "Cambrian explosion". It seems that by that time the progenitors of all the major animal phyla[3] had already evolved.

For a long time, scientists thought that there was very little complex life before that time. However, in recent years, evidence has come to light of many complex soft-bodied organisms' forms of prints in shales and sandstones[4] from the time just before that. To what extent these were the ancestors of the current extant lifeforms is not clear. There is no obvious connection: the Cambrian explosion of life is an outstanding mystery.

Figure 10.1 provides a quick overview of this 541 Myr stretch of time and provides us with some of the basic concepts. That stretch

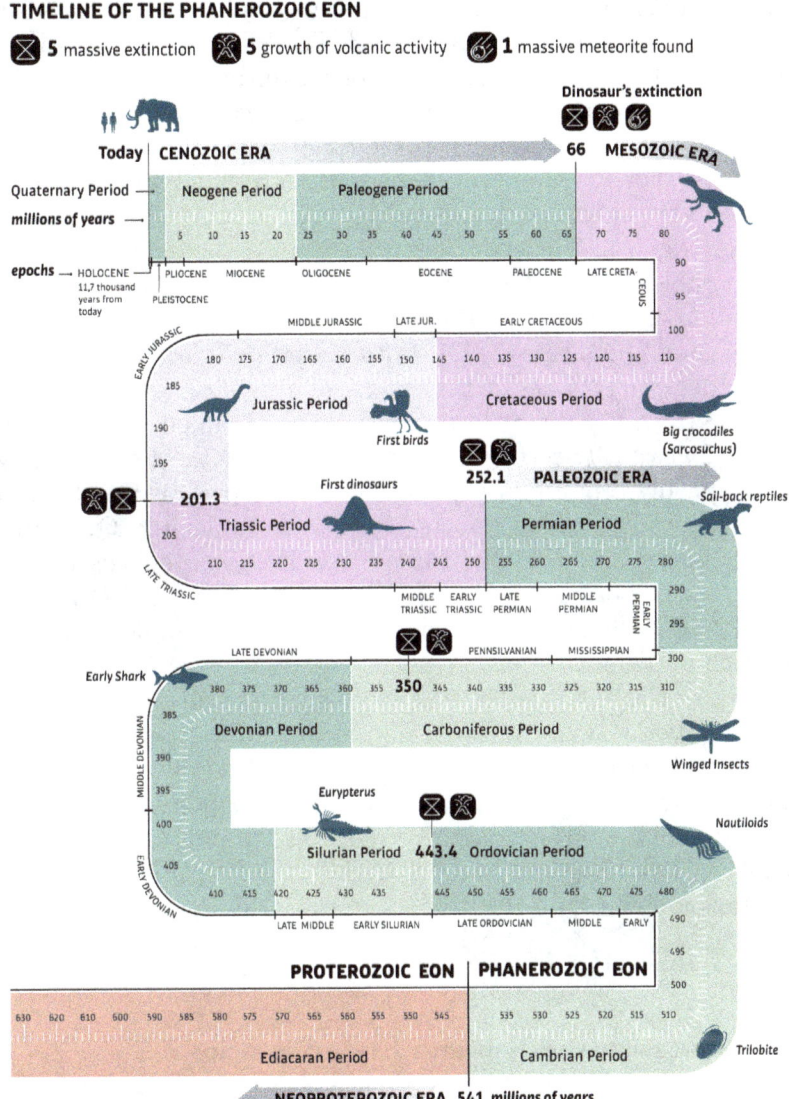

Figure 10.1 The past half billion years represents only 12% of the age of the Earth: the Phanerozoic Eon, which is divided into 12 "periods", each being characterized by its own ecosystem. Credit: Infographic sketched by the authors and elaborated by Javier Pérez Belmonte.

of time is referred to as the "Phanerozoic Eon", the age of "visible animals" or the age of "revealed life".[5] The Phanerozoic was divided into twelve "Periods" defined by geologists during the first half of the 19th century to reflect the times when there were radical changes in rock formations and the fossil content.[6] It is hardly surprising that some of the boundaries between Periods correspond with significant changes to the species of fossils occupying the rocks — this was how they were defined. These Periods were given obscure and somewhat unmemorable names: **C**ambrian, **O**rdovician, **S**ilurian, **D**evonian, **C**arboniferous, **P**ermian, **T**riassic, **J**urassic, **C**retaceous, **P**aleogene, **N**eogene and **Q**uaternary.[7] The word "Cambria", for example, is the Roman name for Wales, where rocks of this kind were first studied in 1831 by Adam Sedgwick while accompanied by the young Charles Darwin. The time spanned by each of the Periods varies, being typically around 45 Myr. Each of these periods is subdivided into geologically defined epochs. So, for example, the Quaternary period divides into the Pleistocene and Holocene epochs and the Carboniferous period divides into the Mississippian and Pennsylvanian epochs each of which divides into Upper, Middle and Lower subepochs.

As shown in Figure 10.1, the Geological Period previous to the Cambrian Period was the Ediacaran Period which consisted mostly of simple lifeforms that had neither skeletons nor shells and which appear to have left no identifiable descendants. The Age preceding the Cambrian Period is simply referred to as the Precambrian, lasting from 4600 Mya until 541 Mya.[8] Together, the Precambrian Age and Phanerozoic Eons span the entire history of the Earth.[9]

10.2 The Rise and Fall and Rise of Species

Now, with the geological timescale established, we can see in more detail which kinds of animals lived at what times. We describe animals biologically by grouping them into *species*, *genus* (plural *genera*) and *family*. Animals of the same *species* can interbreed. A *genus* in biological classification is a group of species that are closely related through common descent. Whereas members of the same species can reproduce and produce fertile offspring, members of the same genus that are not of the same species generally cannot. A *family* is a group of genera that share some attributes such as body form.[10]

A familiar example is the family of present day Great Apes that are typified by a relatively large body frame, hairless faces, protruding lips, flat nails and complex fingertips. The Great Apes also have no tail. They also have rotating shoulders and large and complex brains. The family of Great Apes consists of 12 species which are divided into four groups or genera. The four genera are orangutans, gorillas, chimpanzees and humans.[11] There are three living species of orangutan, two species of gorilla and two of chimpanzee, and one species of human.[12] The humans have 97% of their DNA in common with the orangutans and the gorillas.[13]

This shows that the extinction of a species is biologically less serious than the extinction of an entire genus, which in turn is less catastrophic than the extinction of an entire family of genera.

It was Georges Cuvier who, in 1813, used the word "extinction" to describe the disappearance of fossils from the fossil record. He attributed the extinctions to unknown catastrophes, a hypothesis that was referred to as "catastrophism". An alternative hypothesis was later proposed by Charles Darwin in his 1859 book, *On the Origin of Species*. This was the Theory of Evolution in which older species disappeared because over long periods of time they evolved into new species that differed systematically from their ancestors. Darwin's view was referred to as "gradualism". The battle between the two views has been long and acrimonious, but the Darwinian approach has largely prevailed, becoming an entrenched dogma. The Darwinian dogma prevailed for a century during which the effects of geographic change and climate change on habitat and on evolution were not only ignored but vehemently rejected.

This establishment view was well represented by George Gaylord Simpson of the American Museum of Natural History and Columbia University. He wrote two highly regarded texts on the subject: *Tempo and Mode in Evolution* (1944) and *The Major Features of Evolution* (1953) which set the view on extinctions for the 1960s: the disappearance of species was an inevitable consequence of the cycle of natural selection and the slow emergence of new species to fill in the gaps.

An important step was taken by Norman Newell in the early 1960s when he was a professor of geology at Columbia University. Stephen

Jay Gould was one of Newell's PhD students who described Newell enthusiastically as the scientist who brought biology into the world of palaeontologists who had been mainly trained as geologists with only a smattering of knowledge about invertebrates.[14] By that time, there were numerous articles about the fossil record distributed throughout the literature, but no possibility of integrating them into a single coherent database. Newell was able to collect a sample of 2500 families (as opposed to species) that provided the possibility of binning[15] the data in time-bins of around 10 Myr. The data and its presentation in clear graphical form were presented in an article Newell wrote for the February 1963 edition of the *Scientific American*.[16] With the available data, Newell was able to show that there were several periods of extinction and, importantly, that despite the extinctions there was an increase in species diversity. However, Newell's time resolution and small sample size were not enough to get down to the detail of the extinctions.

In 1982, David Raup and John Sepkoski[17] presented a new compilation of data on 3300 fossil marine families that revealed the detail of five rapid extinction periods, each followed by a rise in diversity. Then, in 1984, Sepkoski produced a detailed view of the growth in diversity in a graph, displayed here as Figure 10.2, which shows that extinctions which take place in a mere million years and are followed by Radiations of species over a period of tens of million years. The survivors adapt to new ecosystems and habitats and thrive, until they too suffer an extinction. It should be noted that this graph is based on counts of biological families[18]: a family consists of genera which are themselves collection of species. The death of a family implies the disappearance of all its constituent genera and their species. The extinction of a family is generally a disaster that has a far more significant impact than the disappearance of a species.[19]

There are some notable dips, short periods during which there is a rapid decline in the numbers followed by a slower rise. These tend to occur at or near the transitions from one period to the next, but this is not a coincidence. It is the way the geological periods were first defined and named almost 200 years ago when geologists recognized large changes in the fossil flora and fauna.

There is a continual rise in the number of families from the start of the Cambrian Period until the first extinction at the transition from the Ordovician Period to the Silurian, 440 Mya. Then, there

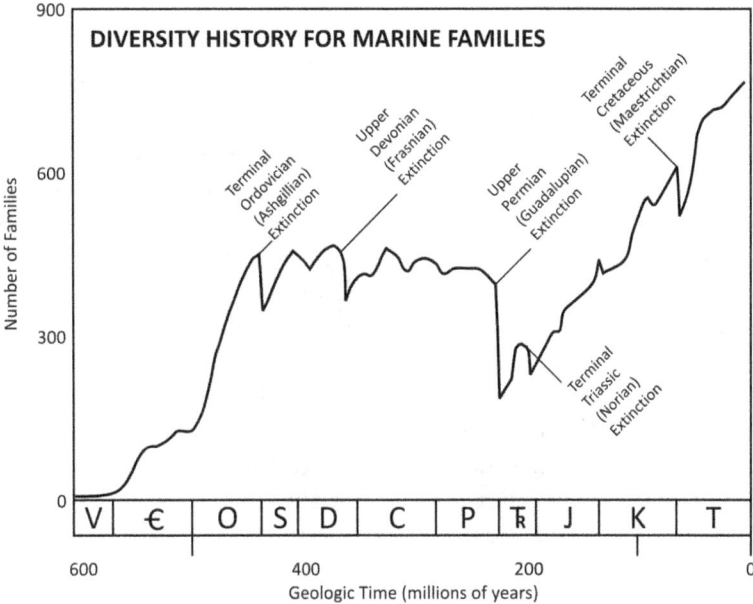

Figure 10.2 Growth of species diversity. Credit: Adapted from Sepkoski 1984 *Paleobiology*, 10, pp. 246–267. The curve is the evolution of the number of all known families of species that ever existed. The sharp dips mark the extinctions. They last a brief million years and are followed by a rapid re-generation, a "radiation", of life over a period of some 10 million years that leads to greater diversity. (Abbreviations for names of geological periods follows the recommendation of the U.S. Geological Survey Geologic Names Committee 2010.)

is another similar rise after the time of the Great Permian extinction when the Permian Period gives way to the Triassic. These rises, or "radiations" as they are called, keep on going despite continual extinctions.

Simple evolution by Darwinian Gradualism was not what was happening. Darwin's friend, Alfred Russel Wallace, had published a paper in 1855, 2 years before Darwin's *On the Origin of Species* was published, arguing that environmental factors were of central importance in evolution. Darwin advocated selection within the same species, whereas Wallace saw the influence of environmental factors as the key driver of evolution. Wallace was in fact correct. Without these external influences, gradual evolution is far too slow, whereas we see very rapid recoveries of species following an extinction event.

The number of species increases rapidly as the survivors adapt and take over the vacated habitats.[20]

The extinctions themselves become clearer if we simply count the number of species, genera or families that were alive before a given time and count the number of those in the sample which have disappeared during the following few million years. We express this as a percentage. The extinctions graph shown in Figure 10.3 is created from data on genera from Sepkoski and others.[21] The original version of this figure was due to Newell in the 1960s and updated by Raup and Sepkoski. Details of these extinctions are summarized in Table 10.1.

There are clearly more than the famous five extinctions. One gets the impression that extinctions are going on all the time at different levels. Counting up the total death toll in the number of species that ever lived shows that only 1% of them survived to the present day. Some 90% of that death toll is due to the five great extinctions.

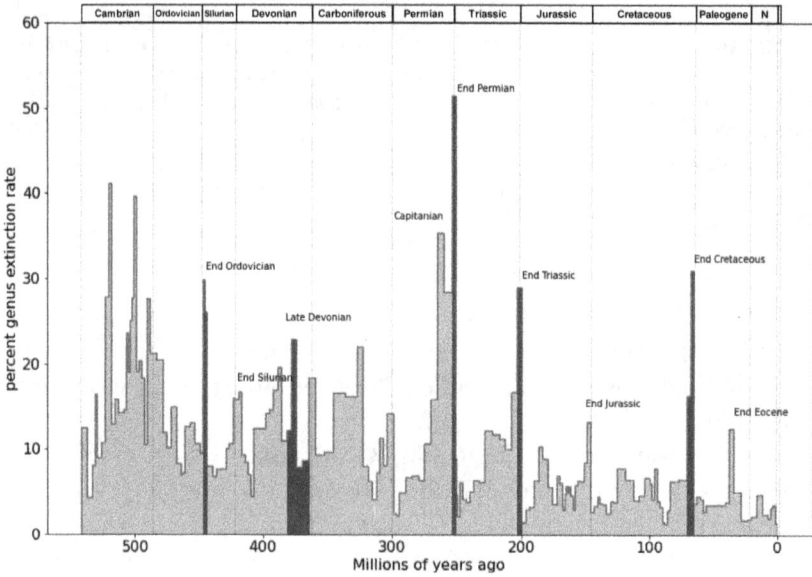

Figure 10.3 Fraction of genera becoming extinct within short intervals of time. The big five extinction events are shown along with a few others, all of which are associated with major volcanic activity in Large Igneous Provinces. Only one, the End Cretaceous extinction, is generally thought to be associated with an impact. Data from R.A. Rohde and R.A. Muller. (2005). *Nature*, 434, 208–210.

Table 10.1 The five great extinctions.

Name	Myr ago	% Extinction			Possible cause of extinction / Examples of what died
		F	G	S	
Ordovician–Silurian	444	26	60	85	Ice age with lower sea levels followed by a rapid melting. Many trilobites, corals, some brachiopods and hard-shell marine invertebrates.
Late Devonian	383–359	22	57	83	Many pulses of climate change and reduced oxygen. Possibly result of volcanism. Armoured fish and trilobites and corals.
Permian–Triassic	252	51	82	95	Volcanic Activity (Siberian Traps). Destruction of forests. Marine species, insects, amphibians and all trilobites.
Triassic–Jurassic	201	22	53	80	Volcanic activity at the then-margins of the Atlantic. Many reptiles and ribbon-like fish.
Cretaceous–Paleogene	66	16	47	76	Bolide impact and volcanic activity (Deccan Traps). Last of the nonavian dinosaurs, ammonites and molluscs.

Notes: The table summarizes the five great extinctions in terms of the percentage of Families (**F**), Genera (**G**) and Species (**S**) of animals (fauna) that went extinct within a period of a few million years around that time. The numbers are from Anthony Hallam's 1998 paper "Mass extinctions in Phanerozoic time". The last column gives the likely cause of the extinction and names a few of the fauna that did not make it through to the next geological period. The numbers do not distinguish between land and sea animals nor geographical location, so they are merely indicative values. As seen in Figure 10.3, extinctions have been going on at a significant rate all the time so much so that 99% of all species that ever lived are now extinct.

DINOSAUR'S DEATH DIAGRAM

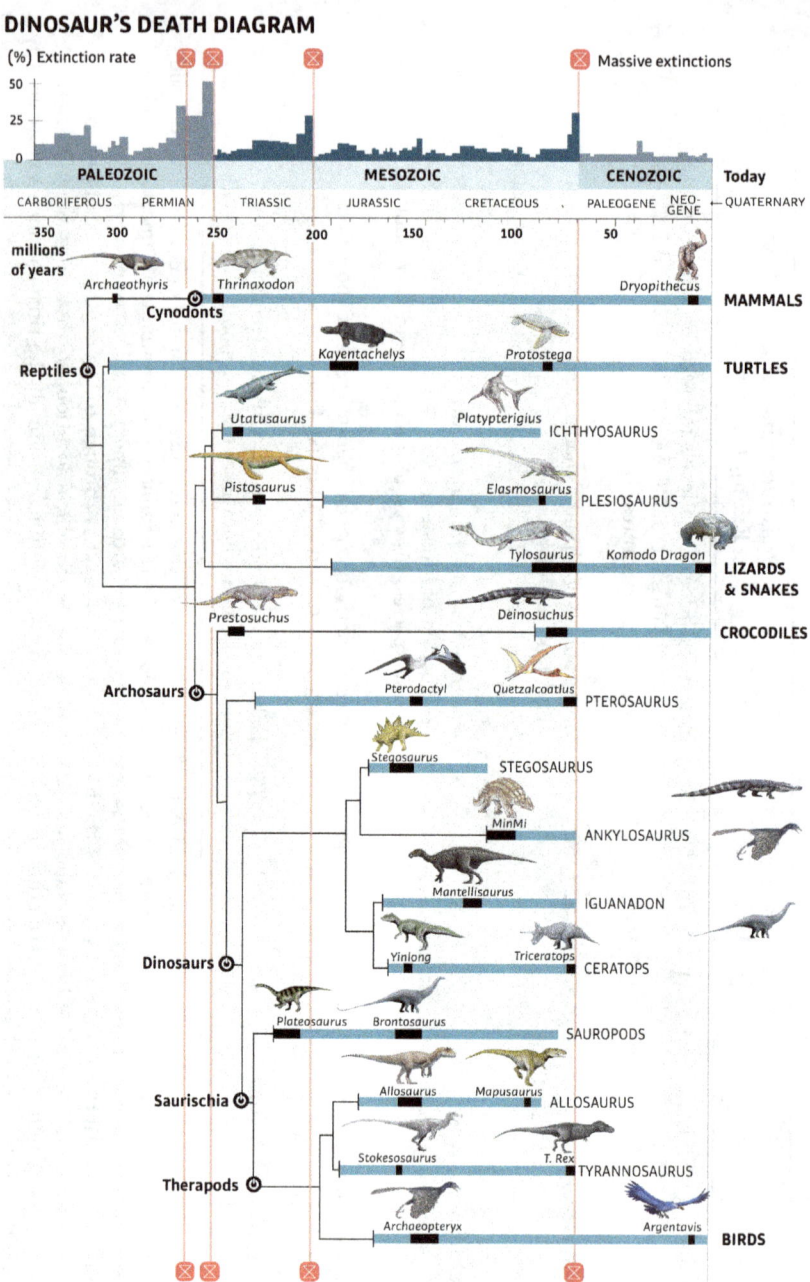

Figure 10.4

A significant factor is that the famous five are worldwide catastrophes, they were not caused by local events.

What, with all these extinctions, went extinct and when? It is easy to think of an animal and look up in *Wikipedia* to find out when it lived, but that is no more useful in giving a historical context to the life than in looking up the dates of birth and death of some obscure monarch. So, here we shall choose groups of biologically related animals, as in "all Tyrannosaurus-like", and chart when they lived. After all, one Tyrannosaur is pretty much like another: big and fierce with little arms. From sufficiently well-preserved fossil remnants, any animal or plant can be biologically grouped with any of its descendants, and we can see when the group first appeared and when it disappeared.

Biology provides a scientific framework which depicts the descent of species, organisms and genes from a common ancestor: the "phylogenetic tree". A small part of the Tree of Life is shown in Figure 10.4. This part of the tree focuses on the descendants of the earliest reptiles. Among those descendants are the dinosaurs. An early off-shoot from this tree is the branch that leads to all the mammals.

In the figure, the horizontal thick gray line segments define groups of species that share a common ancestor via a parent–child relationship. Examples of specific animals are represented by black rectangles on the gray lines. The length of the black rectangle indicates the period between when the animal first appears in the fossil record and when it vanishes. At some point in their past, two of these lines may share a common ancestor. The common ancestor is the place where the descendants branch off in two different directions.

We notice that there are animals which are popularly described as dinosaurs but were not in fact dinosaurs. For example, neither the pterodactyl, which is a pterosaur, nor the pistosaurus, which is a

←───

Figure 10.4 (*see figure on facing page*) The family tree of Reptiles and their descendants. Each thick horizontal line depicts a parent–child related group of species. The black rectangles on the thick lines depict the interval of time during which the displayed animal lived. The thick lines start and end with the birth and death of the group (dates from Wikipedia, 2020). Biologically related groups are connected by thin black lines, as in a family tree. Credit: Infographic sketched by the authors and elaborated by Javier Pérez Belmonte.

plesiosaur, are dinosaurs, but dinosaurs, pterosaurs and plesiosaurs are all archosaurs. Crocodiles and birds are also descended from the same common ancestor, the archosaurus, that lived 240 Myr ago. This was the ancestor of all the dinosaurs. However, other extant reptiles are not descendants of the archosaurus and, hence, perhaps surprisingly, crocodiles are more closely related to birds than they are to lizards, snakes and turtles.

The positions of the four great extinctions that occurred within the last 350 Myr have been marked on the diagram. We see that the dinosaurs emerged out of the great end-Triassic extinction, 200 Mya. Some of our dinosaur-like friends died in an extinction event, but others did not. Some survived the catastrophe only to die out in the changed environment that followed. In particular, the diagram shows that only one line of dinosaur descendants made it through the K-Pg boundary extinction. These were the birds, which had evolved and been flying around in the Cretaceous. The K-Pg boundary was the end of the nonavian dinosaurs.

One curious aspect of these extinctions is that land plants seem to have been largely unaffected by whatever killed off the fauna. The only convincing evidence for an enhanced rate of extinction of land plants is at the Permian–Triassic extinction when the rate of extinction rose to about six times the normal rate of plant attrition.[22] There is marginal evidence for an earlier plant extinction event having occurred at the Carboniferous–Permian boundary. Importantly, the flowering plants ("angiosperms") that appeared around 130 Mya, rapidly grew to dominance during the Cretaceous and seem to have sailed through the Cretaceous–Paleogene transition with no evident ill effects. The previously dominant form of plant life, the "gymnosperms" (flowerless plants that produce cones and naked seeds, as in conifers[23]), had originated in the late Carboniferous and become dominant during the Jurassic and early Cretaceous. They too survived the K-Pg extinction, though with continually decreasing diversity.

10.3 So What Caused the Extinctions?

By the end of the 1970s, there was a greater awareness of the Earth's history, and the events of times long past, both in the scientific community and among the general public. In particular, dinosaurs and their fate were much discussed and speculation about the cause of their extinction was common. The emergence during the preceding decade of plate tectonics as the explanation for continental drift provided a natural framework for the discussion. Plate tectonics provided a basis for understanding the geological and climatic changes that had taken place throughout the last 540 Myr of Earth history.

The strongest case was focused on volcanic activity and its effects on Earth climate. By this time, the anti-drift dogma was being eroded. Mantle convection provided acceptable evidence-based understanding of continental drift: a new chapter in geology. At the same time, it was becoming apparent that this new version of Earth History could also account for extinctions through the climate and habitat change that was a natural consequence of the changing Earth. The hypothesis that large-scale volcanic activity might be the driver of mass extinctions grew more plausible.

The evidence from sea floor spreading showed that the levels of volcanic activity varied considerably over the geological time span. Extreme volcanic activity, with its associated climate change, had always been thought of as the likely cause of large-scale extinctions.

The seafloor spreading rate maps the level of underwater volcanic activity: the greater the amount of magma issuing from the ocean ridges (i.e. volcanic activity), the faster the ocean plates are driven apart. This provides a starting point for making a quantitative study of the relationship between volcanism and species extinction rates. A qualitative relationship between species diversity and continental movements was already being noticed in 1970. In a paper in *Nature*, Kennett and Watkins presented evidence in deep sea sedimentary cores taken from the Antarctic Pacific Waters that some volcanic maxima occurred at times when the magnetic field orientation was changing. In the same year, and also published in *Nature*, James Valentine and Eldridge Moores were able to correlate species diversity with patterns of continent assembly and fragmentation.[24] A year later Valentine published a seminal paper discussing shallow marine diversity in the context of a plate tectonic model.[25] The paper examined the current geography of continents and other land features and related it to marine bio-geographic patterns, especially to patterns of species diversity and provinciality.

An important breakthrough came in 1982 when Raup and Sepkoski published results on extinction events derived from their important compilation of marine fossil data covering the entire Phanerozoic Eon. They found that the pattern of extinction of land-based fauna was reflected in the marine environment. Their substantial data set provided the means to quantify species extinction rates.

In 1984 and 1986, the Raup–Sepkoski data were used by a group in Copenhagen[26] to demonstrate a strong correlation between seafloor spreading rates (tectonism), measured in both the Atlantic and Pacific oceans, and variations in marine fossil diversity. The correlation for dinoflagellates and the calcareous nanoplankton was significant at the 99% level. In their words: "a close relationship exists not only between major rifting events and biological crises but also between relatively small scale tectonic processes and minor evolutionary trends". This provided strong evidence-based support for Valentine and Moores' contention that large-scale volcanic activity was the cause of extinctions.

10.4   Visitations from Space

A possible alternative to volcanism was that the cause of extinctions might be of extraterrestrial origin. The 1950 suggestion by Otto Schindewolf that extinctions might be driven by supernova explosions was not pursued, nor were similar suggestions that the cause could lie in disruptions of the Solar System by passing stars, or oscillations of the Solar System through the spiral arms of our Galaxy.[27] These alternatives all seemed somewhat contrived with no evidential support other than the phenomenon they sought to explain.

In 1973, the astronomer Ernst Öpik and the 1934 Chemistry Nobel Prize winner, Harold Urey, independently published papers within a month of one another looking into the consequences of an impact of a comet or asteroid hitting the Earth. Both Öpik and Urey were born in 1893 and both were 80 years of age in the year when they wrote these articles.

There are many asteroids in the Solar System. Tens of thousands have been identified and cataloged since the day, the very first of the 19th century, when the Italian astronomer Giuseppe Piazzi discovered Ceres, the first known asteroid, with a diameter of

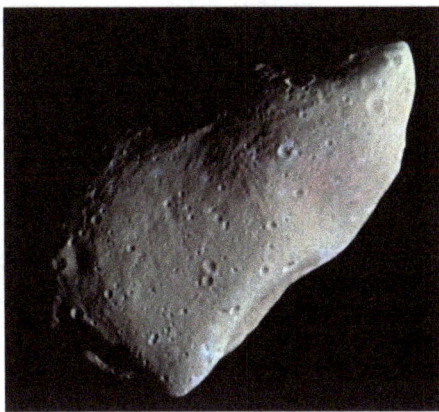

Figure 10.5 The Gaspra asteroid photographed by the Galileo space probe in 1991. It is an irregular body with approximate dimensions of $19 \times 12 \times 11$ km. Craters caused by impacts with other, smaller bodies can very clearly be seen on its surface. In fact, it is very likely that Gaspra itself is actually a piece of a once-larger asteroid that broke up as a result of a major collision. Image courtesy of NASA/JPL/California Institute of Technology.

1,032 kilometers. Like Ceres, most asteroids orbit around the Sun, following orbits located between Mars and Jupiter, the so-called asteroid belt. The asteroid Gaspra is shown in Figure 10.5. The total mass of all these asteroids is less than 5% of the Moon's mass. There are also asteroids closer to Earth which, along with comets that occasionally come near our planet, are known as NEOs (near-Earth objects). The efforts of astronomical observatories and worldwide space agencies detecting and cataloging these objects and the analysis of their orbits are necessary because, although it is unlikely an asteroid will hit Earth soon, it is certain to happen eventually.

There are three types of asteroids in Earth's vicinity. These have been given striking names: Amor, Apollo and Aten. Each type follows a different type of orbit. The orbits of the Amor group come near but do not cross Earth's orbit; those of the Apollo asteroids cross Earth's for periods of over a year, and Aten's asteroids cross it for periods of less than a year. It is thought that there are around a thousand asteroids near Earth that have diameters greater than a kilometer.

The impact of one of these larger asteroids would be devastating. An asteroid with a diameter of a kilometer would, in addition to creating a sizeable crater, cause large-scale destruction on our planet.

On 24 July 2002, the Lincoln Near-Earth Asteroid Research (LINEAR) team published its discovery of (89959) 2002-NT7, an asteroid from the Apollo group with a diameter of about two kilometers. A preliminary calculation involving few data on its orbit led various sensationalist media outlets to announce that the asteroid could quite probably collide with Earth on 1 February 2019.

The press, especially in Britain, reported the development through headlines such as[28] "Space Rock on Collision Course". This caused some level of alarm among the public, even though it was known that the probability of a collision was very low: one in a million. The story prompted a debate on how scientific news of this kind should be communicated. NASA's Tony Philips said at the time that it was as though the newspapers had announced that a person was going to win the lottery just because she had been seen buying a ticket. One of his colleagues, Donald Yeomans, the director of Jet Propulsion Laboratory's NEO program office, made this comment about communicating scientific discoveries to society[29]: "The goal, I suppose, is to be at the same time sober, informative but not too nerdy". Of course, in February 2019, no one remembered the media uproar from 17 years earlier, and the asteroid calmly made its way past Earth at a distance of about 60 million kilometers (40% of Earth's average distance from the Sun).

Öpik estimated the probabilities of impacts and the resultant crater sizes for impacts by "stray bodies" of various sizes and compared these with known crater sizes on Earth and on the Moon. Based on 1970s astronomical knowledge, he estimated 18 impacts of objects bigger that 8.5 km during the 4.5 billion year lifespan of the Earth: one every 250 Myr, guessing that such an object would affect less than half of the Earth.[30] Urey's article, which was only suggestive, considered the probable consequences of an object 1 km in diameter, weighing perhaps 1 gigatonne, hitting the ground at 45 km per second.[31] While this would be insufficient to cause worldwide extinction, he put forward the opinion that such an impact might stimulate wide-area volcanism.

Then, in 1980, a discovery took place that completely changed the tenor of the debate. Walter Alvarez was a young University of California Berkeley geologist working with the nuclear physicists Luis Alvarez (Walter's father), Frank Asaro and Helen Michel. They examined samples of the clay layer boundary between the Cretaceous and Paleogene periods that Walter had brought back from Gubbio in Italy. His father, the winner of the 1968 Nobel Prize for Physics, had suggested examining the samples for element abundances. The samples were found to contain an anomalously high level of the rare chemical element iridium. Iridium excess is sometimes found in meteorites, so they postulated that a large asteroid had hit the Earth at 66 Mya, causing the iridium excess and the mass extinction.

The research was published in *Science* and followed by a conference held in October 1981 at Snowbird in Utah. The conference was well attended by the nuclear physicists who supported the Alvarez thesis, and the paleontologists who were not at all convinced that the proposed impactor was responsible for the demise of the nonavian dinosaurs.

By all accounts, the meeting was at best frosty. The paleo-people were not to be diverted from their research and continued to work as they had always worked, feeling that the great extinctions were associated with times of volcanic activity and a consequence of the ever-changing Earth.

That was the beginning of another story and the rise of a new dogma.

Chapter 11

The Dinosaur Extinction Debate

So comes snow after fire, and even dragons have their endings.

J.R.R. Tolkien, The Hobbit

Science occasionally captures the public imagination. Few people get worked up about the discovery of the Higgs Boson, looking into the face of the Big Bang generates a little more interest, but ask a child of 10 years of age what killed off the dinosaurs and the answer will come back instantly: "a meteorite". Ask what a meteorite is and the answer is less certain, but they all know what a dinosaur is, or was. People make jokes about dinosaurs. "Question: Why did the Archaeopteryx catch the worm? Answer: Because it was an early bird!". Nobody makes jokes about the Higgs Boson.[1] The 20th century animated cartoon Flintstone, the family has a pet dinosaur, called Dino. The 21st century Simpsons children know all about dinosaurs. The cartoonist Gary Larson frequently featured dinosaurs in his "Far Side" images.

The strange thing is that the public view of the extinction of dinosaurs is entirely one-sided: a thing from outer space annihilated them allowing mammals to evolve into us. Children are taught this as a "known fact", and so it goes almost unquestioned. How did this situation come about?

The story starts in 1980 when one Nobel Prize-winning scientist, who was neither a geologist nor a biologist, launched an idea that grabbed media attention and the public imagination. He attributed the cause of the last great extinction to the impact of a large object from space, and this scenario was promoted as "The Death of the Dinosaurs". The essence of the conflict was whether the forces driving evolution were of terrestrial origin, in particular volcanic activity and shifting continents, or extraterrestrial, as in a supernova explosion or impact of a massive body. It was a takeover of a branch of science by an alien community that created a publicity-driven juggernaut and openly derided the existing community.[2]

11.1　The Bolt from the Blue

The takeover began in June 1980 when the premier US journal *Science* published a 14-page paper entitled "Extraterrestrial cause for the cretaceous-tertiary extinction". The authors were Luis Alvarez, Walter Alvarez, Frank Asaro and Helen Michel.

The aftermath of the 1980 paper was enthusiasm on the part of those working in the fields of physical chemistry and nuclear physics, mirrored by reticence on the part of the majority of paleontologists and in particular those who were specialists in vertebrate biology. The relentless avalanche of papers published by the impact enthusiasts far outweighed the slow and deliberate data sifting of the paleontologists who were carefully tracing the lineage of diverse species through the times of environmental crises. Added to this was the often abrasive character of Luis Alvarez[3] which led him to deride researchers outside of Physics and Chemistry as being mere "stamp collectors".[4] The debate descended to one of the all-time lows of scientific discourse.[5]

It was not only the weight of papers being published by the "impactors" but also their rapid organization of and participation in numerous international conferences, notably one at Snowbird in Utah.[6] Press releases were received with glee by the scientific press who used imaginative graphics to depict the event and by the broadcast media at large. The bolide impact was seen as the ultimate scientific drama of our past.[7]

There have been many suggestions that the journal *Science* was discriminating against articles that did not support the impact theory for the extinctions. As it happens, the editor of *Science* in 1980, Philip Abelson, had been Luis Alvarez's graduate student in 1939 and then a long-time colleague. When presented with the Alvarez manuscript, Abelson had responded that it was far too long and that the journal had already published far too many dubious papers on the subject. Abelson bent to Alvarez's pressure: Alvarez halved the length and the paper was accepted despite still being twice as long as the typical lead paper in the journal.

11.2   The Father and Son Team

It had been noticed during the 1970s that some elements present on Earth in very small quantities, platinum, iridium, osmium, and rhodium, were far more abundant in meteorites than on Earth. The idea had arisen that perhaps nearly all of these elements had been deposited by ablation of meteorites throughout the past history of our planet and that the presence of iridium in soil deposits might provide an alternative dating method. Just measure the amount of iridium down to a given depth and you have a date estimate.

In 1977, Walter the geologist had collected samples from Gubbio in the Apennines mountain range of Italy. Gubbio was a place where the geological Cretaceous–Paleogene ("K-Pg") boundary[8] was exposed as a 1 cm bed of clay with the Cretaceous on one side and the Paleogene on the other. His interest was to study the changes in the population of small marine creatures across this boundary. There were questions that needed to be answered: How old is that layer of clay and was how long had it had taken to deposit it? On return from Italy, Walter talked to his father Luis about that. Luis suggested measuring the amount of iridium in the layer: measure the mass of iridium and you have the age of the layer. Luis, neither a geologist nor a paleontologist, was enthused by the subject and so they both returned to Italy in 1978 to gather more sample material.[9]

With the help of Frank Asaro and Helen Michel from the Lawrence Berkeley Laboratory, who had access to the equipment necessary to measure the quantities of rare elements in samples, the abundances of 28 elements were measured and 27 were found to be normal for rocks in the Earth's crust. Only one, iridium, was exceptional with an abundance of 5.5 parts per billion, in comparison with the standard crustal abundance of 0.1 parts per billion. It was known that cliffs in Denmark, just to the south of Copenhagen, exposed the K-Pg boundary. These were shown to have a three times greater excess of iridium compared to the Italian samples. Significantly this excess manifested itself only in that thin layer: there was no abnormal trace of iridium on either side of a thin layer. This looked like instantaneous dumping of iridium: an impact by a meteorite, comet or asteroid. Whatever it was, an estimate for the mass of the impactor suggested that it would have had a diameter of some 10 km and thus be associated with a crater some 150–200 km in diameter.[10]

11.3   The Coup de Grace

The immediate question was as follows: Where is the evidence for the perpetrator of this hypothetical event? Not surprisingly, a number of research groups were on the hunt for the impact crater.

The key evidence had in fact been found in 1978 by two geophysicists: Glen Penfield and Antonio Carmargo. They had been involved in oil exploration around the Yucatan Peninsula on behalf of a petroleum company owned by the Mexican government, Pemex, and had been able to map a 180 km diameter feature at a depth of some 10–20 km near the municipality of Chicxulub (pop. 4,000).[11] They later identified the feature as an impact crater from the Cretaceous–Paleogene era. Apparently, Pemex had known about this feature even before that time but had withheld the information as valuable intellectual property.

Of course, with the publicity surrounding the 1980 paper, Penfield was keen to announce their discovery and so attended a conference of the Society of Exploration Geophysicists where he presented the details of the discovery. It was a poor choice: the conference was not well attended and certainly not by the "right people". Penfield had maps and measurements but was not allowed to show samples as evidence for the impact. His report attracted little attention. Knowing who to talk to and how to grab their attention is a primary requirement in getting listened to.

Alan Hildebrand and William Boynton of the Department of Planetary Sciences at the University of Arizona spent several years from 1984 onwards defining the likely characteristics of such a crater and narrowing down the search until, in 1990, they got information from a *Houston Chronicle* reporter that Glen Penfield had told him a decade before about his earlier discovery of an impact crater. The rest, as they say, is history: Hildebrand contacted Penfield, and they went and got more samples, finding that these samples fulfilled the requirements set out earlier in the work with Boynton. They then found glassy fragments that would be associated with an explosive impact[12] in their samples and dated them to the time of the K-Pg boundary. In 1991, they announced the discovery, with Penfield as the first author.

The discovery of a credible candidate for the impact crater radically changed the scientific attitude to the impact theory. It now made sense to spend effort on understanding the physical properties of the impactor, accurately dating the impact and understanding what the global consequences of such an impact might be. Accurate timing could place the impact event either at the beginning, middle or end of the period of extinction claimed by the paleontologists.

11.4 The Dogma Is Established

After the 1991 announcement of the discovery of the Chicxulub impact Crater, and its identification with the massacre of the beloved dinosaurs, all bets were off as regards proposing an alternative cause for the Cretaceous–Paleogene extinction.

The existence of such a crater had been a necessary requirement of the impact hypothesis, Alvarez himself had stated as much. The discovery was taken as proof of the hypothesis, all other suggestions were made by "impact deniers". The press, the media, in general, and the editors of learned journals were loathed to contemplate any alternative idea as a crank theory by people who could not face the reality of the situation: the evidence for the impact. The learned journals such as *Nature* published a number of papers on the impact theory during the following decade, and only one that ran counter to the newly established narrative.[13]

People immediately began to speculate about the detailed consequences of the impact and how it could lead to the observed worldwide extinction. The impact would certainly have destroyed every living thing in its vicinity. It might have caused serious problems as far as what is now the Dakotas and Montana, the location of the Hell Creek Formation that is so rich in dinosaur fossils laid down at the time of the K-Pg extinction.[14] But the extrapolation of the extinction process to the entire planet depends on models calculated from nuclear explosion simulations.

There were many off-hand assertions about poison gases and dust being generated and blocking out the Sun, causing the famous "nuclear winter" that had been talked about during the cold war. Mere indications, however plausible, are not proof: many paleontologists and paleobiologists, while not doubting that there was an impact, had serious questions about the biological and environmental details of the impact story.

More recently, global climate models have been used to describe the aftermath of the impact, but difficult questions remain. How long would the effects last after the impact, and why should they be so species selective? Fossil evidence suggests that the extinctions took place over an extended period, starting before the impact occurred. While many species went extinct, many others remained. Suggestions that the dinosaurs became extinct because the vegetation on which they fed was destroyed by severe climate disruption do not accord well with the fact that plants did not everywhere suffer an extinction at the K-Pg boundary.[15] However, it is perhaps noteworthy that in the area that is now Montana and Dakota some 80% of plant taxa were wiped out, suggesting that this area bore the brunt of the asteroid impact and suffered a substantially higher level of damage compared with the rest of the world.

The paleontologists and paleobiologists did not go into hibernation: there was plenty more to study during the 500 Myr or so of the Phanerozoic Period, including several major extinctions which were, so far, not attributed to an impact. More and more data were being gathered, the genealogy of large numbers of species was being reorganized in various different ways, and more accurate ways of assigning dates to fossils and analyzing their environment were emerging. Nor did they change their opinions concerning the cause of the K-Pg extinction: on the contrary, there were even more questions that needed answering.

So why, for at least a decade, was there no debate in those important journals? As stated earlier, there might have been reticence on the part of the scientific media to promote what might be seen as "crank science". Had a paper been submitted, it might well have encountered a referee of the opposite persuasion.

11.5 The Role of Public Opinion

A key difference between the impactors, led by Alvarez and his group, and the volcanists, who were more diverse, is that the impactors were very much better at publicizing their work. They had a big advantage for this, because the theory that an asteroid strike caused the death of the dinosaurs is a much more dramatic tale to tell, and leaves a more lasting effect on the public audience. The power of suggestion on people can be seen in other similar situations. For example, when Walt Disney Studios made the film "White Wilderness" in 1958, they included an invented scene of lemming suicide. Although this was a total fabrication, and publicly recanted afterwards, there are many people still today who believe that lemmings commit suicide by jumping off a cliff.

Most scientific research, and especially research which leads to no obvious and short-term financial gain, is funded by tax-payer money. The money is allocated to scientific projects by expert bodies, populated by expert scientists. But does public opinion affect the decision-making process of those august scientific bodies who allocate public funds for scientific research? In the 21st century, the answer is a resounding "yes". Having a good publicity machine is a huge advantage when competing for research money.

In the UK, the tax-payer-funded BBC has long been regarded as a quality presenter of validated information on radio, television, and the internet with its freely accessible, advertising-free, BBC website. Over the decades it has tackled the extinction issue in a number of feature programs.[16] A 1981 *Horizon* documentary program presented the Alvarez hypothesis in a program on the death of the dinosaurs. Later, in 1994, following the discovery of the Chicxulub impact crater, they covered the subject in a program *Hunt for the Doomsday Asteroid*. A decade later, in 2004, the BBC produced and ran a more critical program: *What Really Killed the Dinosaurs?* That program presented some of the inconsistencies inherent in the asteroid impact story. This was later followed in 2010 by another Horizon program, *Asteroids — The Good, the Bad, and the Ugly* blaming an asteroid as the cause of that mass extinction. In terms of the public and the press audience in the UK and beyond, the BBC said it, so it must be correct, even though this view is not shared by all scientists.

11.6　Assessing the Data

Paraphrasing a famous quote that nobody ever made: "For a man with only a hammer, the solution to all problems is a nail". In the years that followed the acceptance of the impact theory for the K-Pg extinction, the impactors have sought to attribute other extinctions to bolide impacts. The hunt was on for craters that could be associated with other extinction events.[17]

A large number of craters have been found in many parts of the world and of many different sizes and ages. Some of them can be associated with extinction events if one allows enough latitude for uncertainties in the dating. However, as an explanation for extinction events in general, there are two big problems.

First, there are many impact craters which are not associated with any recognizable extinction event. Bond and Grasby[18] found that of the 175 impact structures that had been identified on Earth by 2019[19] only the Chicxulub crater is associated with a mass extinction. If impacts were a principal cause of extinctions, it is difficult to explain why they should have such differing consequences, with only one being associated with a dramatic world-changing event and the others having little or no obvious consequences at all. Bond and Grasby concluded that "The absence of convincing temporal links between impacts and extinctions other than end Cretaceous suggests impacts are not the main drivers of extinctions".[20] Second, while several mass extinctions can be loosely linked to some bolide impact somewhere, the greatest mass extinction of all, the end of the Permian, does not correspond to any plausible candidate crater.

The alternative theory is that extinctions are caused by terrestrial geological processes, and in particular by extensive volcanic activity and continental movements. This has been a long-held opinion among paleontologists, and there is a growing amount of data supporting this approach. Alvarez derided paleontologists as mere "stamp collectors", but after many decades of data gathering, a lot of "stamps" have been collected.

Of particular importance is the now clear link between plate tectonics and volcanism. This is revealed in the "Large Igneous Provinces" ("LIPs") that indicate past periods of massive volcanic activity.[21] The geography of the oceans is defined structurally in terms of areas called "Provinces", of which there are three kinds: continental margins, deep ocean basins and mid-ocean ridges. These occupy respectively around 22%, 42% and 31% of the area, the remaining 5% being the deep ocean trenches. Those provinces which are made of igneous rock are referred to as "Igneous Provinces", and the largest of these as "Large Igneous Provinces" or "LIPs" for short. LIPs are found both at the margins of tectonic plates and in their central regions. In general, LIPs are not long lasting structures, as much of their rocky substance becomes consumed by tectonic plate movements. They typically last about 50 Myr before suffering significant subduction. The present-day geographical location of the main LIPs is shown in Figure 11.1.

All but one of the major extinction events can be linked to a visible LIP and the volcanic fields or "Traps" associated with it. The exception is the earliest great extinction, the end-Ordovician extinction, for which there is a strong presumptive link but the LIP itself may well have been subducted and no longer be visible. The Great End-Permian extinction of 250 Mya is associated with the Siberian Traps: a LIP having the area of Australia and currently covering the Siberian Steppes to a depth of almost a kilometer. The end-Cretaceous extinction of 66 Mya, when the dinosaurs died out, is associated with the Deccan Traps, a Large Igneous Province covering much of India and extending out into the Indian Ocean.

Even on a lesser level, volcanic activity is found to correlate in both strength and timing with minor extinction events. Using seafloor spreading rates as a proxy for tectonic activity and associated volcanism, a group in Copenhagen[22] showed in 1986 that there was a strong correlation between seafloor species taxonomic diversity

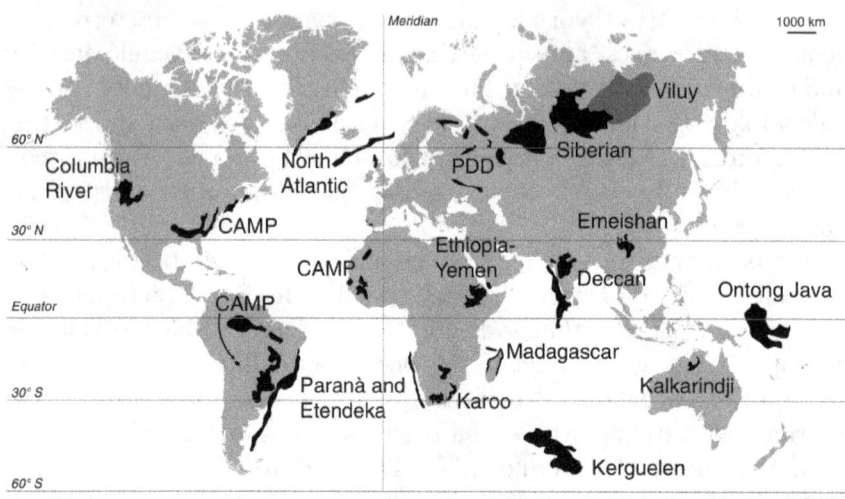

Figure 11.1 Location of the main Large Igneous Provinces (shaded in black). These are areas which in the distant past have experienced exceptional volcanic activity over a period of a few million years. Typically, they are formed from an outpouring of a million or more cubic kilometers of magma over an area greater than 100,000 square kilometers. Antarctica, a host to the Ferrar Traps, is not shown on this map. From David P. G. Bond and Paul B. Wignall, Large igneous provinces and mass extinctions: An update, *Geo. Soc. Amer.*, 2014 (Special paper 505).

and seafloor spreading rates, i.e. volcanism. Their analysis showed a significant correlation in strength and timing with all extinction events, major and minor over the period analyzed. The correlation for dinoflagellates and the calcareous nanoplankton was strongly significant (99%). The sea floor spreading data they used for the North Atlantic and North Pacific oceans was by far the best available at the time. Nonetheless, the publication of the work of that group was strongly impeded. The papers had been submitted to the journal *Science* shortly after the Alvarez bolide impact paper had been published. The Alvarez paper completely deflected interest in anything other than impact causes for extinctions. It was not a good time to talk about evidence for volcanic causation of mass extinctions, and their papers were rejected. However, the work did get published in the American Geophysical Union journal *Paleoceanography*.[23]

The Large Igneous Provinces are associated with bursts of activity involving vast amounts of volcanic activity, dragging up magma from below the crust and dumping it in the massive areas of igneous rock that are the LIPs. During this process, large amounts of carbon and sulfur dioxides are emitted and the protective ozone layer is depleted. Noxious gases like hydrogen sulfide and methane are also released. Those are drivers of climate change and could lead to the death of biological systems from acid rain and excess ultraviolet light when atmospheric ozone is destroyed. The volcanic activity that creates these floods of molten rock is known to have lasted for periods from a few 100,000 years to a million or more years.[24]

Most significantly, there are also increased levels of the element mercury (chemical symbol "Hg"). Mercury is associated with all volcanic activity, particularly with LIPs. Furthermore, it is absent in geological strata which are not associated with major volcanism. Mercury is among the most toxic materials known on Earth and could itself be a driver for extinction.[25] Evidence for excess mercury at extinction sites is found in the rocks and also in abnormal spores of ferns. The correlation between excess mercury and extinction events is extremely strong, as is shown in Figure 11.2.[26]

Just as iridium is taken to mark the occurrence of an asteroid impact, so mercury can be taken to mark the occurrence of extreme volcanic activity. The key difference is that whereas excess iridium is found only in association with a single extinction event, the end-Cretaceous, excess mercury is found in association with all major extinction events, including the end-Cretaceous.

Despite this evidence, the impactors held sway over the volcanists for decades after Alvarez's initial announcement. In the scientific press, and in particular the popular press, the voices of the volcanists were hardly heard. What was missing was a champion, a true expert in the field, who would go all out to challenge the installed views. Such a champion emerged in Professor Gerta Keller of Princeton University.

11.7   Enter Gerta Keller

The Geological Society of London, in January 2004, held a web debate that consisted of written arguments presented by Gerta Keller[27] of Princeton University, opposing the impact rhetoric (the "prosecution"), and by Jan Smit[28] from the Free University of

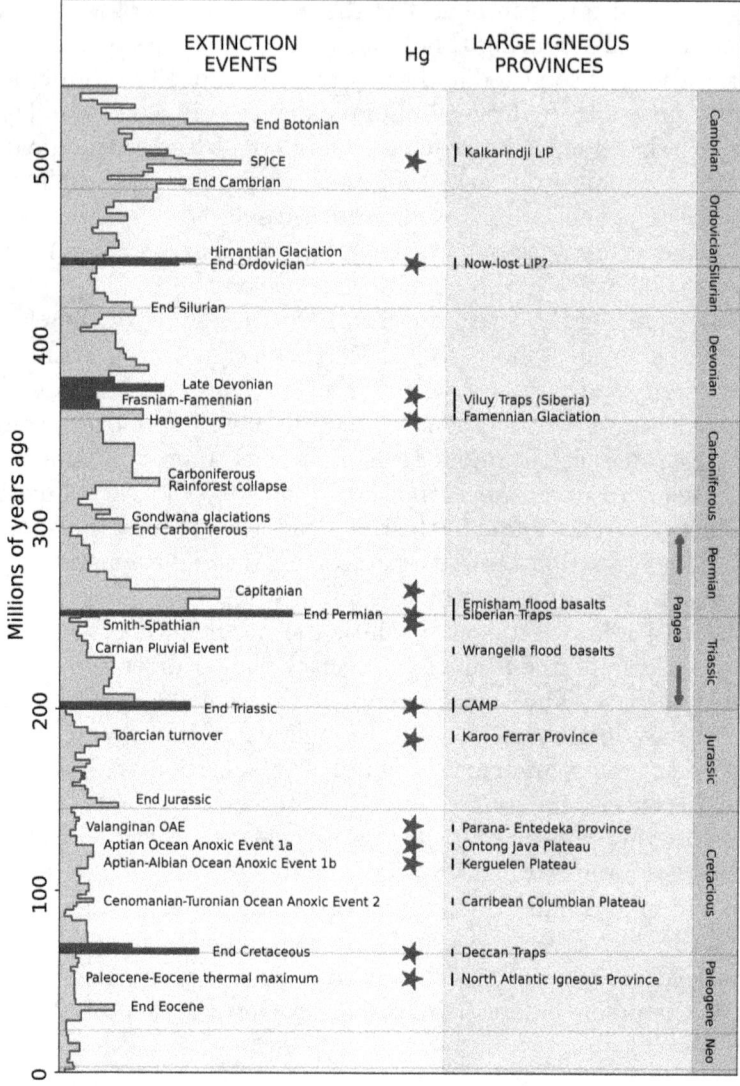

Figure 11.2 The left-hand side of the figure shows extinctions of genera taking place during the Phanerozoic Era. The right-hand side of the graph lists important Large Igneous Provinces and Provinces that have been studied in detail. The stars identify the provinces where mercury (Hg) pollution levels have been measured: they mark the occurrence of extreme volcanic activity. Every star is associated with significant extinction events. This association supports the notion that major volcanic activity is a driver of extinctions, causing changes in climate, atmospheric toxicity and sea-level. Based on Figure 8 of Bond and Grasby, 2017, *loc. cit.*

Amsterdam, supporting the impact theory (the "defence"), albeit a particular version in which the cause of the extinction was a massive worldwide tsunami. There were then two web-discussion sessions[29] in which Keller and Smit separately addressed issues raised by "spectators" about the presentation. This was followed by a final "where next"[30] in which the presenters summed up their views. Not surprisingly, nobody changed their mind. The documents created by the Geological Society are an interesting testament to the schism that was emerging at that point.

Later that year, the BBC aired a *Horizon* production: *What Really Killed the Dinosaurs?* This was a collage of short interviews with proponents of both sides of the argument,[31] including Gerta Keller and Jan Smit. The BBC published a transcript of the video[32] that makes interesting reading. This was clearly no sedate scientific exchange of ideas, devoid of emotive argument:

NARRATOR: Gerta Keller's work provoked a major scientific clash. Defenders of the old impact theory attacked her ideas. The argument quickly turned vicious.

JAN SMIT: Gerta Keller's totally wrong.

GERTA KELLER: Jan Smit has an awful lot at stake.

JAN SMIT: What she is doing with the evidence makes me totally mad.

GERTA KELLER: Jan Smit says the things he does because he is desperate.

JAN SMIT: Sometimes it's not evidence, sometimes it's not fact.

GERTA KELLER: Desperate to rescue his impact Tsunami hypnosis.

NARRATOR: But behind the bitter personal fight was a fundamental scientific dispute about how to interpret the same bits of evidence.

Prior to this point, everything had gone the way of the impact theory. But the opposition had found their champion, Gerta Keller, who emerged both as a leading researcher on the Cretaceous–Paleogene extinction and a powerful spokesperson (see Figure 11.3).

We know of Gerta Keller through her scientific writings: she and her collaborators write long meticulously detailed papers that are beautifully illustrated. She has a minimalist *Wikipedia* page but we learn more about her and her role in the war against "the impactors" in a superb article by Bianca Bosker in the *Atlantic Daily* weekly web magazine.[33] Bosker writes that "Impactors' case-closed confidence

Figure 11.3 Professor Gerta Keller in the atrium of Guyot Hall, with an Allosaurus dinosaur excavated during a 1941 dig. Illustration by Isabel Gálvez (www.isabelgalvez.eu). Inspired by a photograph taken by Denyse Applewhite.

belies decades of vicious infighting, with the two sides trading accusations of slander, sabotage, threats, discrimination, spurious data, and attempts to torpedo careers" and goes on to say that "Keller keeps a running list of insults that other scientists have hurled at her, either behind her back or to her face. She says she's been called a 'bitch' and 'the most dangerous woman in the world,' who 'should be stoned and burned at the stake'".

Since the early 1980s, Keller had become an expert in the study of creatures called "foraminifera" or forams for short. These are single-celled amoeboid protists with external shells. They are found throughout the geological record from early Cambrian times to the present day. They are broadly classed into two groups: the larger group, "benthic", are found in seafloor sediments and a smaller group, "planktonic" that float in the water at different depths.

Keller's focus on forams is very relevant. They play a key role in identifying and quantifying mass extinction events. Recognizing and defining extinctions of lifeforms has always been difficult because of the rarity of most fossils — particularly large ones like dinosaurs. Moreover, it is not just the loss of one individual species that defines an extinction — lots of species become extinct all the way through the geological record. What counts is the loss of many families and genera of species over a short period of time (see Table 10.1). However, in order to do meaningful statistics across narrow time boundaries, it is necessary to have large numbers of identifiable fossils throughout all the substrata. Forams are the most useful markers for this purpose. There are over 50,000 recognized species of foram of which 40,000 are now extinct. Each species of foram is also characteristic of the conditions at the time it was living — such as ocean salinity, temperature, oxygen levels and so on. They flourish and die in synchrony with the other species of marine and terrestrial animals.

In 1987, Keller published two papers following her investigations at one of the finest Cretaceous–Paleogene sites in the world: El Kef in Tunisia.[34] Each paper focused on one of two types of foram. The punchline is that "the K/T boundary extinctions of planktic foraminifera extend over an interval from 25 cm below the geochemical boundary (the Ir anomaly) to 7 cm above. Species extinctions appear sequential, with complex, large, ornate forms disappearing first and smaller, less ornate, forms surviving longer. The 14 species extinctions below the boundary appear unrelated to an impact event."[35]

That 25 cm corresponds to more than 100,000 years before the impact occurred. The simple logic that the deeper the layers, the older they are is almost irrefutable evidence that fossils found below the iridium line are older. Extinctions were already going on prior to the impact![36]

More results were presented at the "Topical Conference on Global Catastrophes in Earth History", held in October 1988 at Snowbird in Utah,[37] the same venue as in 1981 when Alvarez presented results of his research on the iridium anomaly. The title of her talk was "Extended period of K/T boundary mass extinction in the marine realm". Her conclusion was that "There appears to be a non-random selectivity in the planktic foraminiferal extinctions with large complex species going extinct earlier than smaller more primitive morphologies. ... This pattern of extinction is unlikely due to random extinctions, but implies a progressive systematic disruption of habitats with pre-K/T boundary extinctions unrelated to the boundary event."

Bianca Bosker, in her *Atlantic* article, reports that Keller said she barely got through her introduction before members of the audience tore into her: "Stupid." "You don't know what you're doing." "Totally wrong." "Nonsense." Keller's approach to this is simple: "Normally, when people get attacked and given a hard time, they leave the field. For me, it's just the opposite. The more people attack me, the more I want to find out what's the real story behind it."

11.8   The Battleground — The Deccan Traps

So, was the last great extinction at the K-Pg boundary due to volcanism or to an impact? Everyone agrees that a massive bolide hit the Earth at the time of the K-Pg extinction and that this would have had a significant effect on the life on Earth at that moment. The question is as follows: How significant?

If there had been no bolide impact, would the extinction have happened anyway? If so, then the impact was not a necessary cause. If there had not been extensive volcanism occurring at that time, would the bolide strike alone have been enough to cause the extinction? If not, then the impact was not a sufficient cause unless the volcanism was itself caused by the impact. If the impact was neither necessary nor sufficient, then the K-Pg extinction would be effectively caused by volcanism, not by the impact.

The academic community seems to be divided in two by this question. The division is acrimonious, taking place at conferences, in learned journals and on social media. That has become a deeply personal issue. The current reality is most probably that the data are not yet good enough to make definitive statements. There are strong systematics and selection effects coming into play in the data analysis and, inevitably, personal bias.

For both theories — impact or volcanism — the ultimate causative mechanism for the extinction is environmental change: the climate, the ocean levels and the atmosphere. The key difference is in the timing. An impact is a very rapid event, but volcanism occurs over a much longer period. To give a definitive answer, we need accurate dating of both extinction rates and lava flows, and the timing of the bolide strike.

The uncertainties in the relative timing of events have recently taken a considerable step forward through use of *cyclostratigraphy* (also called "cycle timing"). This is a timing mechanism in which the ticks of the clock are cyclic variations in the Earth's orbit around the Sun. These variations cause climate change and evidence for these variations can be seen in the geological record.

The idea of using the Solar System as an astronomical clock for use in geology goes back as far as 1895 to the American geologist Grove K. Gilbert (1843–1918). Gilbert recognized that the Earth's climate would be affected by two astronomical phenomena[38]: the 26,000 yr precession of the equinoxes and the change in the direction of the perihelion of the Earth's orbit in the opposite direction, yielding a cycle of 21,000 years. Gilbert recognized a regular alternation of layers of shale and limestone in Cretaceous rocks in Colorado. He could not attribute this regularity to any known geological phenomenon and so he turned to an astronomical explanation: the 26,000 yr precession of the equinoxes and the change in the direction of the perihelion of the Earth's orbit. The direction of the perihelion moves in the opposite sense to the precession, yielding a cycle of 21,000 years. He was also aware of the fact that the Earth's orbital eccentricity varied on a longer timescale, roughly 91,000 yr, but he chose to ignore that. Gilbert remarked that this was not the first time that scientists had speculated that there might be an effect of these orbital variations on Earth's climate, but he, being a geologist, saw the evidence for the connection.

The issue of these orbital variations and Earth's climate was fully addressed by the Serbian scientist Milutin Milankovitch (1879–1958) in a large collection of research articles written during the period 1912 to 1939 when he collected all of his works into a single volume, *Canon of Insolation of the Earth and Its Application to the Problem of the Ice Ages*.[39] His 626-page "Canon" was submitted to the Royal Serbian Academy in 1941, two days before Nazi Germany attacked the Kingdom of Yugoslavia. These orbital variations are now referred to as "Milankovitch cycles". Milankovich showed that the cycles are caused by the varying influence of other planetary bodies on the Earth's orbit about the Sun and applied this to Ice Ages. Confirmation of his theory came in 1976 with a thorough study by James Hays, John Imbrie and Nicholas Shackelton published in *Science*.[40]

The best evidence for the dating of extinctions comes from the study of foraminifera, those single-celled creatures with hard shells. The fact that forams have existed throughout most of the time life has been on Earth makes them paleobiological time machines. Cyclostratigraphy potentially offers an accurate clock for that time machine.

The sizes of lava flows are also key data: the timing and rates of flow of the lava are critical for examining the cause of the K-Pg extinction. Pulsed elevated concentrations of the element mercury (Hg) have been found to correlate with intense volcanic activity[41] and therefore provide valuable evidence in favor of volcanic activity. All of the major mass extinctions have shown evidence of significant volcanic activity (see Figure 11.2).

These data come only from the painstaking fieldwork of gathering, measuring and dating samples. Alongside these data are computer models of climate and atmospheric dispersal of material. However, the models are not a substitute for accurate observations, and should never be allowed to usurp them. How often do we have grounds to repeat that adage: "A wonderful theory, pity about the facts"?

11.9   The Current Status

It is important to know how the rapid transition from the Cretaceous period to the Paleogene is recognized. It would be absurd to define the K-Pg boundary as the place of the iridium anomaly — that would just give a circular definition. In reality, and in keeping with the criteria of other geological boundaries, there are several criteria. (1) Mass extinction of life forms is the primary criterion: the planktonic foraminifera are the only marine life form that suffered a near-total extinction and so reflect the extinction history. (2) The appearance of new species following the transition. (3) The presence of a red clay layer. (4) A strong change in the abundance of the ^{13}C (Carbon-13) isotope of carbon relative to the ^{12}C isotope. This is indicative of strong climate change, though alone it is not sufficient to define the transition.[42] To these, we can add (5) the presence of mercury, Hg, anomalies that mark worldwide atmospheric fallout from the Deccan eruptions.[43] Finally, (6) the presence of an anomalous abundance of iridium is a useful marker for identifying the boundary layer, even though it cannot be used to define it.

Where Gerta Keller has led us now is to a view that is a refinement of her early work: more data from more places and working at a much higher time resolution using the much improved radioactive indicators for rock layers and our understanding of low-level variations in the Earth's orbit. As we have already said these variations are referred to as "Milankovitch cycles",[44] they are caused by the variable influence of other planetary bodies on the Earth's orbit about the Sun. The consequence of these cycles can be seen and mapped as variations in the structure of the layers of rock over periods of tens of thousands of years.

The focus is on the Deccan Traps, a Large Igneous Province that covers an area the size of Mongolia surrounding Mumbai and Pune in India. At the time of the K-Pg transition, millions of cubic kilometers of magma were spewed out of the Earth in three pulses over a period of about 300,000 years. Keller's team has obtained high-quality data from the Elles site in Tunisia and used Hor Hahar in Israel for corroborative evidence of this scenario. Keller et al. put the K-Pg boundary at 66.016 ± 0.050 Mya,[45] which is around 30,000 years after the largest volcanic pulse occurring at about 66.05 Mya.[46]

The Reinvention of Science

A MILLION YEARS OF VOLCANISM, POLLUTION AND EXTINCTION

Reconstructed volcanic pulses and volcanic mercury 66 million years ago

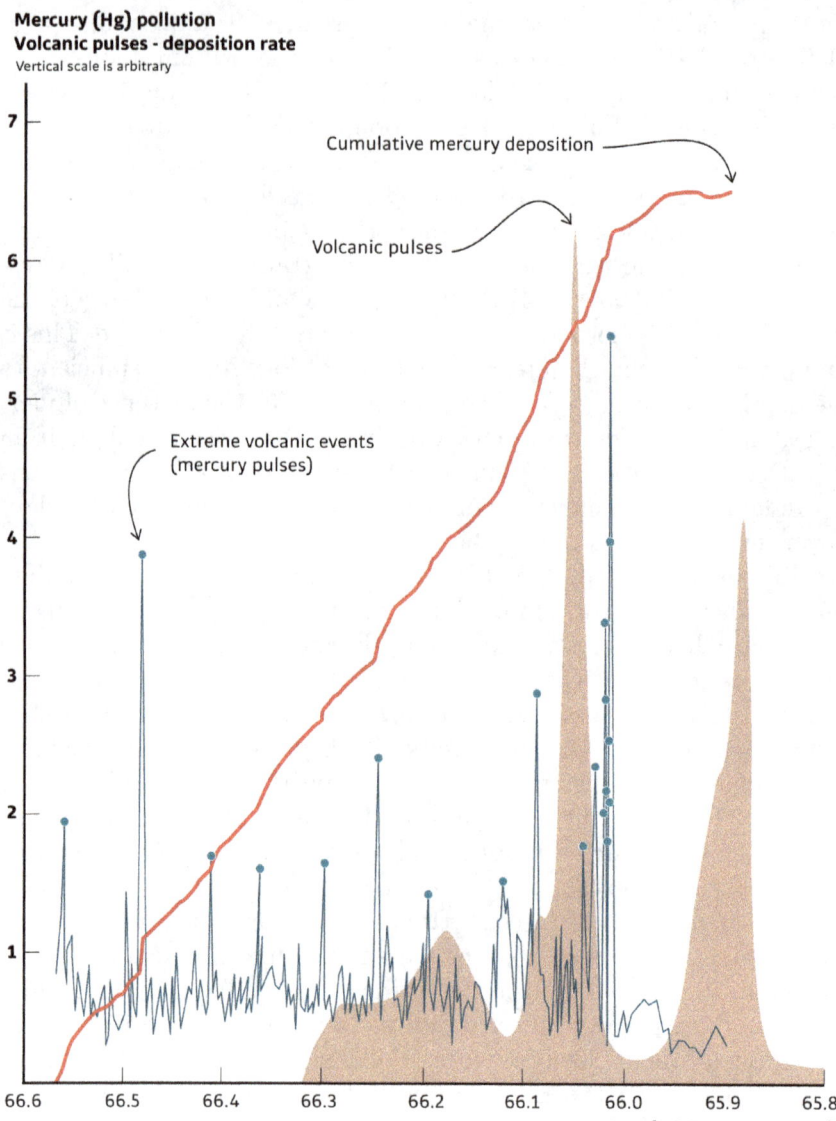

Figure 11.4

Data from the Elles site reveal that the final thousand years before the K-Pg transition was a brief period of massive extinction. The data displayed in Figure 11.4[47] show almost a million years from 66.6 Mya to 65.8 Mya, a period of 800,000 years. The extreme volcanic events are marked by peaks in the amount of mercury deposited and we notice a crowding up in the last thousand or so years before the boundary is reached where we see nine almost overlapping extreme volcanic events. The K-Pg boundary at 66.016 Myr is clearly visible as a crescendo of volcanic activity. A further 100,000 years after the boundary, at 65.9 Mya, there was another big pulse of volcanism, followed by a relatively quiet period of 300,000 years before the final pulse occurred at 65.6 Mya (not shown in the figure). The solid line shows the total toxic loading of accumulated mercury throughout this period. Life on Earth was never the same again.

So, 350,000 years prior to the K-Pg boundary, the Deccan Traps started to become highly active, emitting copious amounts of noxious gases, transforming the atmosphere and the climate. Species were dying out faster than others could adapt to fill the gap. The extinction was already on its way.[48] This had been a period of substantial climate warming that ended 25,000 years before the K-Pg boundary when, as revealed by the higher time resolution Milankovitch dating of sediments having mercury concentrations, there were numerous massive pulses of volcanism.

The great extinction debate continues to rage, but there is now a noticeable change in the arguments. Some of the impactors are beginning to accept that volcanism has an important role to play in mass extinctions. The fact that a large bolide struck the Earth at Chicxulub around 66 Mya is undeniable. The evidence in terms of the crater, the glass spherules and the iridium excesses cannot

---

Figure 11.4 (*see figure on facing page*) Volcanism before and after the K-Pg boundary. The 300,000 years prior to the boundary show substantial evidence for significant climate disruption throughout this period. The white area displays the intensity of volcanic pulses, lava megaflows, at the Deccan traps. The thin line represents the measurements of the abundance of mercury emitted during this period, while the dots identify the Extreme Events of volcanic mercury emission . The solid fat line shows the growth curve for the total amount of volcanic mercury emitted as time passes — the "toxic load". Credit: Infographic sketched by the authors and elaborated by Javier Pérez Belmonte.

be seriously disputed and is not disputed. Only the accuracy of the timing is controversial.

Equally, the fact that mass extinctions of life, both major and minor, are linked to extreme volcanism is also undeniable. The evidence, in terms of substantial data correlations and mercury measurements extending over hundreds of millions of years, cannot be seriously disputed. These include the end-Cretaceous extinction when the dinosaurs finally died out.

Extreme volcanic activity, for periods extending over thousands of years, provides a credible mechanism for causing mass extinctions, through the agencies of noxious gases, mercury poisoning, dust and greenhouse gases affecting climate and sea changes. A bolide strike would be locally devastating but its effects would be much more short-lived than extended volcanic action. As a direct cause for the end-Cretaceous extinctions volcanism provides a better explanation than the direct effects of a bolide strike, although this is denied by the die-hard impactors.

Nevertheless, it remains an extraordinary coincidence that a bolide should strike the Earth right in the middle of an ongoing LIP and associated mass extinction. This has prompted a number of people to look for a causal link between the two. Coincidences as close as this are always suspicious and we naturally look for a causative agent. But not all close coincidences in nature are causal. For example, it is a remarkable coincidence that the size of the Moon and the Sun when viewed from Earth are almost identical — allowing the phenomenon of a solar eclipse, total in which the Solar corona is visible. There is no causation in that coincidence, and no one tries to invent one. However, there have been several attempts to attribute the Deccan volcanism to the effects of the bolide.[49] This is a difficult argument to sustain, as there is no easy route for plate tectonics to allow for such a connection.

The argument is strongly affected by the most critical question: Did the bolide arrive before the volcanism started, or not? Evidence from the mercury measurements and the extinction data suggests that volcanic activity was well under way before the bolide struck. It may be reasonable to propose that the impact could enhance the strength of the ongoing LIP, but how, and to what extent, remains unanswered.

It all comes down to measuring the timing. The advantage of the study at the pristine Tunisian Elles site is that the strata are sufficiently clear and distinct to allow highly accurate dating to be achieved. The same benefit does not apply to the other two principal sites — Chicxulub and the Deccan Traps — where there has been some mixing of the geological deposits. It is not so much a question of error in the data analysis but a matter of what time resolution is possible. Dating the time of impact by Argon (^{40}A-^{39}A) and U-Pb decays has significant error bars, as well as systematic errors that are hard to calculate.[50] Dating by using the Milankovitch cycles, as Keller does, is far more accurate. It is remarkable that we can time events taking place almost 65 Myr in the past with an accuracy of around 20,000 years.[51] Keller's opponents have to argue that her timing estimates are wrong and that her errors are much larger than she thinks.

The two current best estimates for the time of the Chicxulub impact are 66.0± 0.05 Mya from Uranium decay dating and 66.06 ± 0.20 Mya from Argon decay dating. This would place the timing of the impact in the middle of the active phase of the Deccan Traps.

The great extinction debate is still open. New scientific evidence is needed to improve the dating and thus make it possible to determine the precise times at which the different devastating events occurred, and the part they played in killing the dinosaurs. New calculations about bolide impacts and new understanding of what affects the behavior of LIPs are needed to determine whether a causal link between a bolide impact and extreme volcanic activity is possible.

Given the intensity of feeling between the two sides, science may have to wait for another generation before a resolution is found.

Part IV

The Dragon's Treasure
Awards and Extreme Stars

Chapter 12

The Not-Invented Here Syndrome

Imagination is more important than knowledge.

Albert Einstein, What Life Means to Einstein:
An Interview by George Sylvester Viereck
26 October 1929, The Saturday Evening Post

12.1   The Homunculus

At this point, it is important to clarify that postulating unobserved entities was and is a common approach in many disciplines. On occasion, these entities were associated with new experimental or observational devices. When the first microscopes entered circulation, one particular unobserved entity in early physiology was postulated: the homunculus.[1]

Science historian Owen Gingerich states that Galileo,[2] "after being only a timid Copernican, was converted by his astronomical findings, using the telescope, into a flaming Copernican." Galileo did not invent the telescope nor was he the first to build one. It appears to have been a Dutch invention. When Galileo learned about its lens-based operation in 1609, he was able to build one in a few months.

The Dutch cloth merchant Antonie van Leeuwenhoek (1632–1733), based in Delft, was neither the inventor of the microscope nor the first person to use one (just as Galileo did not invent telescopes). But van Leeuwenhoek went on and built more than five hundred microscopes using one lens each, which he polished with a refined technique that he never revealed. Looking through these instruments, he carried out systematic observations of microbial life, and today he is recognized as the father of microbiology. Van Leeuwenhoek did not belong to the academic world — he spoke and read only Dutch — but he was not some mere dilettante. As the biochemist and writer Nick Lane acknowledges,[3] he was a scientist of the highest caliber. Two people are central to van Leeuwenhoek's story: his friend and renowned Dutch physician and anatomist Regnier de Graaf (1641–1673), who also settled in Delft, and the first Secretary of the Royal Society of London, Henry Oldenburg (1619–1677), who was born in Bremen. The former encouraged him to write down the details of his observations and sent them to the Royal Society; the latter took on the task of translating his scientific correspondence from Dutch to English, so it could be read by other members of the Royal Society. This was apparently the main reason why Oldenburg learned Dutch. In his translation, he introduced the term "animalcules" to refer to the bacteria and protozoa that van Leeuwenhoek described in his letters. In 1677, van Leeuwenhoek observed spermatozoa for the first time. Clinical anatomist, professor and bestselling science writer Alice Roberts explains this beautifully in her book[4] *The Incredible Unlikeliness of Being: Evolution and the Making of Us*:

> You know that semen somehow helps to produce babies, so you get hold of some (I'll leave those details to your imagination) and you look at a small drop of that milky fluid using your microscopy. You stare down, astounded by the sight that greets your eyes. The whole field of view is buzzing with movement. You can make out individual, tadpole-like cells, thrashing their tails furiously. They seem to be micro-organisms, like the protists you've already discovered (and written letters to the Royal Society about). But these "animalcules" came from a human.

Years later, in 1694, a compatriot of van Leeuwenhoek, the lens polisher, mathematician, and physicist Nicolaas Hartsoeker (1656–1725) also observed sperm and made a drawing in which a tiny human figure can be seen inside the sperm (see Figure 12.1). It is unclear whether he actually drew what he claimed to have seen with his microscope (Hartsoeker was the inventor of the screw-barrel microscope) or did the drawing solely to defend his spermist version of the performist theory of conception. According to this theory, each new being exists, fully formed but extraordinarily small, in the father's

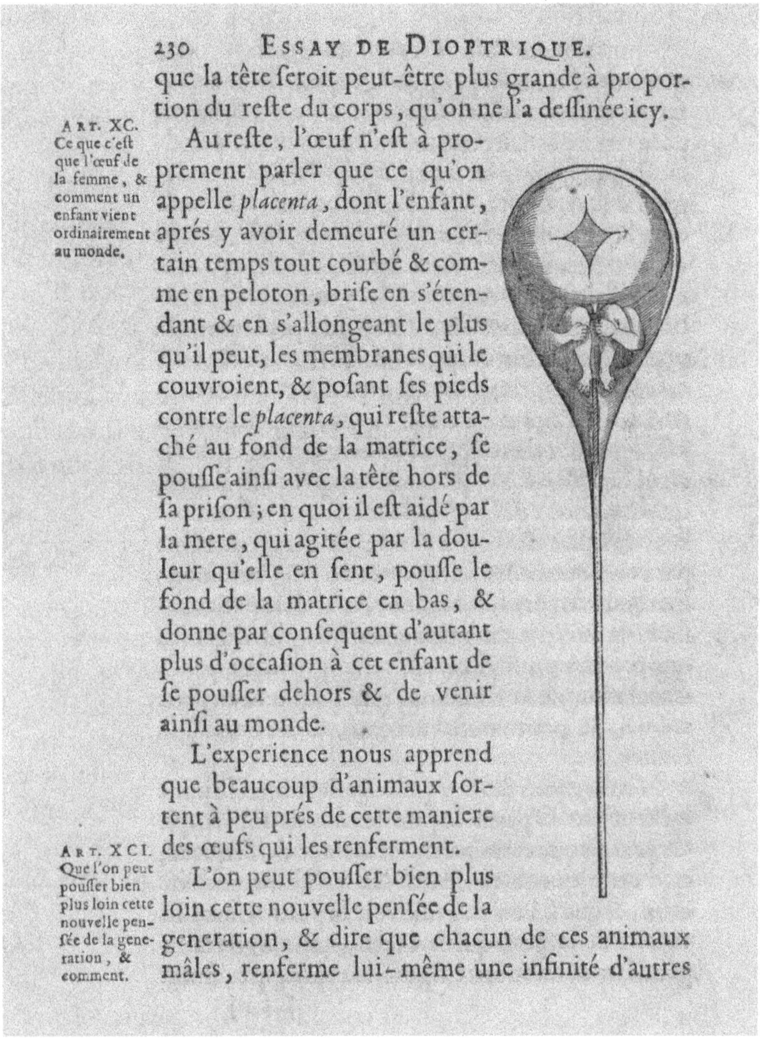

230   ESSAY DE DIOPTRIQUE.

que la tête feroit peut-être plus grande à propor-
tion du refte du corps, qu'on ne l'a deffinée icy.

A R T. XC.
Ce que c'eft
que l'œuf de
la femme, &
comment un
enfant vient
ordinairement
au monde.

Au refte, l'œuf n'eft à pro-
prement parler que ce qu'on
appelle *placenta*, dont l'enfant,
aprés y avoir demeuré un certain temps tout courbé & com-
me en peloton, brife en s'éten-
dant & en s'allongeant le plus
qu'il peut, les membranes qui le
couvroient, & pofant fes pieds
contre le *placenta*, qui refte atta-
ché au fond de la matrice, fe
pouffe ainfi avec la tête hors de
fa prifon ; en quoi il eft aidé par
la mere, qui agitée par la dou-
leur qu'elle en fent, pouffe le
fond de la matrice en bas, &
donne par confequent d'autant
plus d'occafion à cet enfant de
fe pouffer dehors & de venir
ainfi au monde.

L'experience nous apprend
que beaucoup d'animaux for-
tent à peu prés de cette maniere

A R T. X C I.
Que l'on peut
pouffer bien
plus loin cette
nouvelle pen-
fée de la gene-
ration, &
comment.

des œufs qui les renferment.
L'on peut pouffer bien plus
loin cette nouvelle penfée de la
generation, & dire que chacun de ces animaux
mâles, renferme lui-même une infinité d'autres

Figure 12.1 The homunculus depicted by N. Hartsoeker in 1694. *Essay de Dioptrique*, Paris: Jean Anisson. Image courtesy History of Science Collections, University of Oklahoma Libraries.

sperm as a "homunculus". This term was never used by this theory's advocates; they referred to it as an "animalcule"[5] or used French terms: *le petit animal* (the little animal) or *l'enfant* (the child). According to this theory, the role of the female ovum was limited to providing the nutrients and the place in which the new being would grow.[6] The enormous abundance of sperm in semen, however, served to trigger discussions in scientific circles about the wasteful nature of the reproductive process (and the Creator himself), since millions of preformed animals and human beings would never be born.

In 1696, the French aristocrat and astronomer François de Plantade (1670–1741) published, under the pseudonym Dalenpatius, a drawing of his own microscopic observations of the sperm, in which the sperm's tail came out of the homunculus's head, as if it were a cap, contradicting Hartsoeker's drawing. Some argue that the intent behind Dalenpatius's drawing was to mock those who claimed that they saw little men in sperm. Van Leeuwenhoek did not understand it that way. He refuted the image in a letter sent to the Royal Society along with Dalenpatius's drawing, which caused the drawing to be wrongly attributed to van Leeuwenhoek for a long time.

The theological doctrines of pre-formation and original sin, plus the relative unimportance of females, were all still alive when microscopes were first turned on human cells. Thus it is not surprising that at least a few late 17th century natural philosophers reported seeing or imaging some sort of mannequin in sperm, though, as we have seen, Hartsoeker and Dalempatius differed on whether the tail of the sperm as the head or the tail of the "little man". Those images were not confirmed by the next generation of microscopes and microscopists, but the associated idea of "ontogeny recapitulates phylogeny" survived until Darwin's time and beyond. If you feel so inclined, you might even trace the postulated entity down to the present time as "nature vs. nurture" in determining human characteristics as deeply discussed by Steven Pinker in his best-selling 2002 book *The Blank Slate: The Modern Denial of Human Nature.*

Although van Leeuwenhoek is considered the father of microbiology today, neither he nor Hartsoeker nor Dalenpatius was part of the circles that focused on physiology and anatomy, so their contributions, in which they explained what they had observed with the first microscopes, were met with skepticism in more academic settings. After all, they were not students of the field and did not have adequate university training or previous knowledge (although there are those who view this last aspect as an advantage since it allowed them to observe without too many preconceived ideas[7]). This attitude of rejecting scientific contributions to a particular field from people who do not belong to that scientific community is called "NIH" (not-invented here) syndrome, which we will talk about throughout this chapter in different contexts and times.

New ideas and results can also be rejected for perfectly rational, unbiased reasons. Arthur Eddington, who co-organized the expeditions that found Einstein bending of starlight by the Sun during a 1919 eclipse (see Chapter 14), said that he refused to believe an observation until it was confirmed by theory. He meant, we think, until he understood the observation, a perfectly sensible reaction to something that seems impossible. It is, however, most likely to arise when an idea or result relevant to the interests of a particular scientific community comes from someone outside that community, whose methods and vocabulary do not instantly encourage credence. Neither the first nor the last proposal of this type, but one very well known is the idea of continental drift proposed by Alfred Wegener, as described in Chapter 9. We look here at three more examples of the phenomenon: (1) the discovery of helium as a new element from its emission line in the solar chromosphere spectrum by Norman Lockyer (*et al.* perhaps) in 1868, (2) the suggestion of a partly ionized layer in the Earth's upper atmosphere in 1902 by Oliver Heaviside (now eponomized) and (3) the different suggestions put forward by the astronomer Fred Hoyle, including panspermia.

12.2   Helium

In November 1859, one of the most important scientific works of all time was released: *On the Origin of Species*, by the English naturalist Charles Darwin. Seen in perspective, this work completely revolutionized Biology. Only a month earlier, two German scientists who had been working together for a couple of years in Heidelberg released findings that revolutionized physics and chemistry. The protagonists of this story are physicist Robert Wilhelm Bunsen (1811–1899) and chemist Gustav Robert Kirchhoff (1824–1887). Please hang here with us while we try to explain the new technology and ideas that led up to the 1868 announcement of what became helium. Bunsen is known for his burner. The Bunsen burner consists of a metal tube in whose mouth a flammable gas is mixed with controlled amounts of air. When the burner is lit, the mixture produces two flames: a primary, pale-blue one, forming an internal cone, and a practically colorless one forming a larger external cone. This instrument allowed Bunsen to study how different substances burned and, in particular, the color of the flame they produced. Evaluating these colors is quite subjective, and different substances can produce flames with similar colors, so determining a substance from the color of its flame is not easy. This is where his colleague Kirchhoff came in. His idea was to observe the flame through a prism and analyze the spectrum it produced.

Newton had already observed that white light decomposed into a rainbow of colors when it passed through a prism, but it was the German astronomer and telescope maker Joseph von Fraunhofer (1787–1826) who in 1814 used an instrument that he called a spectroscope[8] to observe sunlight. He was surprised to note that dark lines distributed across the entire continuous spectrum were superimposed on the rainbow of colors. The lines were of different widths and intensities: some were lighter, others darker. Fraunhofer died of tuberculosis in 1826 without finding an explanation for the mystery of these dark lines in the solar spectrum. More than 40 years after they had been observed, a satisfactory explanation had not yet been found. In 1859, Bunsen and Kirchhoff finally found one.[9] A fire in the city of Mannheim, 18 kilometers from Heidelberg but perfectly visible from their laboratory, prompted both scientists to point their spectroscope in that direction. By analyzing the lines of the spectrum, they were able to determine the substances that were burning in the fire. This event made them think that they could take a similar approach to determining the elements that make up the Sun and its atmosphere based on the dark lines in its spectrum that Fraunhofer had observed. At the end of October of that same year, Kirchhoff announced his discovery at the Berlin Academy of Sciences[10]: "I made some observations which disclose an unexpected explanation of Fraunhofer's lines, and authorize conclusions therefrom respecting the material constitution of the atmosphere of the Sun, and perhaps also of the brighter fixed stars." Astrophysics was born.

Spectroscopy — spreading light out into rainbows and looking for wavelengths with extra or missing flux — was an exciting new tool in the 1860s, for both astronomers and chemists. Astronomers were learning that previously known elements on Earth were also to be found in the Sun and stars, and chemists were finding new elements from their colors.

Bunsen and Kirchhoff identified ruby-red rubidium in 1861 and sky-blue cesium in 1860. William Crookes contributed twig-green thallium in 1861. A comparable number of false discoveries were also reported by users of both spectroscopic and chemical methods. In 1861, Friedrich Wilhelm Dupré (1834–1908) and his brother August (1835–1907) announced "the existence of a fourth member of the cadmium group of metals." The evidence was a previously unobserved blue line in the spectrum of London's water, which if they were to venture a name would surely be called *Effluvium*, but was in fact internal reflection from the surfaces of prisms and lenses in their spectroscope. Those were cases where the scientists had samples available for experimentation purposes.

The helium story has become somewhat confused because the Frenchman Pierre Jules Janssen (1824–1907) and the Englishman Norman Lockyer (1836–1920) by chance wrote letters to the French Academy that arrived on the same day and announced that each had discovered a way of studying solar prominences without waiting for an eclipse through the judicious use of spectroscopes. This is the discovery (not that of helium) for which the French Ministry of Education issued a joint Gold Medal in 1872. Their contribution was described on the medal as an *Analyse des Protuberances Solaires*.

Janssen and Lockyer gladly agreed to share the award. They were both aware that the first to apply the method had been Janssen. He did so just one day after the eclipse of 18 August 1868 seen from Guntur (India), which he had successfully observed. In a letter to his mother written on 6 September, Janssen said,[11] "Thus the true eclipse took place for me on the 19th and not on the 18th. Since then I have been able to trace day by day the shape, the place, the composition of solar prominences visible up to now only during eclipses." A few days later, on 19 September, he would send a letter from India to the Secretary of the Paris Academy of Sciences to explain his method. It would take more than a month for the letter to reach its destination. Its arrival coincided with that of a letter containing a similar description sent from London by Norman Lockyer, who had successfully carried out similar observations on 20 October. This remarkable coincidence likely helped to associate the announcement of this method, which in Lockyer's words allowed artificial eclipses to be generated, with the discovery of helium.[12] The proposed technique would undoubtedly contribute to the discovery of Helium, but it would take considerably longer to come about.

The fingerprint of helium in the spectrum of the Sun's atmosphere is an emission line produced by this chemical element, which is especially visible in light from the upper atmospheric layer called the chromosphere which is hotter than the photosphere we normally see and faint enough that an eclipse or the method of Janssen and Lockyer were necessary for detection. But seeing the line is not the same as concluding that its presence in the spectrum is due to a previously unreported chemical element. Reaching this conclusion would take time and would be a story full of doubts and uncertainties.

Lockyer actually saw the relevant emission line — a yellow-green line that is close to but not coincident with the emission lines of sodium — back in London a couple of months after not only Janssen but also Norman Pogson (of magnitudes fame), Colonel John Herschel (grandson of the discoverer of Uranus) and a few others had seen but not drawn attention to it during the August 1868 eclipse, which was best seen from India. Lockyer, mindful of his own background as a solar observer, turned to his chemist acquaintance Edward Frankland (1825–1899, to whom we owe much of the concept of valences). Together they coined the name helium[13] for the source of the aberrant line, after going to great lengths to account for it as hydrogen under either low or high pressure. This new element did not meet the fate of jargonium, nigrium, norium or asterium.[14]

But Lockyer and Frankland had sporadic doubts, and helium was not accepted into Mendeleev's then-new periodic table until after Lord Rayleigh had found argon in air (1895) and the Scottish chemist William Ramsay (1852–1916) had seen that bright yellow-green line in a terrestrial sample that same year, soon after finding krypton, neon and xenon.[15] For these discoveries, Rayleigh received the 1904 Nobel Prize for Physics and Ramsay the Nobel Prize for Chemistry. The periodic table thus gained an 8th column, which was nearly complete within 5 years.

In another sense, helium is part of a different story, one that began with the 1864 discovery of nebulium by the English astronomer William Huggins (1824–1910) and ended with the 1869 discovery of coronium[16] by the American astronomers Charles Young

(1834–1908) and William Harkness (1837–1903, born in Scotland), independently, during another solar eclipse, this one visible across most of the United States. You will not find either nebulium or coronium on your modern periodic table, even though Mendeleev and others made room for them in the 1870s: each turned out to be the product of familiar elements (oxygen and nitrogen for nebulium and iron for coronium) under unfamiliar conditions. But the three — helium, nebulium and coronium — had three aspects in common: discovery in astronomical objects by spectroscopic means just after spectroscopy began to be applied to the heavens; wavelengths in the yellow-green part of the spectrum, where the human eye is most sensitive; and identification during the time just before the application of photography to the study of astronomical spectra. Since early photographic emulsions had very little sensitivity at those wavelengths (capturing mostly blue and violet into ultraviolet light), observations made even a decade later would probably not have found the relevant emission lines in nebulae and the outer layers of the Sun.

12.3   The Heaviside Layer

The mystery was the reception of wireless telegraphy (Morse code) signals sent by Guillermo Marconi (1874–1937) in 1901 from Cornwall UK into St. Johns Newfoundland (Canada) and at various other points up and down the American Atlantic seaboard. If the radio (wireless) waves were to travel in a straight line across more than 2,000 km of the Earth's surface, the capturing antenna would have had to be hundreds of kilometers high rather than a few tens of meters. By chance that year, Oliver Heaviside (1850–1925, a self-taught electrical engineer, who left school at 16 and was never regularly employed except by the Great Northern Telegraph Company) was commissioned to write an article on telegraphy for the 1902 Encyclopedia Britannica.[17] He considered primary telegraphy by wire but added a short section on wireless. In this, he considered the possibility of ocean water (which contains dissolved salts and is therefore a pretty good electrical conductor) having acted as a sort of waveguide, bending the radio waves around the Earth's surface, but opined that a better bet would be a conducting layer of atmosphere above the Earth's surface which could act as a mirror via total internal reflection.

The direction of bending of the radio waves meant that the index of refraction of that layer (at a distance above ground roughly half the height of that Brobdingnagian antenna) would have to be negative and the speed of light therein larger than c (the speed of light or radio waves in a vacuum). You can imagine how that went over with a physics community that was just getting ready to welcome Einstein's special relativity, saying that nothing is going to carry information at speeds greater than c. Heaviside was already fairly unpopular, in so far as he was then known, for having used imaginary and complex numbers to describe complicated electrical circuits. Don't worry. Electromagnetic waves in media other than vacua have two characteristic speeds: a group velocity (which carries information and indeed is c or smaller) and a phase velocity, which does not carry information and is c or larger.

Meanwhile, as it were, Arthur E. Kennelly (1861–1939) who was also largely educated outside formal universities, was just starting a professorship at Harvard in 1902. That year, he put forward independently[18] nearly the same idea to account for the Marconi experiments. But he was a clubbable sort, recipient of many honors in his career, with a wife and child, in contrast to Heaviside who spent most of his life living with his parents, later a sister-in-law, and finally alone and reclusive, with nearly all his recognition posthumous.

The name Heaviside layer was proposed for the hypothetical entity by the British physicist William Eccles (1875–1966) as early as 1910, but it remained hypothetical until 1923/24 when Edward Appleton (1892–1965, physics Nobel 1947 precisely for the discovery of this ionized layer) thought about radio reception at distances of tens of kilometers, which seemed to show the effects of interference between two channels of transmission, one in a straight line and one reflected from above. He set out to study the interference, using an existing BBC transmission tower and receivers he could move around places like Oxford. Sure enough, reflection was occurring at 100 km or so above the ground. Details were mapped out a few years later by the Russian-born American physicist Gregory Breit (1899–1981) and the American geophysicist Merle Tuve (1901–1982) in the United States. If Appleton's name rings a bell, it could be for the higher, Appleton (150–800 km) layer in the atmosphere, but more probably because he made vital contributions to the World War II British effort to develop radar for defense purposes.

We have it on good authority that his friends called him "Skip"' Appleton.

The Kennelly–Heaviside and Appleton layers of the upper atmosphere are now called unromantically the E and F layers and known to harbor ionized atoms, responsible for the reflection.

12.4 An Hour with Hoyle

A 1957 paper,[19] called "Synthesis of elements in stars", by E. Margaret Burbidge (1919–2020), Geoffrey R. Burbidge (1925–2010), William A. Fowler (1911–1995) and Fred Hoyle (1915–2001) systematized an enormous number of observations, experiments and ideas from astronomy, nuclear physics and related fields to account both for the production of (nearly) all the elements and isotopes found on Earth and in the Sun and stars, and the relative amounts of them. The paper was of sufficient importance to future work in astrophysics that it was frequently abbreviated as B^2FH.

The American–Canadian astrophysicist Alastair G. W. Cameron (1925–2005) produced a very similar synthesis in the same time frame, but it was largely published in classified documents, and he was not recognized for what he had done until considerably later.

When a Nobel Prize came (Physics in 1983), it went to William A. Fowler (shared with Subrahmanyan Chandrasekhar (1910–1995) for very different contributions to astrophysics), with no glory for the Burbidges or Hoyle, and people have been asking "Why?" ever since. Well, at least, from early on, "Why not Hoyle?" and, more recently, "Why not Margaret Burbidge?" now that thinkers about history and structure of science and its rewards have begun to focus on achievements by women. Some of Cameron's own students occasionally asked "Why not Al?" but we don't think anyone ever asked "Why not Geoff Burbidge?". Fowler himself was particularly distressed by the non-recognition of Hoyle, who had been both a friend and colleague from the mid 1950s until their deaths.

Well, why not Hoyle? Was he an outsider to the community for which his contributions were most important, in the way that Lockyer, the discover of helium in the Sun; Heaviside of his layer; and Wegener of continental drift were (see Chapter 9), and so part of this story? Perhaps and partially.

Hoyle was born in rural Yorkshire in 1915 and kept firmly his non-Oxbridge accent throughout his life. Throughout his career, before, during, and after writing the seminal paper with the Burbidges and Fowler he had very firm allies and very firm opponents among his colleagues in England, though he came in due course to a Cambridge degree in mathematics (the normal preparation for a career in astronomy, there, then) and to a fellowship at St. Johns College (one of the more prestigious ones). His most profound disagreement with the majority of the astrophysical community was over what is called the steady-state model of the universe. In 1948, Hoyle, Hermann Bondi (1919–2005), and Thomas Gold (1920–2004) put forward an alternative to a universe of decreasing density and finite age, in which new matter appears to keep the density, number of galaxies, and everything else is constant forever.

The theory, initially called new cosmology and then steady-state theory, quickly gathered modest support on both sides of the Atlantic. In this theory, the universe has no beginning. It is therefore infinitely old and has always had the same appearance, even though it is expanding. This universe that remains unchanged over time is based on the so-called perfect cosmological principle, which generalizes the cosmological principle by adding that homogeneity exists not only in space but also in time. That is, all observers no matter where and when would see a universe with the same characteristics. As a result of the observed cosmic expansion, the density of the universe should gradually decrease, but if it has to remain constant as the steady-state theory requires, then continuous creation to maintain constant density is needed. That creation has never been observed either in space or in the laboratory, but the amount of matter that would have to be created is so small (an atom of hydrogen for every cubic meter in 10 billion years) that it would be extremely difficult to detect. Years later, some of the proposers of the steady-state theory were involved in bitter disputes with Big Bang defenders. The controversy made a significant contribution to changing cosmology, which went from being a highly mathematical discipline to being a branch of physics in which the confrontation between theoretical models and observations would take on a leading role, as both alternative theories could be compared with astronomical observations. In fact, various observational programs were designed deliberately to "test" steady state in a way that requires developing new observation techniques.

In 1948, there were no data that disagreed with the steady-state idea. Indeed it was a better fit to the multi-billion year ages of some stars, at a time when it seemed as if an evolutionary universe could be only about 2 billion years old. That discrepancy was later resolved, but Hoyle stuck by steady-state, while Bondi and Gold (even more outsiders to the British establishment, as Viennese Jewish refugees) went on to other tasks and careers.

Radio astronomy played a fundamental role here. Using radio masts as astronomical detectors was tried by US engineer Karl Jansky (1905–1950) in 1932. He was the first mariner to scan the sky outside the visible spectrum and to detect radiation from the galaxy. In subsequent years, radio astronomy developed significantly out of World War II radar meteorology and soon there was a great deal to say in the cosmological debate. Martin Ryle (1918–1984) was a Cambridge professor in 1950, the first to hold a chair of radio astronomy. He oversaw the preparation of the Cambridge catalogs of astronomical radio sources and quasars, apparently stellar objects that emit a huge amount of energy some in the form of radio waves. The study of these catalogs showed that the distant universe (the one observed as it was thousands of million years ago) is very different from our local environment (that is its actual state). This meant that the universe evolves, as the relativist Big Bang model predicted, an observation contrary to the steady-state hypothesis. The fact that there are types of astronomical objects more commonly found at vast distances contradicts the perfect cosmological principle: the content of the universe has changed greatly over time.

Hoyle and Ryle were in the same department at Cambridge University and their close proximity only served to fan the flames of scientific debate. When in 1963 Maarten Schmidt (1929–2022) measured the redshift of the first quasar to be discovered (3C 273), many supporters of the steady-state theory, like Dennis Sciama (1926–1999) and his student Martin Rees (now English Astronomer Royal), were convinced that it was strong proof of an evolving universe. The definitive boost would come only 2 years later, also by means of radio astronomy, with the discovery of the Cosmic Microwave Background (CMB) radiation.

There are at least four biographies of Hoyle,[20] which fill in many details, without entirely agreeing with each other. As time went on, he espoused additional nonstandard ideas that made him no friends, for instance, that archaeopteryx was a forged fossil, made up of two parts, one bird and one reptile. His loud protest when a Nobel went to Antony Hewish (1924–2021) for the discovery of pulsars, leaving out the graduate student, Susan Jocelyn Bell (b. 1943, later Burnell, see next chapter), who had first recognized the anomaly in radio telescope records was perhaps justified, but not calculated to endear him to the committees selecting prize winners. Hewish and Martin Ryle won in 1974 and Fowler in 1983.

Another source of considerable discord with most of the scientific community was Hoyle's opinion that terrestrial life did not originate on Earth but had been carried through space to us (the concept is called panspermia) and is continuously being replenished by viruses, bacteria, and so forth coming from outside our atmosphere. The concept was satirized as "plagues come from comet tails", or, sometimes, "Flu comes from comet tails". There is anecdotal evidence that some of the Nobel Committee had asked "how can we give the Prize to someone who believes that the origin of life on Earth is extraterrestrial?" The anecdote also shows that Hoyle was a great scientist who was always somewhat out of the mainstream. He loved a good argument and challenging the establishment.

We know a bit more about the origins of life on Earth and quite a lot more about comet tails than did Hoyle in 1978, but at least a few of his younger colleagues still adhere to a panspermic point of view. It allows longer for the origin and evolution of life than the 4.5 Gyr possible on Earth. Indeed, if we live in some sort of a steady-state universe (to which still also a few younger Hoyle colleagues adhere, in modified form), there is an infinity of time for these processes. Again, as earlier in his career, Hoyle had a few very firm allies and many adamant opponents. If, when, and by whom he might have been nominated for a Nobel prize (perhaps chemistry rather than physics?) will be known only when records are unsealed in future years.

If this story seems sadder to your authors than those of Lockyer, Heaviside and Wegener, it is probably because we knew Fred Hoyle and many of his associates fairly well and, most of the time, very much liked him.

Chapter 13

World Wars and Nobel Prizes

Yet why not say what happened?

Robert Lowell, "Epilogue" from Day by Day

Coming into the 20th century, mainstream science's views have often changed so quickly that it has become possible for one person to end up as the rear guard of the ideas or experiments for which he was once the advance guard. The best-known example is Albert Einstein, whose early work laid some of the foundations of quantum mechanics, demonstrated the equivalence of mass and energy, showed in his theory of general relativity that matter and energy bend space and time, and much else. But he lived another 25 years after all of these were essentially complete, and in those years was, though highly honored, increasingly out of step with the rest of the physics and mathematics communities.

The two items he declined to leave aside were causality and a geometric interpretation of the forces of physics. Modern quantum mechanics (ideas and an enormous number of real events that it describes) says that atoms, particles, and other small things "decide" at random what to do, decaying or not, passing through one slit or another, and so forth. Einstein, sometimes quoted as saying "the Old One does not play dice",[1] was sure that there must be some underlying more basic theory that would permit definite calculations of which particle would go which way. "Causality" of this kind can actually be shown not to exist for some experiments (though not all).

As for the second, geometric, preference, Einstein thought it must be possible to express all the known forces (meaning just gravity and electromagnetism, when he first started working on the problem) in terms of tensors[2] like those of general relativity but more complex. He mentions this in a letter to David Hilbert (1862–1943), the German mathematician, in November 1915, even before all of the classic general relativity papers have been published. And he struggled the rest of his life to find a unified theory or theory of everything. Physicists are trying for a similar goal today, but the foundation is the quantum mechanical description of electromagnetism, the weak, and strong (nuclear) forces, not the geometry of general relativity, which cannot, even in principle, be put into quantum mechanical form.

Studying the universe in isolation from the world around us has never been possible, least of all in the 20th century. Einstein wrote the core of his general relativity theory in Berlin in 1915, when Germany was already very much at war with much of the rest of Europe. His letters from the period are a remarkable mix of highly mathematical discussions with Willem de Sitter (1872–1934), Hendrik Lorentz in Holland and Hilbert elsewhere in Germany interspersed with worries about money, food, attempting to cross into Switzerland to see his sons, and plans for documents or committees that might somehow oppose the war. By 1917–1918, major food shortages in Germany and a variety of stomach and digestive ailments had forced Einstein to subsist largely on rice and milk (prepared for him by the cousin who became his second wife). One cannot help but wonder how much these privations affected his later thinking about science and everything else. That his postwar contributions (summarized as "Einstein, Infeld, Hoffmann" and "Einstein, Podolsky, Rosen") were not the equal of his earlier work is probably true. The murder of Walther Rathenau (1867–1922) also contributed to driving him out of Germany, first temporarily in 1922 and then permanently.

On 6 April 1922, Einstein met in Paris the French philosopher, Henri Bergson (1859–1941), to debate the nature of time. Einstein maintained there were only two sorts: the time of physics and that of psychology. Bergson strongly supported the existence of what he called "philosophical time". It had taken heroic efforts by the French physicist Paul Langevin (1872–1946) and others to arrange the event.

Rathenau, shortly before his death, had urged Einstein to accept the invitation, while Max Planck (1858–1947) discouraged it. And Einstein actually needed permission from the Prussian Academy of Sciences to travel to the country that Germans then felt was deeply oppressing them in the wake of the Treaty of Versailles. Einstein and Bergson would never have agreed in any case, the former stating unequivocally "Il n'y a donc pas un temps des philosophes", which means as Jimena Canales[3] explains that "the time of the philosophers did not exist." But the intellectual climate had been deeply darkened by the war and its immediate aftermath, and certainly the April 1922 debate held in Paris had the protagonists' nationalities among its poisons.

13.1 Who Got Moseley's Prize?

Another curious "last to lay the old aside" story also carries major Word War I scars: a death at Gallipoli, and Swedish attempts to carry on the traditions of the Nobel Prize. We call it "Who got Moseley's prize?" And the answer is Charles G. Barkla (1877–1944; no, you are not expected to have heard of him before, nor probably Moseley), was another, extreme "last to lay the old aside".

The 15 years 1895–1910 saw an explosion of new laboratory data and ideas concerning the structure of matter, radioactivity and X- and gamma-rays. Barkla, a northerner from Lancaster, arrived at the Cavendish Laboratory, Cambridge in 1899, just in time to work with the discoverer of the electron, J.J. Thomson (1856–1940), and be a part of it all. His territory was the scattering of X-rays by different solid materials. He was the first to show that X-rays can be polarized. Next, he demonstrated that samples of any particular element zapped with high energy electrons or a range of energies in X- or gamma-rays will respond by emitting both a whole X-ray rainbow and one or more features at discrete energies that differ from one element to another. He labeled some of those discrete energies K and L, so as to allow for others of both higher and lower energy. M and N were indeed found. And Barkla thought he had found J, the ones with the highest energy. His subsequent career — some time in Liverpool; a professorship in London; and finally the Chair of Natural Philosophy in Edinburgh (1913–1944) — was largely devoted to what he called the "J phenomenon", whose properties, as he reported, became stranger and stranger with time, though the students he mentored continued to get jobs.

Barkla also clung firmly to the way of measuring X-ray energy that he had first used in which the higher the energy, the thicker a sheet of aluminum foil they will penetrate (don't try this at home!).

Meanwhile, to Oxford came Henry G.J. Moseley (1887–1915), who received his first degrees in mathematics and physics in 1910 (see Figure 13.1), and made enough of an impression on his surroundings that Ernest Rutherford (1871–1937), the discoverer of atomic nuclei, was glad to take him on as an advanced student at the University of Manchester. Moseley started out doing some of the projects on radioactivity that interested Rutherford but soon moved on to X-ray scattering. With really very "little help from his friends", he developed a much better way of measuring energies, scattering the X-rays off crystals like salt. The angles at which the X-rays came off depended uniquely on their energies (highest energies get bent least), and Moseley could, therefore, let the scattered X-rays hit photographic film, and where they hit provided an accurate value for energy. Compared to what Barkla had done, this was something like the advance from using a color filter to spreading light in a photographable spectrum from red to violet and measuring how much the various wavelengths had been bent by a prism or grating.

Within just a couple of years, Moseley showed that each element that he zapped not only yielded a unique signature but that the signature was closely correlated with the positions of the element on a periodic table of the elements. Moseley's method enabled him to sort out a couple of places in that table where physicists and chemists disagreed about the order of the elements (the chemists were right) and to show that there were only 3 or 4 elements missing between aluminum and gold, at a time when discovering a new element was the goal of many scientists and many of them had reported successes, not all of which could be true. Moseley's gaps, elements 43 and 61 (radioactive and found at accelerators, not in chemistry labs) and 72 and 75 (rare earths very difficult to separate from their neighbors, found in the 1920s), would surely have entitled him to say "I told you so!" and march off to Stockholm to collect his Nobel Prize. Whether physics or chemistry could be argued, and Rutherford had actually

Figure 13.1 Henry Moseley at the University of Oxford's Balliol-Trinity Laboratories, 1910. Illustration by Isabel Gálvez (www.isabelgalvez.eu) inspired by a picture from the History of Science Museum, University of Oxford.

nominated him for both in 1915. Other distinguished scientists soon said that they would gladly have joined in the parade.

But the Great War broke out in 1914, and instead of Stockholm, Moseley marched off to war. He volunteered and insisted on a commission on active duty, rather than some kind of science-based war work that occupied most of his fellow physical scientists, both contemporary and older. William L. Bragg (1890–1971), the pioneer of X-ray crystal scattering, was working on sound ranging in Belgium when a telegram informed him of the 1915 Nobel he shared with his father, William H. Bragg (1862–1942), who was doing war work back in England, as were Rutherford and many others. Moseley was deployed to Gallipoli, where the British (with help from Australian and New Zealand troops) were attempting to force their way through the Dardanelles into the Black Sea and take the Turks from behind. He was shot and died instantly on 10 August 1915, when the Nobel Committee had not yet completed its deliberations.

13.2   The Winners and the Losers

Barkla, meanwhile, was up in Edinburgh, plugging away on his J phenomenon and not volunteering for anything in conflict with his somewhat Germanic sympathies (but all three of his sons served in the Second World War, and one was killed). The 1916 Nobels in chemistry and physics went unawarded, as did the 1917 chemistry prize. But 1917 physics went to Charles Glover Barkla "for his discovery of the characteristic X-rays of the elements."

Meanwhile, Barkla received a 1916 prize from the Royal Society (London), and in his prize lecture, it is already clear that he has no use for Moseley's results or for Niels Bohr's model of the atom, or indeed most of what later becomes known as quantum mechanics (a point he shared with Einstein who was only 2 years younger). He engaged in published spats with Bragg, Rutherford, and Arthur Holly Compton (1892–1962, another Nobelist) never on what we would now think of as the winning side.

International politics surely influenced the Nobel Awards and non-awards 1914–1920, perhaps the Swedish Academy recognizing that it would have to live with both the winners and the losers, however the war turned out. But by 1920, when the war-time prizes were actually presented, Barkla was the only British scientist who chose to attend.

It is easy to imagine several counter-factual histories: one in which Barkla moved with the times; many in which Moseley survived the

Great War (and perhaps even shared a prize with Barkla); but also, of course, other sad ones in which Moseley hung up on some idea he had worked on in 1913–1914, for instance, using energies of M and N series X-rays to measure the numbers of electrons in outer Bohr orbits and hence to predict chemical properties of heavy atoms. But given his early collaborations on the basis of equality with several colleagues (including Darwin, but George, not Charles), it seems much more likely that he would have remained part of the mainstream, retiring somewhere around 1957 from one of the professorships he had intended to apply for in 1915 (Oxford and Birmingham).

About "relocations", we should also mention here that the forced emigration, after the *Anschluss*[4] in 1938, of the brilliant Jewish woman physicist Lise Meitner (1878–1968) affected her subsequent career and contributed to her exclusion from the 1944 Nobel Prize in Chemistry for nuclear fission that was awarded to her collaborator Otto Hahn (1879–1968). A similar series of events happened to Marietta Blau (1894–1970), an Austrian Jewish physicist who also had to leave her country in 1938, for the same reason. She probably deserved to share the Nobel Prize for the use of photographic emulsions for particle detection (cosmic rays, radioactivity and accelerators) — a prize that later went to the British physicist Cecil Powell (1903–1969).

Historians tell us that the main cause of World War II was World War I and its impossible peace treaties or even that they should really be regarded as a continuous conflict (partisans of the 100 years' war sometimes also laid down their long and crossbows). But by 1939, England, Germany, the United States, the Soviet Union and all the rest had learned not to deliberately send their most scientifically gifted young men onto the battlefield. Instead, they variously worked on radar, rockets, atomic bomb development, ballistics and so forth. An obvious exception included the young men attending the service academies in their various countries. One of these, Joseph Weber (1919–2000), brings us back to aspects of the general theory of relativity, and in particular, gravitational waves.[5] This will be the topic of the next chapter. He used to say that Alfred Nobel (1833–1896) did more damage with his prizes than ever he did with his explosives. In his recent book, Keating (2018) agrees with this thought arguing that the Nobel Prize, instead of increasing scientific progress, may actually disrupt it. Nearly all the scientists in the world surely have their own lists of colleagues who could have and should have won one.

"Could have" means alive when the topic they had worked on was recognized. This excludes both Henry Moseley killed at Gallipoli in 1915 and Rosalind Franklin (1920–1958) who imaged the X-ray diffraction patterns of crystalline DNA that led Watson and Crick to the double helix structure but died of cancer well before the 1962 Chemistry Nobel went to James D. Watson, Francis H.C. Crick (1916–2004) and Maurice H.F. Wilkins (1916–2004) who had been Franklin's boss.

"Should have" means that the person made a major, perhaps irreplaceable, contribution to an invention or discovery that did earn a Nobel Prize for someone else. Many physicists would put near the tops of their lists Lise Meitner, and perhaps Fritz Strassmann or Otto Robert Frisch (1904–1979) for the discovery of nuclear fission. That 1944 Chemistry prize went exclusively to her contemporary, Berlin chemist Otto Hahn, who remained in Berlin when she necessarily left, because her Austrian citizenship was no longer protecting her from German actions against Jews after *Anschluss* in 1938.

For many astronomers, the list is topped by S. Jocelyn Bell. A near tie for top ranking is Vera Cooper Rubin (1918–2016) for her role in the discovery of dark matter (see Chapter 16), a topic which has not yet been distinguished by a Nobel. Vera sadly died in 2016 and so is no longer in the "could have" category, while Jocelyn (b. in 1943) is still alive and well.

13.3   Jocelyn Bell and the Prize She Didn't Get (Yet)

When first-year Cambridge graduate student Jocelyn Bell joined a small team to build a new sort of telescope to detect cosmic radio waves in 1965, little did she know that she was embarking on a path that would lead her to the confirmation of outrageous predictions from the 1930s, thereby scooping some more senior scientists, and, eventually, receiving worldwide renown in the astronomical and other communities. The new radio telescope was the idea of her supervisor, Dr. Antony Hewish (1924–2021; Physics Nobel Prize, 1974, and thereby hangs part of our tale). Look though it might be like a field of chicken coops, the new radio array was well matched to its intended task of finding new examples of radio sources whose observed brightnesses varied rapidly because they were twinkling (called scintillation by radio astronomers) in the varying density of hot gas in and around the Solar System, the way starlight twinkles in varying densities and temperatures of the Earth's night air.

The radio sources already known to scintillate were quasars, a word apparently derived from Hong-Yee Chiu's effort to pronounce the acronym, QSRS, for quasi-stellar radio sources. These are now known to be giant black holes at the centers of a few galaxies that are gobbling up gas from disks around themselves: Hewish (and therefore his students) wanted to find lots of additional quasars so as to understand better how they work.

One of the many fruits of Antony Hewish's quest for ever-better angular resolution at radio wavelengths was his work with student Sam Okoye which was actually a prediscovery of the pulsar in the Crab Nebula supernova remnant. But Hewish wanted to find and study scintillating sources, that is, compact ones (including quasars) whose radio flux would twinkle like stars on a crisp winter night. Thus, in 1965, he set out to build a large array, working at low frequency (where scintillation is most conspicuous). Bell arrived just in time to be one of about five people (plus summer volunteers) who built the array.

Susan Jocelyn Bell[6] (always called Jocelyn and later Bell Burnell) was the student given the job of looking at the raw data coming from the array, to find scintillating sources. The data took the form of black ink tracings on rolls of paper, and she has said that at one point she was behind a million feel of chart paper. Co-author VT sympathizes, having early in her married life helped a bit in analyzing data from Joseph Weber's gravitational wave detectors (see the next chapter), which took the form of red ink traces on Esterline Angus chart paper, though she was never that far behind! Jocelyn indeed found scintillating sources, and they formed the backbone of her PhD dissertation.

But along the way, she found something else. Ordinary matter makes up most stars, planets and us. Compress it enough and you get white dwarfs, known since the 1920s and modeled since 1930. A few years later, German-American Walter Baade and Swiss–American Fritz Zwicky proposed that matter could be compressed much further, under strong enough gravity, until it was made almost entirely of neutrons. Formation of neutron stars, they hypothesized, could provide the energy for supernovae and far cosmic ray acceleration as well. A neutron star of one or two solar masses would fit inside the ring roads of most European cities. On the ever of World War II, Americans J. Robert Oppenheimer (soon to head the Manhattan project and make explosions of his own) and his student George Volkoff calculated models for neutron stars, more or less agreeing with what Baade and Zwicky had guessed. And then there was a war.

Astronomers started thinking about neutron stars again around 1965, and some young, prominent theorists, including Austrian–British–American Thomas Gold, Italian Franco Pacini and mostly European Lodewijk Woltjer suggested that, if such things existed, they might rotate very rapidly and have very strong magnetic fields, compared to other sorts of stars. In the Soviet Union, Azerbaijani student Okhtay Guseinov suggested to his (much better known) Russian–Jewish advisor Yakov Borisevich Zeldovich that there was a way to find such things (as well as what were soon to be known as black holes) if they orbited normal stars. They do, and "the orderly progress of science" would soon have led to the discovery of such collapsed objects. But they were scooped by Jocelyn!

In her careful examination of those million feet of chart paper, she had found a small number of patches of data that looked like fuzz, and which, when the time resolution of the data system was improved, turned out to show regularly spaced blips of radio energy coming from one place in the sky. At first, she and her colleagues worried that they might be recording noise from malfunctioning electrical equipment nearby. Their site was in the country, but cars with faulty ignition systems and vacuum cleaners (hoovers to the English) that spark were found even there. But the source passed overhead 4 minutes earlier every day, meaning a fixed location in the sky, not on Earth or on the Sun. Three more such sources turned up before Hewish and other members of the Cavendish radio astronomy group announced the result.

Bell quickly learned to recognize the patterns on the chart paper coming from scintillating sources and those likely to be due to interference (a sparking vacuum cleaner in the next village, for instance). A third pattern, which she called "scruff", didn't seem to be either of these. She spotted it first on 6 August 1967, and scruff was resolved into periodic pulses on 28 November. And the scruff occurred most days when a certain part of the sky was overhead, shining into the telescope. Since it wasn't always there, it took her a while to realize that the source was passing overhead four minutes earlier each day, keeping time by the stars and not by the Sun, and somewhat longer to persuade her advisor that it was perhaps important and they should modify the data recording to spread out the time structure of the scruff. Lo and behold! The pattern was actually a series of pulses, 1.3 seconds apart, not always there (scintillation at work, meaning the source was compact), and shifted back and forth in time by the Earth's motion around the Sun (at a speed of about 30 km/sec). There was no sign of any other time shift (Doppler effect) in the data, meaning the source was not a planet orbiting another star. Extraterrestrial signals was an early thought, and Jocelyn coined the names LGM 1,2,3,4 for Little Green Man 1,2,3,4 when she found three more similar sources, the shortest period one clocking in at only 0.25 second. Finally, the name "pulsars", by analogy with quasars, was coined before there was complete agreement about the nature of the sources. White dwarfs, neutron stars and black holes were all mentioned, with pulsation, orbits and rotation as timing mechanisms. But the winner was rotating neutron stars, because nothing else could match the shortest pulse periods of less than a second, and only rotation would slow down as the sources lost energy. The discovery of a pulsar in the Crab Nebula supernova remnant (subject of VTs doctoral dissertation) in 1969 clinched the deal.

We interrupt this story to address "why weren't pulsars found earlier?" since the closer ones are actually quite bright radio sources. Two attributes were lacking in the apparatus used in previous surveys: a short response time (rather than automatically smoothing over time to eliminate assorted noise) and a repetitive observing routine, to ensure that apparently sporadic signals were in fact from permanent celestial sources.

Bell had first noticed the scruff in August 1967. She was certain of a cosmic source by December, when she found the second, and found the next two sources before the group announced and published the first one, a very short note in *Nature*, in February, 1968. The paper had as its authors Hewish, Bell and a couple of other guys.[7]

Cambridge in the summer of 1968 (see Figure 13.2) was a very exciting place to be, both for Jocelyn Bell and for co-author Trimble who was at Fred Hoyle's Institute of Theoretical Astronomy as a half-pay postdoc at the time. Co-author Jones arrived in Cambridge 2 years later.

Figure 13.2 Jocelyn Bell, in 1968, in front of the Mullard Radio Astronomy Observatory, at Cambridge University. Illustration by Isabel Gálvez (www.isabelgalvez.eu) inspired by a picture taken for the *Daily Herald* newspaper.

It was the 1974 Nobel Prize in Physics that was shared by Antony Hewish and Sir Martin Ryle[8] (head of the Cavendish radio group, whose very important contributions to radio astronomy had little to do with pulsars). Confusingly, that same year the first binary pulsar (meaning one in a short-period orbit with another neutron star) was discovered by Russell Hulse[9] and Joseph Taylor, who shared the 1993 Physics Nobel, and it will be 2043 or some such before the details of those negotiations can be read. But our guess is that student Hulse was the single most significant beneficiary of Jocelyn's exclusion from the 1974 award. It is worth mentioning that in both cases it was the student who unambiguously recognized the unusual pattern and had some difficulty in persuading the supervisor that it was important.

There is no Nobel Prize for astronomy (or mathematics) essentially because Alfred Nobel thought they were unlikely to contribute to human betterment. The believability of any more exciting stories founders on his never having been married. The physics prize has, however, gradually extended its territory into geosciences — Appleton for the ionosphere (1947) and space science or astronomy (Hess for the discovery of cosmic rays 1936) and others. It has been written many places that the 1974 Nobel to Ryle and Hewish was the first in astronomy, but, by then, this was really a fairly modest expansion of the permitted territory called "physics". Real constraints have continued to include, first, at most three winners per prize (though the will said "the person") and, second, that winners must be able to come to Stockholm to receive the prize, as judged at the time the decision was made or announced. There is also supposedly an informal constraint against awarding a later prize for something already recognized, though the case of "transuranic element" (Fermi, Physics, 1938, and Seaborg, Chemistry, 1951) is on the ragged edge of this rule.

Since Sweden remained neutral during both the world wars, one might expect, at various times, her scientists to have held biases either for or against the various belligerents. Other possible biases, against, may have included, or still include (1) young scientists, (2) women and (3) Jews.

On the issue of youth, Bell Burnell has said that since a supervisor will be blamed if a project comes out badly, the supervisor should also receive credit when a student carries it out to great success. Cesare Lattes (1924–2005) who made very major contributions to the discovery for which Cecil Powell received the 1951 Physics Nobel (discovery of pions among the cosmic rays, using nuclear emulsion techniques) feels that he was left out because of not being the senior member of the collaboration and has sounded less accepting of this than Bell Burnell.

A look at statistics of physics winners, from Michelson (1907) to Ginzburg (2003) makes it hard to believe that Jews have been strongly disfavored, except perhaps when Einstein was, for some years, a strong contender but not a winner until 1921. Jewish women are perhaps another story, as has been suggested by Ruth L. Sime in the cases of Lise Meitner and Marietta Blau. Blau's work was an essential prelude to Powell's, and she was nominated and briefly thought about after Powell won, but then fell into the "No rethinking" pit.

As for Bell, as a young woman, somehow the relevant Swedish academics were able to re-evaluate "young student" between 1974 and 1993, though the details will not be openly available until 2043 (the year of her 100th birthday!). As for the combination, none of the very few scientific female Nobelists could readily be described as young.

Every scientist who has a favorite "should have, could have" winner will surely agree that there have been, and perhaps still are, prejudiced decision-makers in Stockholm, whether their bias is age, gender, nationally or culturally based. The only advice we have to offer is if you are asked to nominate (very many people now are each year) try for a nomination that will enhance "diversity" and the fairness of the process.

The person who flew most disastrously off the handle of Bell not being recognized was Sir (as of 1/1/1972) Fred Hoyle, as described in the previous chapter. Speaking to a reporter in Montreal, he said (perhaps not realizing he would be quoted in extenso) "Yes, Jocelyn Bell was the actual discoverer, not Hewish, who was her supervisor, so she should have been included... When Dr. Bell confirmed her initial findings, her directors kept it a secret for months — no one in the group was allowed to speak to anyone else outside." Hewish's response, as Mitton pointed out, was "What Hoyle has said is untrue, absolutely untrue. It makes me so angry. It is ridiculous to suggest that the results were stolen." and "Hoyle had made an astonishing fabrication for which he could see no explanation."

At the time, Jocelyn, by then Bell Burnell, said "It was Professor Hewish who did the preparatory work; I was the individual who analyzed the data — doing the spade work." She expanded on this on later occasions, saying that a Nobel should go to someone who had been a graduate student when the work was done only under truly exceptional circumstances, and she did not feel that her recognition of LGM1 and the others met this criterion. But it is important also to understand the circumstances in which Jocelyn made those statements at that time: she was in the final year of her PhD when she made the discovery. She was going to need academic references from the Cambridge Radio astronomy group. Therefore she couldn't afford to be critical. During the following years, as a postdoctoral fellow in a vulnerable position, she had also to be cautious with her public statements regarding this matter.

A more measured response was that of Tom Gold (of Bondi, Gold and Hoyle steady-state cosmology). He spoke after dinner at the December 1974 Texas Symposium on Relativist Astrophysics (also the occasion on which the discovery of the first pulsar in a binary system was announced), and suggested that the organizers of the next, 1976 event should invite Bell Burnell to give the evening talk. This in due course happened. She called the published version "petit four" (Bell Burnell, 1977) and had the audience hanging on her every word. Many were memorable, but two stick in the mind. First, she regarded the microphone, describing it as a "male chauvinist pig" microphone, meant to have one part clipped to a gentleman's tie and the other part inserted in a gentleman's jacket pocket. Her very elegant dinner dress had neither, and the remark struck a chord with other women scientists in the audience. The second was a sort of warning to devotees of oral histories: "There comes a time when you don't remember what actually happened. You only remember how you told the story the last time." She has now, in response to very many requests, told the story often.

On a later occasion, she said "I think there are still a number of inbuilt structural disadvantages for women", as Coroniti and Williams remarked in their chapter about Jocelyn Bell in *Out of the Shadows* (2006, edited by Nina Byers and Gary Williams, CUP).

Prof. Dame S. Jocelyn Bell Burnell has had a rather peripatetic career and has a much longer WIKI than Hewish. For some years after their marriage, she followed her husband, Martin, back and forth across Britain, picking up research and teaching jobs where she could. Thus, she was a gamma-ray astronomer for a while, then an X-ray astronomer, working particularly with the British satellite Ariel V and participating in the discovery of X-ray pulsating and bursting sources.

Later Bell Burnell became an infrared astronomer and project manager for the James Clerk Maxwell submillimeter telescope in Hawaii (1986–1990), then on to a professorship at the Open University in Milton Keynes. She was Dean of Sciences at the University of Bath (2001–2004) and in 2018 was appointed Chancellor of the University of Dundee in Scotland.

Meanwhile, as it were, Jocelyn has received prizes and lectureships associated with the names of J. Robert Oppenheimer, Beatrice Tinsley, William Herschel, Karl Jansky, Grote Reber, John Bolton (three of the founders of radioastronomy), Magellanic and Gordon. She has served as president of the Royal Astronomical Society (2002–2004) and was the first female president of the Institute of Physics (2008–2010) and of the Royal Society of Edinburgh (2014–2016). She was "promoted" to Dame Commander of the Order of the British Empire in 2007. Her various accolades mention not only pulsars but also her extended contributions to education and the promotion of the status of women in science. Bell shares with co-author Trimble the *Honoris Causa* doctorate by the University of Valencia: Trimble (2010) and Bell (2017).

Jocelyn Bell was awarded a Breakthrough prize in 2018 (3 million US dollars) for the discovery of pulsars. We agree with Brian Keating, who has said that this Breakthrough award "rights past injustices and properly honors the pioneering and pivotal contributions of a scientist who opened a new window on the cosmos."

Chapter 14

Black Holes
and Gravitational Waves

If there should really exist in nature any bodies,
whose density is not less than that of the Sun,
and whose diameters are more than 500 times
the diameter of the Sun ... their light could not
arrive at us.

John Michell, Dark Stars, 1783

The 18th century Cambridge academic John Michell[1] was possibly the first astronomer/mathematician to have considered the effect of Newton's theory of gravitation on light propagation. In his paper, published in 1783,[2] he discussed the concept of a "dark star": a stellar object so massive, or so compact, that the corpuscles making up the light ray could not escape the pull of gravity. Michell had embraced Newton's corpuscular view of light and so viewed this problem as analogous to that of calculating the speed that a ball (or a light corpuscle) launched vertically must have in order to escape the gravity of the Earth.[3]

Michell was exploiting the consequences of Ole Rømer's 1676 discovery that light traveled at a finite velocity, using Bradley's 1729 precise measurement of the velocity of light (see Chapter 7). It should be noted that Michell, like many of his contemporaries, did not believe that the speed of light was a universal constant.

The same idea, that objects massive enough and small enough that the escape velocity was greater than the speed of light would be dark stars, was later proposed by the great French polymath Pierre-Simon Laplace in 1798 in his *Exposition of the System of the World*, with a detailed calculation published in 1799.[4]

The idea that gravity could trap light in this manner was considered bizarre at the time when these calculations were performed: adherence to the notion that "Light travels in straight lines" certainly makes trapping light seem impossible. Moreover, many scientists of the 18th century were not at all convinced that the velocity of light was finite, notwithstanding Bradley's measurements. "Dark Stars" only became of interest 150 years later with the rise of black hole physics in the 1960s. Those "dark stars" were an example of postulated but unseen entities, the "dragons" of our metaphor.

Nonetheless, in 1804, Johann Soldner (1776–1883), an astronomer working in Berlin, published an article taking the work of Michell and Laplace an important step further: he calculated the gravitational deflection of light rays passing close by a star.[5] This was a remarkably prescient piece of work that was totally ignored during the following 117 years. In 1911, Einstein also calculated the Newtonian deflection of light by the Sun coming to a very similar value as that given by Soldner. But that too was ignored until Einstein tackled the problem in 1915 using his theory of general relativity, coming up with a value that was twice the value he and Soldner had given for Newtonian gravity.

Although Newton had introduced the world to both gravitation and the corpuscular theory of light, he did not discuss the action of gravity on light. There is the famous statement in his *Optiks* that "Do not Bodies act upon Light at a distance, and by their action bend its Rays; and is not this action (*ceteris paribus*) strongest at the least distance?",[6] but the context suggests that he was thinking of media causing refraction rather than the action of gravity. He was inclined to believe the outcome of Rømer's experiment and had repeated Rømer's calculation for publications in his 1704 work, *Optiks*. It is interesting that Halley had also revisited the analysis of Rømer's data in 1696 and derived a value of 300,000 kilometers per second, before Newton's *Optiks* appeared, and in doing so set the Earth at a distance of around 8 light-minutes from the Sun.[7]

14.1 Gravity and Light

The publication of Einstein's theory of gravitation came in 1916. This was his *General Theory of Relativity*: his theory for how gravity works, putting light rays at the forefront of the observational and measurement processes. The experimental verification of the prediction for the gravitational bending of light by the Sun became one of the first successful tests of the theory and finally convinced the science community at large that Einstein's theory was the way to go with gravity, in general, and strong gravitational fields, in particular. The anomalous rotation, or precession, of the orbit of the planet Mercury was an observed effect that could not be explained within the terms of Newtonian gravity. General Relativity did not provide a prediction. It simply gave a quantitatively correct explanation for the phenomenon and that was all that was needed: it was an important *test* of the theory.

The deflection of light by gravitational fields was a real prediction and a great triumph for the theory.[8] In the framework of General Relativity, the Sun could be viewed as a *weak gravitational lens* that bent the rays of starlight passing close by the Sun on their way to the Earth. The distortion of the image of the sky by the focusing action of the gravitational lens could be observed and measured only at the time of a Solar Eclipse. The advent of Radio Astronomy in the 1950s and the discovery of radio sources all over the sky meant that the experiment could later be done without waiting for an eclipse using compact sources. However, that was not without an additional complication, that the radio waves passing close by the Sun were affected by the Solar atmosphere in a way that visible light was not.

The big event for the deflection of light was the discovery of *strong gravitational lensing* caused by the passage of light past objects having strong gravitational fields. This is a purely relativistic effect, predicted by Einstein long before it was observed, and so provides one of the key tests of Einstein's theory. The Hubble Space Telescope has provided many beautiful images of strong lensing. An example is shown in Figure 14.1.

Figure 14.1 An example of the focusing of light by the gravitational field of the central galaxies in a cluster: a strong gravitational lens. The ring displays four bright knots on its circumference and two in the middle. The four knots are separate images of a single quasar (2M1310-1714) lying directly behind the two central cluster galaxies. (There are three other such knots that are not visible in this picture.) The quasar is at a light travel distance that is about three times the distance to the intervening cluster. The existence of strong gravitational lenses was foreseen by Albert Einstein (1936) and by Fritz Zwicky (1937), and the first was discovered in 1979 by Dennis Walsh, Bob Carswell and Ray Weymann. Original HST image by ESA/Hubble & NASA (license: CC4.0), https://esahub ble.org/images/potw2132a/.

14.2 Black Holes

In 1915, Karl Schwarzschild (1873–1916) came up with the first solution of Einstein's formidable equations: the relativistic solution describing the gravitational field due to a point mass. The solution was unquestionably correct, it now bears his name, the Schwarzschild Metric, but there were difficulties with its interpretation. The solution seemed to divide the space around the mass into two regions, an exterior region extending out to infinity and an interior region, a domain from which nothing escapes. This was poorly understood for a number of decades during which some people were arguing that the interface between the regions was fictitious and while others argued that it was real. The notion of trapping light by the force of gravity was not fully appreciated. The Newtonian papers of Michell and Laplace had long been forgotten.

It was in the late 1960s that John Archibald Wheeler of Princeton University popularized the term "Black Hole". It seems, however, that the term could have been first used by Robert Dicke (1916–1997, also from Princeton).[10] They both saw the black hole as the ultimate fate of a massive star that had exhausted all of its nuclear fuel: with no source of energy to maintain the pressure that holds it up, its eventual fate must be to collapse.[11] Wheeler was one of the most inspiring of post World War II scientists: he was a brilliant and imaginative physicist, politically conservative, but scientifically willing to go where others dared not even think. During the 1960s and beyond, Wheeler inspired many young mathematicians and physicists around the world to study diverse aspects of General Relativity. He might reasonably be regarded as the father of both Black Hole Physics and the study of gravitational waves.[12] The term "Black Hole" is quite appropriate: it generally refers to "a deep, immeasurable space into which people or things vanish; an abyss".[13]

With his students, he drove forward the study of black holes as astronomical objects. Groups of researchers around the world started exploring the details of black holes, including rotating black holes, within the framework of Einstein's theory of gravity and, in particular, studying the trajectories of light rays in their vicinity. Wheeler also coined the term *Geometrodynamics* as a description of general relativity that emphasizes the inseparable connection between space–time geometry and the dynamics of gravity.

The physical interpretation of the Schwarzschild Black Hole solution was mostly completed during the 1950s. The main facts were little different from what Michell and Laplace had described using Newtonian gravity, except that now there was no need for an *ad hoc* description of light propagation. Einstein had built that into his theory. A few round numbers are of help here to set the scene. If the Sun were to shrink to a radius of 3 kilometers, the escape velocity from its surface would be the speed of light. If we shrink the Sun even further, this 3 km value of the radius is still important: it is still the place where the speed of light equals the escape velocity. For a body of 1 solar mass, this 3 km radius sphere divides the space into two regions between which there can be no two-way communication: light, or anything else for that matter, can get into the inner region, but can never get out again.

This boundary is called the *Event Horizon* of the black hole and the radius is called the *Schwarschild radius*. The Event Horizon is just an imaginary surface; there is nothing tangible about it: it is possible to cross that surface from the outside to the inside (apart from getting perhaps torn apart by gravitational tidal fields in the case of low-mass black holes produced by collapsing stars), but once crossed, there is no getting out.[14] For a 1 million solar mass black hole, this radius is 3 million kilometers, and for a 100 million solar mass black hole the radius is about 300 million kilometers, about twice the Earth–Sun distance.

In the case of a rotating black hole, the event horizon is no longer spherical: it is an oblate spheroid with its minor axis coinciding with the axis of rotation of the body inside. The physics of the rotating black hole, or *Kerr black hole* named for New Zealand mathematician Roy Kerr who found the appropriate solution to the Einstein equations,[15] is very complex but turns out to be of great importance since most observed black holes appear to be rotating. It is important to recognize that neither of the idealized mathematical solutions, the Schwarzschild and Kerr black holes, can by themselves be sources of gravitational radiation. They radiate gravitational waves only when they are accreting material, eating entire stars or interacting with something else. Moreover, almost all objects we see that could potentially collapse to form a black hole have rotation. As the collapse proceeds the rotation speed increases,[16] getting ever closer to the speed of light. We expect that most black holes we see are Kerr Black Holes. This accretion process is often the source of high-energy X-rays and jets.

14.3 The M87 Black Hole

Ultra-dense compact objects have been known to exist at the centers of galaxies since the work of Wallace Sargent (1935–2012) and his collaborators at Caltech (the California Institute of Technology) in the late 1970s.[17] They presented evidence for a supermassive object at the center of the galaxy M87,[18] which is at a distance of 55 million light-years, and situated at the center of the Virgo cluster of galaxies.[19] They concluded that "the presence of a central, supermassive black hole must be a serious possibility". The galaxy is famous for its remarkable jet emerging from the very center of the galaxy. The jet is *observed* to be moving away from the black hole at 5–6 times the speed of light: the jet is in fact moving at speeds close to, but less than, the speed of light. The *illusion* of super-luminal motion is merely a consequence of the finiteness of the speed of light. This arises when very fast proper motions at small angles to the line of sight are converted to apparent speeds. The phenomenon of super-luminal motion is fully explained within the framework of Einstein's Theory of Special Relativity. The mass they assigned to the central object, 5 billion solar masses, is consistent with the mass determined from the Event Horizon modeling of the very center of M87. It is currently believed that almost all massive galaxies host a central black hole.[20]

An important aspect of the EHT project is the modeling of the light emitted from the environment of black holes accreting material from their surroundings. The light that is observed may have come from almost anywhere in the space surrounding the black hole, even the backside. This is because light emitted from the vicinity of a black hole can orbit the black hole several times before leaving the area and traveling to the observer. If the black hole is rotating the situation is even more complex.[21] Matching the models to what is observed provides the details of the black hole and its environment. Typically, such models comprise a black hole, possibly rotating, surrounded by a spinning disk of material that is spiraling into the black hole under the force of its gravity and strong magnetic fields that are amplified by the rotation of the disk. There are numerous physical processes going on which conspire to heat the accreted material and radiate energy in the form of light at all wavelengths, γ-rays, X-rays, ultraviolet, optical and infrared light and radio waves.

The *Event Horizon Telescope* team in 2019 produced the first image of the black hole and its environment at the center of the giant elliptical galaxy known as Messier 87 (M87). The image shows the black hole itself and its immediate surroundings as a kind of dark "silhouette", or "shadow",[22] against the surrounding glowing swirling gases orbiting the hole (Figure 14.2). The dark area at the center of the luminous region contains the black hole. The black hole itself is a little smaller than the central dark region but cannot be seen directly because no light can leave it. The luminous material is orbiting around the black hole. One side of the ring is brighter because the gas on that side is approaching us while the other side is darker because it is receding from us; this is another effect of relativity which is seen only when velocities approach the speed of light.

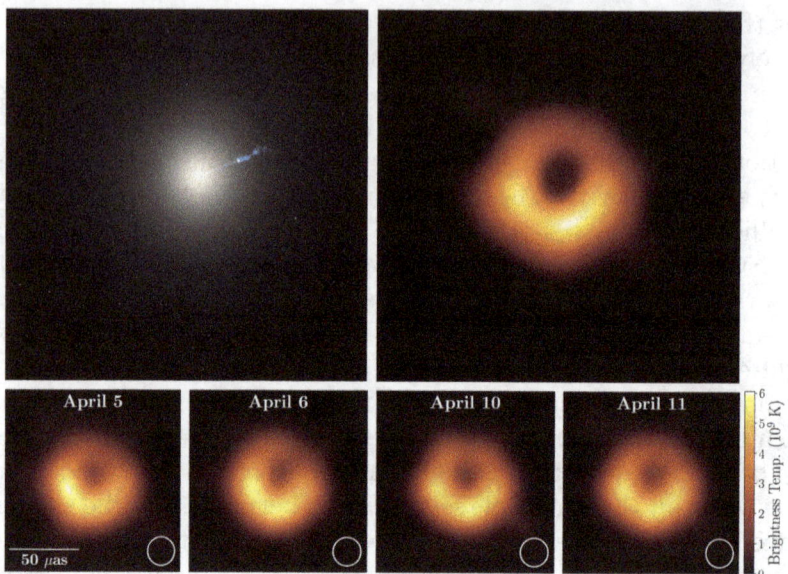

Figure 14.2   Images of the M87 Black Hole from the Event Horizon Telescope (EHT).[9] Credit: © EHT Collaboration. The central part of the galaxy M87 with its jet is shown at the top left. The figure at the top right was the first publicly released image from 10 April 2019, showing the silhouette of the hole against the background of glowing gas deep in the center of the galaxy. The gas is the central part of the accretion flow onto the black hole. The bottom four panels are a time sequence of four snapshots showing that the gas is in motion. The horizontal white line in the first panel shows a scale of 50 micro arc seconds (about ¹⁄₁₀₀th of a light-year at the distance of the source), and the circles depict the amount of blurring of the image detail. Adapted from Figure 15 of The Event Horizon Telescope Collaboration *et al.* 2019 ApJL 875 L4, *First M87 Event Horizon Telescope Results. IV. Imaging the Central Supermassive Black Hole.* The image of the central regions of M87 and its jet is from the Hubble Space Telescope archive.

The four smaller panels are images of the black hole on four different days showing the dynamic nature of the black hole surroundings.

Subsequently, the EHT team published a picture (Figure 14.3) of the M87 black hole taken in polarized light. The lines on the figure indicate the direction of polarization which tells us about the strong magnetic field structure that controls the flow of gas into the hole. The size indicates a black hole mass of 6.5 billion times the mass of our own Sun.

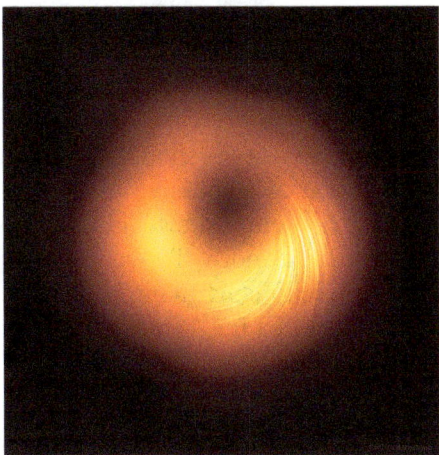

Figure 14.3 The six billion solar mass Black Hole at the center of the M87 giant elliptical galaxy. This is an image made using data from Event Horizon Telescope, a worldwide collaboration of radio telescopes. The hole is seen located in the midst of swirling gas. The lines are indicative of light polarization, caused by the presence of ordered magnetic fields that drive the disc dynamics. Accretion of material onto this black hole is the source of energy for the famous jet seen emerging from galaxy's center. Credit: © Event Horizon Telescope Collaboration.

The relatively nearby galaxy known as "Centaurus A" lies at a distance of 13 million light-years from us. It was discovered in 1826 by the Scottish astronomer James Dunlop (1793–1848) who at the time was living in Australia.[23] The object is a large elliptical galaxy into which a smaller spiral galaxy has fallen. It is an ongoing merger in which we see a lot of star formation and it is also a powerful radio source characterized by two emerging radio jets. At the center is a modest supermassive black hole of 55 million solar masses. Note that this is million, not billion, so the radius of its event horizon is a thousand times smaller that of the black hole in M87 and so cannot be imaged directly in the way the M87 black hole has been imaged. Nevertheless, the Event Horizon Telescope has produced detailed images of the emergence of the jets from the center of the galaxy where the black hole lives.

Another supermassive black hole on the Event Horizon Telescope list is the one that lies at the center of own Galaxy. The story goes back to the early studies of the galactic center, starting around 1979 and culminating in the important review article written in 1987 by the German astronomer Reinhardt Genzel, who went on to win the 2020 Nobel Prize, and the 1964 Nobel Prize winner Charles Townes (1915–2015) where they stated that taking the evidence accumulated by studies in all wavebands, "the central mass concentration could be in the form of a supermassive star (should such an object be stable) or, more likely, in the form of a massive black hole". They put the mass contained within the central few light-years as something on the order of a million Suns or more. They stressed that the evidence was "substantial but not fully convincing".

Where did the fully convincing evidence eventually come from? Improvements in the technology of optical telescopes were the principal drivers. The age of 4 meter class telescopes was coming to an end, giving way to telescopes of up to 10 meters in aperture. The revolution in aperture started in 1993 with the opening of the 10 meter Keck 1 telescope, followed in 1996 with the Keck 2 telescope, both at the Mauna Kea Observatory in Hawaii. The European Southern Observatory's New Technology Telescope "NTT" came on line at a European Southern Observatory site in Chile in 1989. This was an experimental 3.6 meter telescope with a flexible mirror that could be adjusted in real time to compensate for atmospheric blurring of images, an important feature referred to as "adaptive optics". It was the first telescope to have adaptive optics, a key development that was adopted then in most of the large telescopes that were built subsequently. During the period 1998–2001 ESO put up the "VLT" at the Paranal Observatory in Northern Chile: the Very Large Telescope with four 8.2 meter diameter main mirrors. The center of our Galaxy was an ideal target for this technology.

Two groups, working independently, were involved. Reinhard Genzel exploited the NTT to find stars that were very close to the center of the Galaxy and to measure their individual positions and velocities on a regular basis. He was following the stars in their orbits about whatever lay at the very center of the Galaxy. In 1996, Andreas Eckart and Genzel published a paper in *Nature* presenting the space velocities of 39 stars orbiting between 1/10th and 1 light-year from the central object. With these data they concluded that there was a total mass of 2½ million Suns squashed into a region of around 1/20th of a light-year, an inconceivable density "suggesting that it is most probably occupied by a massive black hole". Meanwhile, in the

USA, Andrea Ghez[24] from the University of California in Los Angeles (UCLA) was using the Keck telescopes to do the same thing. When Andrea Ghez presented the work of her group at the Calcutta conference on *Observational Evidence for Black Holes*, in January 1998,[25] it was widely agreed that "In terms of the mass density of the Compact object [at the center of the Milky Way], this is the strongest evidence of a black hole so far".

By 1999, both groups, working independently with different instruments, were able to produce strong evidence that the orbits of these stars were taking them close to some ultra-small and very massive object. The evidence would inevitably get even better. As their work progressed year on year from the early 1990s the stars were seen to be tracing their orbits about this central mass. One star in particular, designated "S2", orbits the central mass once every 16 years and so by 2018 the orbit had been traced 1½ times. This allowed a computation of the orbit using both Newtonian and Einsteinian gravity. Not surprisingly but importantly, Newton failed while Einstein was right on the mark. Consequently, we have a new value for the distance of the galactic center: 25,900 light-years with an accuracy of better than 1%.[26]

Reinhard Genzel and Andrea Ghez shared half of the 2020 Nobel Prize in Physics for "discovering the supermassive black hole that lurks at the center of the Milky Way". The other half of the Prize was awarded to the British mathematician and physicist Roger Penrose for "the discovery that black hole formation is a robust prediction of the general theory of relativity".[27] This was important because even Einstein had expressed serious doubts about whether black holes could exist.

These supermassive black holes are impressive. Black holes of stellar mass size, which were first recognized in X-ray emitting binary stars, are now being discovered by gravitational wave observatories: our next subject for discussion.

14.4   Gravitational Wave Detection

A key moment in astronomy came in 1974 when Russell Hulse and Joseph Taylor announced the discovery of a binary pulsar: a rotating neutron star orbiting about another star. The pulsar was like a clock beeping every 59 milliseconds. The beep period in turn varied with a period of 7.75 hours, indicating that the pulsar was orbiting another star. The orbital period was found to be decreasing by 40 seconds over a period of 30 years. Moreover, the elliptical orbit precesses at an astonishing rate of about 4° per year. That is some 35,000 times faster than the precession of the orbit of Mercury.

The data on the pulsar were entirely consistent with the hypothesis that the orbit was shrinking due to loss of energy through gravitational radiation. Hulse and Taylor shared the 1993 Nobel Prize in Physics for "the discovery of a new type of pulsar, a discovery that has opened up new possibilities for the study of gravitation".

This was the first evidence, albeit indirect, that "gravitational radiation" existed. This stimulated a renewed interest in gravitational wave detection at just the time when Joseph Weber's pioneering and controversial attempts to detect gravitational waves were fading from the public consciousness. Several groups had built "Weber Bars" as the detectors were called, but after a few years, they could not achieve anything like the level of detection that Weber had reported. The time had come to try a different technique.

Four decades following Joe Weber's pioneering work, in 2015, the two installations of the *Laser Interferometric Gravitational-Wave Observatory* (LIGO) simultaneously detected their first gravitational wave signal. The event took place in a distant galaxy some 1.4 billion light-years away.[28] The disturbance at that great distance caused a ripple in space–time that crossed space and time and reached the Solar System where Earth-bound scientists detected, a submicroscopic vibration of less than the size of a proton in a tube 4 kilometers in length. Gravitational waves had been detected and there were prizes all around to the LIGO team leaders, including in the one year 2016 the Kavli, Shaw, Gruber and Breakthrough prizes and the Nobel Prize in 2017.

In order to appreciate the importance of this discovery, it is necessary to address the following question: "What is a gravitational wave?" There is no such concept in Newton's theory of gravity where the force of gravity propagates at infinite speed. We have to turn to Einstein's view of gravity as expressed in his General Theory of Relativity, published in its final form in 1916. The waves appear as a solution to Einstein's equations that describes propagating ripples in the geometry of space–time. However, the very notion of "ripples

Figure 14.4 The coordinate ambiguity experienced in general relativity: Is the floor of this hallway flat or is it what it looks like — an undulating surface? How can you possibly know?[30] This shows that using a coordinate grid can be misleading, which is the problem experienced when interpreting solutions of the Einstein equations in arbitrary coordinate systems. Credit: https://i.imgur.com/V4WqenN.jpg, unattributed.

in the geometry of space time" puzzled most mathematicians and physicists working on the theory.

Einstein himself wrote several papers on the subject changing his mind each time about whether the gravitational wave solutions of his equations represented a real physical phenomenon or were simply an artifact of the choice of coordinate system. He also expressed doubts about the existence of black holes. His skepticism on both counts was shared by many others.

The source of the ambiguity which caused this uncertainty and debate lies in the interpretation of valid solutions to the Einstein equations: Was the solution "real", or was it merely an artifact of the way the situation was described?[29] This is simply illustrated in Figure 14.4 where we see what, at first glance, is an undulating surface. When we realize that this is a carpet running down a hallway, we think we know we are being misled by the grid of lines drawn on the carpet. The illusion of a wavy surface is caused by that wavy coordinate grid, not by any physical agent. Nonetheless, it could in fact be a bumpy carpet, however unreasonable that might seem in light of our experience. The carpet was designed to dissuade people from running in hotel hallways.

It was not until the mid 1950s that progress was made when mathematicians started to work with and interpret general relativity in terms of measurable physical quantities: if you could measure it, do an experiment to make the measurement.[31] The solution to the carpet illusion in Figure 14.4 would be to color the carpet with a measure of the *curvature* of the carpet at each point: a flat carpet would be monochrome (as is the carpet background in the image) and without that grid of lines we would perceive it as flat.[32]

14.5   A Challenge for Brave Experimentalists

In 1946, at the end of World War II, Joseph Weber[33] then aged 27 was a distinguished naval veteran of World War II who went back to university in order to get a PhD. He got his PhD in 1951 and in 1952, at a conference in Ottawa, gave the first-ever public talk on the principles of the maser and laser. He went on to be nominated twice for the Nobel Prize in 1962 and 1963, but in 1964, that Prize went to Charles Townes and to Nikolay Basov (1922–2001) and Aleksandr Prokhorov (1916–2002) of the USSR who independently succeeded in building the first working prototypes. In 1956, he won both Guggenheim and Fulbright Fellowships that took him to Princeton to work with Robert Oppenheimer (1904–1967) and John Wheeler as his advisors. That was a life-changing experience during which he wrote an important paper with Wheeler on the reality of a particular type of gravitational wave[34] and he was able to accompany Wheeler on an extended visit to the Lorentz Institute of the University of Leiden.[35]

At that time, when the mathematical existence of gravitational waves was not yet established, this must have seemed like a very brave if not impossible challenge.[36] Weber (see Figure 14.5), an experienced experimentalist by then, met this head-on and wrote two essays that

Figure 14.5 Joe Weber in 1969 working in his laboratory at the University of Maryland, with one of his aluminum bars. Illustration by Isabel Gálvez (www.isab elgalvez.eu) inspired by a picture from Joseph Weber Papers, *Special Collections*, University of Maryland Libraries.

he submitted for the annual "Gravity Essay Prize" in 1958, when he came third, and 1959, when he won it.[37] That was followed by an article in the prestigious *Physical Review* in 1960 and a book in 1961 on the subject of general relativity and gravitational waves.

By 1966, he had built his first prototype detector, a one-and-a-half-ton cylindrical bar of aluminum, one and a half meters in length, suspended in a vacuum. This was the start of an almost science-fiction concept: he showed that he could detect a vibration in this bar that was less than the diameter of an electron. By 1967, he reported some ten possible detections. By building a second bar later the same year and placing it two kilometers from the first, he could confirm three of those ten events were simultaneously detected by both bars. This was inspiring stuff: detection by two well-separated bars eliminated excitation by local sources of vibration such as heavy trucks running down the nearby highways or people jumping around in the corridors. By 1968, he was reporting frequent coincidences.

He obtained funding to build a number of detectors, five being placed in Maryland and one a thousand kilometers away at the Argonne National Laboratory. Detecting coincidences over this range would be pretty convincing evidence for gravitational waves, and if not gravitational waves then it would be necessary to identify some source of a kind that was hitherto completely unknown. In 1969, he was reporting coincident detections of vibrations in multiple detectors.[38] It was at this point that some skepticism was voiced, mainly on theoretical grounds,[39] questioning what, if anything, might be the astrophysical source of these detections. Such a critique might seem rather odd given that we had no idea what might be expected, but Weber continued and by 1970 had reported some 311 coincident detections.[40] Moreover, because of their bar-like shape, the detectors had some crude directional sensitivity. Examination of all these coincident detections revealed a tendency for the sources to cluster towards the center of the Galaxy. Prior to publication, Joe had claimed that the excess of sources was revealed by the observation that most sources were detected around the same time of day, i.e. once every 24 hours. This was a crucial point for his opposition: if the sources were clustering around the galaxy center they should have been seen *twice* a day when the center of the Galaxy was aligned with the equipment, the point being that the Earth is transparent to gravitational waves. When the paper appeared in print, however, it showed the 12-hourly period, which further raised suspicions.

Weber's claims that gravitational waves had been discovered were strongly contested, to the point where at a conference in 1974 Joe almost came to blows with one of his antagonists who publicly denounced him as a fraud. This was heady stuff. Other leading astrophysicists were more sanguine and some felt strongly inclined to believe that Weber might have found something important. William Press and Kip Thorne, in an article about this, concluded that "We (the authors) find Weber's experimental evidence for gravitational waves fairly convincing. But we also recognize that there is as yet no plausible theoretical explanation of the waves' source and observed strength". The time had come for independent verification of Weber's claims. Several groups emerged to build their own gravitational wave "Weber Bars" with improved technology and by 1973 were able to report that, even with greater sensitivity they could not confirm Weber's claim. Among these was Ronald Drever (1931–2017) of the University of Glasgow, who would later become one of the three founding fathers of the LIGO project.[41] Drever had already started working on interferometers, and it was Drever's interferometer architecture that was adopted for LIGO. The hunt for the elusive waves was on, Joe Weber had set the entire process in motion.

In the mid-1990s, J. Weber and B. Radak[42] found positive correlations between gamma-ray bursts detected by the NASA's space observatory Compton Gamma-Ray Observatory (CGRO) with gravitational radiation antenna pulses. As Marcia Bartusiak explains in her book[43] *Einstein's Unfinished Symphony*, this work "fueled Weber's hope that he would eventually be vindicated."

14.6 Interferometers on a Gigantic Scale

The 1980s heralded substantial advances in laser technology and in optics, and it became viable to consider the alternative to Joe Weber's bars of using an interferometer as the detector. The idea was to build two interferometers, like the one built by Michelson and Morley almost a century earlier, but on a gigantic scale. The passing gravitational waves would cause changes in the relative lengths of the arms. This was the concept dreamed up by Kip Thorne,

Rainer Weiss and Ronald Drever: it became the Laser Interferometric Gravitational-wave Observatory (LIGO). It is not just a matter of scale: the use of a special configuration of mirrors, technically known as Fabry-Perot Optical Cavities, increases enormously the effective path length of the light in the interferometers. Without these devices, the LIGO experiment would not have been able to detect gravitational waves.

On the morning of 14 September, 2015, Marco Drago walked into his office at the Albert Einstein Institute in Hanover, Germany, and turned on his computer. Marco, an Italian post-doctoral researcher, was then working on data analysis for the LIGO project to detect gravitational waves. LIGO, located at two sites in the USA, had been running in "engineering mode" for five months, generating data for final checks on the system before going live in four days' time. The engineers in the USA had gone off duty an hour earlier and left no messages of significance in the log of their shift. Looking at the overnight data Marco saw an event that had not been logged. This was the first-ever signal caused by a gravitational wave. The wave was the consequence of the merger of two black holes some 1.2 billion light-years away.[44]

Kip Thorne, Rainer Weiss and Ronald Drever were awarded the prestigious and valuable 2016 Gruber Prize, the same year as the announcement of the discovery. The 2016 Special Breakthrough Prize was awarded to the three of them and to the whole LIGO team. In 2017, Kip Thorne, Rainer Weiss and Barry Barish (with the whole LIGO collaboration) were awarded the Princess of Asturias Award for Technical and Scientific Research. Later that year Thorne, Weiss and Barish were awarded the 2017 Nobel Prize. Ronald Drever had died seven months earlier. It is widely agreed that without Barish LIGO would probably not have happened. Nonetheless, one is left wondering what would have happened had Drever not met with such an untimely death.

One of the LIGO goals was to detect[45] gravitational waves from merging stellar mass black holes. This meant that they would be looking for gravitational waves in the range of 10 Hertz–10 kiloHertz, close to the audio frequency range of the standard piano (30 Hz–4 kHz). The design of LIGO provides maximum sensitivity at a frequency of around 100 Hz. Joe Weber's aluminum bars and disk were sensitive to a frequency of around 1,660 Hz.

If we look at some round numbers, we can understand why this frequency range is relevant. Take the hypothetical case of two 10 solar mass black holes that are spiraling in towards one another. Again, in round numbers, each has a Schwarzschild (gravitational) radius of 30 kilometers so we can consider them spiraling into one another from a distance of 1,000 kilometers to the point of touching at roughly 100 kilometers. At 1,000 kilometers, they orbit 25 times per second and as they spiral closer to 100 kilometers they are orbiting at around 750 orbits per second, so we expect the radiation to start at 50 Hertz and increase to 1,500 Hertz and above as the merger takes place. The signature of the radiation is a rapid rise in pitch, a "chirp" in the language of audio science (see Figure 14.6). It is the detailed analysis of this chirp that allows us to diagnose the nature of the collision. The amount of gravitational radiation depends on the masses of the black holes. The spiraling-in time is shorter for greater masses. That two massive objects could orbit one another so fast is one of those almost inconceivable things about black holes. Another wow-number comes from the fact that the result of the merger will be an object of mass smaller than the combined mass of the stars. This accounts for the large amount of energy released by this catastrophic event in the brief time it takes for the merger to occur. This is why, with a sufficiently sensitive detector, we can sense these explosions billions of light-years away.

To achieve the highest sensitivity, we need to use the greatest possible number of photons of light in the interferometer beam: the longer the arm, i.e. the more photons, the greater the interferometer sensitivity. The Michelson and Morley instrument did not have a laser light source and had arms of around 1.3 meters.[46] In the case of LIGO, the longest practical arm length was around 4 kilometers[47] and the path length of the light beams within the interferometer was extended by using hundreds of reflections of the light within the instrument (see Figure 14.7). If we recall the back-of-the-envelope calculation about the frequency range, our two 10 solar mass black holes were orbiting at 750 orbits per second when at 100 kilometers separation. In the time of one orbit, ¹⁄₇₅₀th of a second, the gravitational wave can travel a little over 400 kilometers and light can make 100 traverses of an interferometer arm.

The reflectance of the mirrors becomes of prime importance. When Isaac Newton built his reflecting telescope, he made his 33

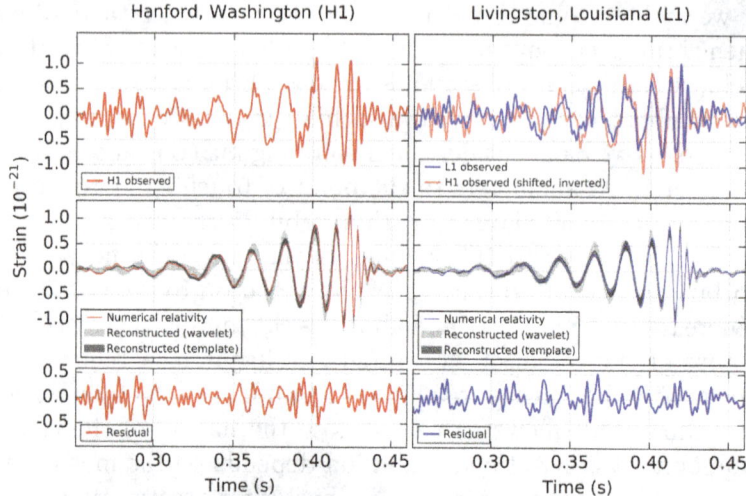

Figure 14.6 The filtered LIGO signal detected on 14 September 2015 in both observatories: Hanford and Livingston. The coincidence in time between the two signals confirms their extra-terrestrial origin. The middle panels show the numerical relativity simulation waveform of the collision of the two black holes. Credit: Caltech/MIT/LIGO Lab.

millimeter main mirror of a metal alloy called *speculum*, a mixture of copper and tin which could be polished to "high" reflectivity.[48] Such mirrors reflected only 67% of the incident light, so Newton's two-mirror system provided only 45% of the available light. The James Webb Space Telescope mirrors achieve 99% reflectance with their special gold coatings. LIGO's mirrors absorb only one out of every 3.3 million photons, and the one it absorbs heats up the mirror ever so slightly. Effects like this are critical considerations when you are trying to measure a change in length that is ¹⁄₁₀,₀₀₀th of the size of a proton.[49]

14.7   Building LIGO

LIGO received funding from the US National Science Foundation ("NSF"). It started in 1980 with a feasibility study and continued through to 2016, a period during which nobody expected to detect anything. The NSF was the sole funding body until the end of the 1990s. As with any large project, there was political wrangling,

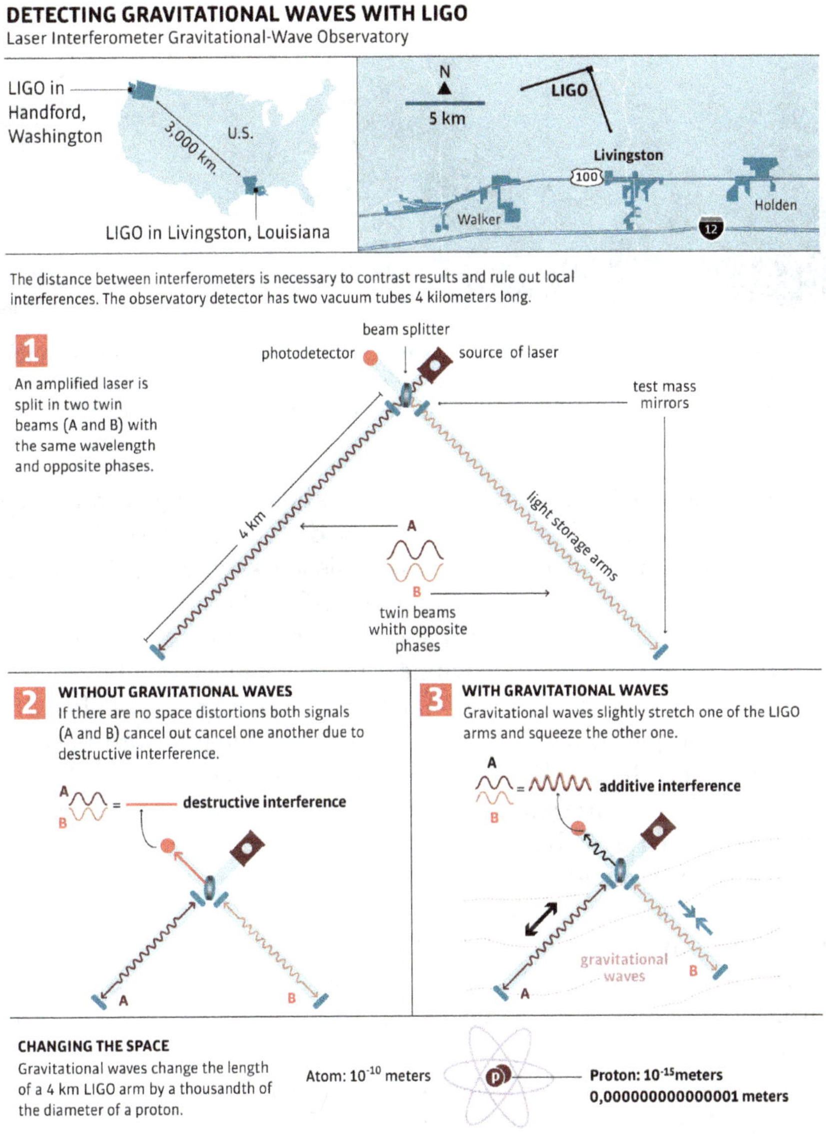

DETECTING GRAVITATIONAL WAVES WITH LIGO
Laser Interferometer Gravitational-Wave Observatory

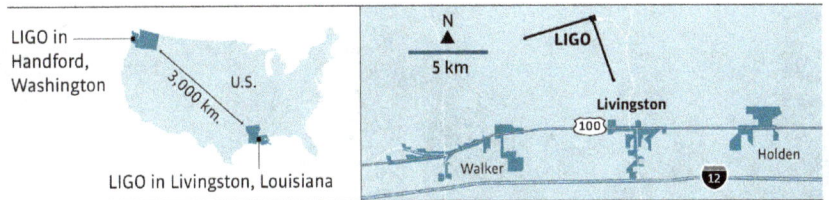

LIGO in Handford, Washington

3,000 km.

U.S.

LIGO in Livingston, Louisiana

N

LIGO

5 km

Livingston

100

Walker

Holden

12

The distance between interferometers is necessary to contrast results and rule out local interferences. The observatory detector has two vacuum tubes 4 kilometers long.

1 An amplified laser is split in two twin beams (A and B) with the same wavelength and opposite phases.

beam splitter

photodetector

source of laser

test mass mirrors

light storage arms

4 km

A

B

twin beams whith opposite phases

2 **WITHOUT GRAVITATIONAL WAVES**
If there are no space distortions both signals (A and B) cancel out cancel one another due to destructive interference.

A
B
= ——— destructive interference

A

B

3 **WITH GRAVITATIONAL WAVES**
Gravitational waves slightly stretch one of the LIGO arms and squeeze the other one.

A
B
= additive interference

gravitational waves

A

B

CHANGING THE SPACE
Gravitational waves change the length of a 4 km LIGO arm by a thousandth of the diameter of a proton.

Atom: 10^{-10} meters

p

Proton: 10^{-15} meters
0,000000000000001 meters

Figure 14.7 Schematic of the LIGO interferometer. The interferometer arms are an ultra-high vacuum environment in which are suspended two mirrors which can oscillate along the direction of the light beam when under the influence of a gravitational wave. Credit: Infographic sketched by the authors and elaborated by Javier Pérez Belmonte.

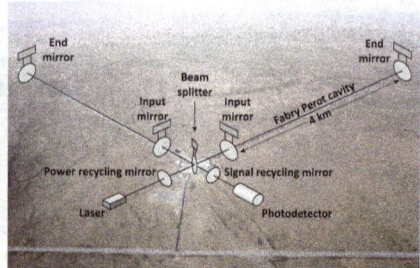

Figure 14.8 On the left panel, we see a picture of the Hanford Observatory with the two 4-km long arms. Credit: Caltech/MIT/LIGO Lab. On the right panel is a sketch done by Fernando Ballesteros of the interferometer.

acrimonious debate and questions raised about funding such a "blue sky" venture at this level.[50] The project director from 1987 to 1994 was Rochus ("Robbie") Vogt, a distinguished scientist and manager of major projects prior to his taking on LIGO. Vogt had been called in essentially to shake up the three team leaders and move them from the small-science model of instrumental development to a big-science model which would achieve the manifestly big science goals of LIGO. Moreover, he joined a project at a time of growing discord between Rai Weiss and Ronald Drever. The main difference between the two was what kind of interferometer should be used. Drever advocated a system in which light bounced up and down the interferometer tube between two suspended mirrors, while Weiss had a different idea. Drever's system was adopted but he found it difficult to work in a big science environment. In 1994, Ron was removed from the project. That was the time when Barry Barish took over as principal Investigator for the project and later, in 1997, as director of LIGO in place of Vogt.[51]

In 1991, the US Congress approved a budget of \$395M. By 1994, it was possible to start building prototypes at two sites: one at Hanford in Washington State and the other 3,002 kilometers away at Livingston in Louisiana (see Figure 14.8). By 2002, the prototype was ready and ran until 2010 without making any detections, but that was as had been expected.

Advanced LIGO started in 2008, still funded by NSF but with additional contributions from international collaborators, and was scheduled to start operations on 18 September 2015. As luck would have it, the first unequivocal gravitational wave signal was detected on 14 September 2015 at 09:50:45 UTC, four days before the intended start of operations! The signal had the characteristic shape expected from the merger of two black holes. Matching numerical simulations to the data showed that this was most likely the merger of a 36-solar mass black hole with a 29-solar mass one, leaving behind a remnant of around 62 solar masses. The distance of the event follows from these numbers: 3 solar masses radiated and the measured strength of the signal. This yields an estimate of around a billion light-years, which puts the black hole pair in a distant galaxy.

By the time Advanced LIGO became fully operational in September 2015, billions of dollars had been invested over a period of some 40 years with nothing to show and no certainty that anything would ever be detected. This was funding and support at a level that no other organization in the world would have put forward. The direct detection of gravitational waves is transformational: all other objects detected in the universe have been detected by virtue of the light they emit, across the electromagnetic spectrum, from radio to gamma-rays, plus neutrinos and cosmic ray particles. We can expect the technology to improve as more detectors are built. Putting interferometers in space would yield fewer nuisance signals ("noise"), and wider antenna separation would give greater sensitivity at the lower frequencies. LIGO has opened up a new window on the Universe.

The near-term future is to have more LIGO-like interferometers come online. The European Virgo interferometer with 3 km arms is located near Pisa in Italy and became operational on 1 August 2017, working as a part of what became the LIGO–Virgo collaboration.[52] Virgo's first detection occurred in 2017 on 14 August 2017, designated GW170814, and the second detection came a few days later, GW170817. An important milestone had been reached for Virgo.

The Japanese Kamioka Gravitational-Wave Detector, KAGRA, became operational in early 2020.[53] KAGRA has arms that are 3 km in length and is located in a deep mine at Kamioka in Japan. Its novel feature is that the system is cryogenically cooled, thereby reducing interfering thermal noise by a large factor. This brings us to the LIGO–Virgo–KAGRA worldwide gravitational interferometer, with more to come.[54]

14.8   Multi-Messenger Astronomy

Almost 100 gravitational wave events were announced during the period up to 2022. Most were identified as being the result of the in-spiraling and merging of two black holes in a rapidly decaying orbit around their common center of mass. Since these events release tremendous amounts of energy in the form of gravitational waves, the events could be recorded from distances measured in billions of light-years. The black holes in the 1 May 2019 event, labeled GW190521, were around 85 and 66 solar masses. This was a colossal event, amounting to the complete conversion to radiation of six solar masses, and took place 17 billion light-years away. The smallest black hole masses detected with confidence so far are around 11 and 8 solar masses in the merger GW170608, which radiated a whole solar mass.

A notable exception to this was the detection on 17 August 2017 of a gravitational wave chirp that lasted about 100 seconds, as opposed to the subsecond chirps that are seen when two black holes merge. The event, GW170817, was detected by both LIGO detectors and by the newly upgraded Virgo interferometer and was ascribed to the merger of two neutron stars.[55] The space-based Fermi Gamma-Ray Observatory (GRO, formerly GLAST) and the International Gamma-Ray Astrophysics Laboratory (INTEGRAL) also reported the detection of a short burst of gamma rays occurring 1.7 seconds after the gravitational wave event. The burst came from a source in the galaxy known as NGC 4993, 140 million light-years away in the direction of the constellation Hydra. Over the following days, weeks and months, the afterglow of the event was studied at all frequencies of light from X-rays to the radio by observatories around the world. At the time of discovery, this was the strongest gravitational wave signal detection and the one closest to us.

The neutron stars had masses near to 1.46 and 1.27 solar masses, and it is estimated that the mass equivalent energy converted to gravitational radiation was at most $\frac{1}{25}$th of a solar mass.[56] The immediate consequence of the merger would have been an ultra-dense object emitting a fireball of gamma rays, referred to by astronomers as a "kilonova", leaving behind either another neutron star or a black hole.

The simultaneous detection of gravitational waves and light waves showed that the speed of gravity and light waves are the same within one part in around 10^{15} (0.00...001 with 14 zeroes).[57] This defined what is now referred to as "Multi-messenger astronomy": observations in all accessible bands of the electromagnetic spectrum and in gravitational waves in addition to particle messengers such as neutrinos and cosmic ray particles.

Shortly after Virgo came online, in January 2020, two events were recorded revealing a new kind of event: both events were ascribed to a black hole devouring a neutron star. The first, GW200105, was detected by the LIGO interferometer at Livingston and Virgo, which had just become operational and was a 9 solar mass black hole devouring a compact object of 1.9 Solar masses, most likely a neutron star.[58] This happened 900 million light-years away. The other LIGO interferometer at Hanford was offline. The other event, GW200115, was detected 10 days after the first by both LIGO interferometers and by the fully operational advanced Virgo. This time it was a 6 solar mass black hole devouring a 1.5 solar mass neutron star at a distance of around a billion light-years.[59]

Within a few years of the birth of gravitational wave science, it has become an important astronomical tool with which discoveries are being made of hitherto unseen events. The next major evolution in gravitational wave astronomy is planned for the mid-2030s: LISA — the "Laser Interferometer Space Antenna". Already, there has been proof of concept with the launch by the European Space Agency of LISA Pathfinder to the L1 Lagrangian point between the Sun and the Earth. This tested various aspects of LISA using a small interferometer and opened up further development towards LISA. The arm length of LISA will be 2.5 million kilometers, the ends marked by three satellites whose distance apart can be varied. Meanwhile, the sensitivity of these instruments is continually improving with advances in technology. Moreover, as other interferometers join the LIGO collaboration, the accuracy of locating events on the sky will improve, allowing more effective verification of the events in electromagnetic wavebands.

14.9 Aftermath

Gravitational-wave research has come a long way since Joe Weber's first pioneering investigations in the mid-1950s. It took 60 years to detect those feeble signals and to start gathering data. Long-term ambitions are now high: more interferometers, Weber bars are back with tens of thousands of times the sensitivity of his 1970s bars, we look forward to LISA in space, and we will search for black holes formed during the first moments of the big bang. Joe Weber worked at a time during which important physics experiments were conducted by one person and a few helpers in a basement (or rooftop) laboratory. Ronald Drever, one of the three founders of the LIGO project, was another such lone experimentalist. The late 1960s and early 1970s was the period of the transition to "big science" done by huge teams engaged in "billion dollar science". It was also the time of transition from the world of analogue electronics to digital electronics: Weber's initial data appeared on long rolls of paper with pen and ink tracings that had to be examined by hand. It was not easy being a lone pioneer. Virginia Trimble is reported as saying[60] "Weber's goal was to bring the Einstein equations into the lab. He felt, and I think it is fair to say, he succeeded in doing that."

Part V

The Dragon's Pride
Light and Darkness

Chapter 15

Measuring the Universe

*Never say you know the last word about any
human heart.*

Henry James

In Chapter 2, we explained how Thomas Digges proposed an infinite
universe within the framework of the Copernican heliocentric model
for the first time. But this idea terrified other astronomers from that
period, as well as their successors. For example, in his *Conversation
with the Starry Messenger*, which he wrote in 1610 in response to
Galileo's manuscript, Johannes Kepler states, "You do not hesitate
to declare that there are over 10,000 stars. The more there are and the
more crowded they are, the stronger becomes my argument against
the infinity of the universe. This world of ours does not belong to an
undifferentiated swarm of countless others!"

Kepler thus aligned himself with the Greek tradition of the Stoics,
a school that was founded in Athens by Zeno, and argued that the
cosmos was finite. The idea of a finite cosmos of stars was backed
in the 20th century by astronomers such as Harlow Shapley. In his
view, everything that we see in the sky formed part of a large galaxy
with a diameter of around 300,000 light years — an immense lone
island in an infinitely large, oceanic void.

Shapley's picture of the cosmos was erased when on 1 January 1925 Henry N. Russell (1877–1957), professor of Astronomy at Princeton University, read out the communication that Edwin P. Hubble had sent to the joint meeting of the American Astronomical Society and the American Association for the Advancement of Science, in which he clearly demonstrated that Andromeda (M31) is a galaxy, though smaller than the Milky Way, and that it is far beyond the limits suggested by Shapley's Great Galaxy. This discovery expanded the universe enormously in the minds of astronomers. Within a few years, hundreds of galaxies in our local environment had been recognized. Successive mappings of the cosmos carried out with increasingly powerful telescopes revealed to us an observable universe whose building bricks are hundreds of billions of galaxies. This represented a huge leap forward in our understanding of the cosmos. The title of Hubble's paper, "Cepheids in spiral nebulae," was relatively technical for a conference with such a general scope. However, it revealed a fundamental result: many of the nebulous objects that astronomers had been observing and cataloging for centuries were, in fact, distant galaxies that were like the Milky Way. The story of that day is magnificently told in Marcia Bartusiak's book *The Day We Found the Universe*.[1] To understand its importance and how the text read by Russell changed our image of the Cosmos forever, it is useful to go back to another important event in the history of astronomy — one that took place nearly 5 years earlier.

15.1   The Great Debate

On 26 April 1920, a debate between Harlow Shapley and Heber Curtis was held in the auditorium of the United States National Academy of Science in Washington DC. The title given to the debate was "The Distance Scale in the Universe." More than a 100 years have passed since that day. Commemorating the 75th anniversary of the debate, one of the coauthors of this book, Virginia Trimble, wrote a paper[2] that tells the story of the debate, its background and its consequences. The suggestion for this academic discussion came from George Ellerly Hale (1868–1938), an American solar astronomer who discovered magnetic fields in sunspots and founded Yerkes, Mount Wilson and Palomar Mountain Observatories.

Heber Doust Curtis and Harlow Shapley were astronomers working at the Lick observatory and at the Mt. Wilson observatory in California, respectively. Curtis, an astronomer who had experience in communicating scientific findings to the public, wanted an open discussion about the extragalactic nature of spiral nebulae (which we now know as spiral galaxies). Curtis was also a notable observer and an excellent photographer of novae in spirals. Shapley, on the other hand, thought the Milky Way was the whole Universe and the Solar System was far from the center, while Curtis held the traditional view that we were near the Galactic center. Months before the debate, Shapley had applied for the position of director of the Harvard College Observatory. This position had been vacated by the death, just over a year ago, of Edward Pickering, its director for more than 40 years. Knowing that the debate audience at the National Academy would include some of the astronomers who would be responsible for choosing the new director, Shapley wanted to make a good impression. He was aware that Harvard university officials were looking for a promising young astronomer who had made relevant contributions to astrophysics. Taking up the position he aspired to would require him to leave Mt. Wilson, where he had had access to the best telescopes in the world. However, it would let him escape a working environment that he had found unpleasant.

Harlow Shapley had always said that his involvement in astronomy had come about by chance. When he was 22, he had at first tried to enroll in the University of Missouri's journalism course. As a younger man, he had worked as a crime reporter for a local newspaper. Upon his arrival at the university, he found out that the journalism faculty would not begin to teach courses until the next academic year. Due to his age, he was not prepared to waste a year of his life. Harlow looked at the courses available that academic year. They were listed in alphabetical order. He ruled out archaeology because he thought that he would never be able to pronounce its name correctly. So, he chose the next discipline on the list: astronomy. After he graduated, he won a fellowship at the prestigious Princeton University to complete a PhD under the supervision of Henry Norris Russell. Harlow worked hard, and his research, which explained variations in the brightness of Cepheid variable stars based on internal pulsations, had a big impact on the scientific community. Shapley had also made a significant contribution to continuing the Copernican program, since, until that time, most astronomers thought that the Sun occupied a central place in our galaxy, the Milky Way. Shapley had realized that the Sun and the Solar System were actually about 2/3 of the way from the center of the galaxy to the fuzzy edge of the stellar disc.

Each of the participants in the debate in some sense adopted a role that was the opposite of his personality. Curtis, who would later become the director of the Allegheny Observatory, was cautious and conservative, and he was not given to drawing hasty conclusions from astronomical observations. Shapley was more innovative and tried to extrapolate, with some risk, the results of his astronomical observations to reach conclusions that might not be sufficiently substantiated. But in this case, he preferred to stay away from more controversial discussions. The title of the debate gave him plenty of room. Both scientists had shared some of their arguments previously, so the debate actually became two consecutive talks. The first to speak was Shapley, who chose to give an informative talk explaining what was known about the Milky Way as well as concepts for understanding cosmic scales, such as the light-year. He focused on his model of the Great Galaxy, the enormous size of which apparently came close to a diameter of 300,000 light-years. Shapley highlighted his important discovery that the Sun and the Solar System were located not in the center of the galaxy but in its outer parts. As already said, Shapley did not talk about spiral nebulae, making it clear that Curtis would expound on this subject, which he did. In papers by both debate participants containing their arguments, published in the *Bulletin of the National Research Council*, Shapley highlighted the findings of his Mt. Wilson colleague and friend Adriaan van Maanen (1884–1946). Van Maanen's results indicated that spiral nebulae have an internal rotational motion. His analysis was based on comparing images of these nebulae taken over two or three decades. The angular velocities that he claimed to measure for spiral galaxies such as M33, M81 and M101 using the Observatory's instruments suggested that these objects could not be at enormous distances, since if that were the case, the tangential velocities of their components in km/s would be inconceivably large. Later it would be established that the rotational movements measured by van Maanen were spurious (and we still don't know why!).

Curtis gave a much more technical talk and made use of slides. He did what the *Washington Post* had announced the day before:

"Dr. Herber D. Curtis will defend the old theory that there are possibly numerous universes similar to our own." He was certainly not anticipating today's discussion about the possible existence of multiverses[3]; the *Post's* description references the idea that spiral nebulae are actually galaxies like ours, which is what Curtis would argue. Why did the newspaper refer to an "old theory"? The idea had certainly been around since the time of the German philosopher Immanuel Kant, to whom the expression "island universes" is attributed, probably wrongly, in reference to these nebulae. This expression was used by Curtis, at least in the written version of his contribution to the debate.

The dominant position, which in the context of this book we might consider the dogma or paradigm accepted by most of the scientific community, was that defended by Shapley. The novelty of his contribution was the enormous size he gave to the Milky Way — 300,000 light-years — and our position far from the galaxy's center. In both respects, he was right. Although the diameter of the Milky Way's disc is only 100,000 light-years, the halo of dark matter that surrounds it is much larger. And the Sun is 27,000 light-years from the galaxy's center. Ultimately, Harlow Shapley had championed the conservative position so robustly put forward by the British science historian Agnes Mary Clerke in her book *The System of Stars*, which had been republished a few years earlier:

> The question of whether nebulae are external galaxies hardly any longer needs discussion. It has been answered by the progress of research. No competent thinker, with the whole of the available evidence before him, can now, it is safe to say, maintain any single nebula to be a star system of coordinate rank with the Milky Way. A practical certainty has been attained that the entire contents, stellar and nebular, of the sphere belong to one mighty aggregation.

As we have said, Shapley had approached the debate with the main aim of impressing senior figures from Harvard University who were in Washington. The interim director who replaced Pickering was the observatory's most senior astronomer, the 67-year-old Solon I. Bailey, but university management was convinced that the center should be led by a young and promising astronomer. At 35 years of age, Harlow Shapley was a good candidate. At the end of that year, he took up the position, which he would hold for more than

30 years. Probably, this was the most important long-term outcome of the debate. Shapley's Milky Way was too big. Curtis' Milky Way was too small. The current scale is close to the geometric mean of the two. And certainly, Curtis was right about the existence of other galaxies.

15.2   The Women Who Invented Astrophysics

When he arrived at Harvard, Shapley met the women astronomers that his predecessor Pickering had been recruiting over the previous three decades to carry out routine calculations. They worked under his direct supervision on photographic plates and stellar spectra. Some of these astronomers made important astronomical contributions, but because of social constraints, women were barred from the academic and scientific progression that most men would have enjoyed. They had to settle for what was clearly poorly paid and even more poorly acknowledged work. Some called the group Pickering's harem (see Figure 15.1). A modern account of their stories can be found in the excellent book written by Dava Sobel *The Glass Universe*.[4]

In this era, "computer" was the label given to a person performing a supposedly routine numerical algorithmic task. Shapley was deeply grateful to one of them, Henrietta Swan Leavitt (1868–1921). The original work that this Harvard computer had carried out had formed the basis for the measurements of distances to the stars that had led him to produce an extraordinary map of the galaxy, in which the Sun had been moved from near the center and placed near the edge. When he arrived at Harvard, Harlow Shapley wanted to repay Miss Leavitt by naming her head of the observatory's photometry section. Sadly, Henrietta died of cancer just a few months after his appointment. She was 52. Her premature death was a tragedy for many of her colleagues, not only because of their recognition of her scientific discoveries but also because of her extraordinary character and humanity. Solon Bailey, the previous director under whom she worked, wrote in her obituary: "Henrietta had the happy faculty of appreciating all that was worthy and lovable in others and was possessed of a nature so full of sunshine that, to her, all of life became beautiful and full of meaning."

Figure 15.1 The women "computers" working in Harvard in 1891. Williamina Fleming is standing in the center of the image and Edward Pickering is standing on the left. The two ladies with magnifying glasses are Annie Jump Cannon (in the center of the image) and Henrietta Leavitt (in front of Pickering). Credit: HUV 1210 (9-4), olvwork289689. Harvard University Archives.

Henrietta Leavitt made a very important personal contribution to astronomy. In her case, that contribution was decisive in understanding the universe's scales. Fortunately, a message by Pickering published on 3 March 1912 in the Harvard College Observatory Bulletin made it clear in its first sentence who was responsible for this important scientific work: "The following statement regarding the periods of 25 variable stars in the Small Magellanic Cloud has been prepared by Miss Leavitt." What followed were the results of a detailed study of variable-luminosity stars in this small satellite galaxy of the Milky Way. Incidentally, the first 1908 paper on variable stars in the Large Magellanic Cloud has Leavitt's name as the author.

The light emitted by variable stars is not constant, hence the name given to them. In 1784, the young English astronomer John Goodricke was the first person to observe that the apparent brightness of some stars varied periodically as we have explained in Chapter 6. His first variable star was Algol, whose light curve (see

Figure 6.2) is rather symmetric. The light curve of Delta Cephei was clearly asymmetric: it increases to a peak and then subsequently decreases more slowly until it reached a low point, only for this pattern of behavior to then repeat itself. Goodricke was made a member of the Royal Society for this discovery at the age of 22. Unfortunately, he died just 14 days after his appointment. Goodricke was working with Edward Pigott, who outlived him many years. The first papers were published with both names.

Henrietta Leavitt was an expert when it came to measuring variations in the brightness of these stars using the photographic plates taken at Harvard's observation station in Peru (see Figure 15.2). One afternoon in October 1907, Leavitt wrote in neat handwriting in her personal notebook: "It would seem that the brightest variable stars have longer periods of variability." It was a notion that her mind mulled over for several years, and by 1912, she had enough evidence to conclude that there was a direct relationship between period duration and average intrinsic brightness, the amount of light emitted by the star. Leavitt had just provided to all of the astronomers across the world the cornerstone that would support the entire cosmic architecture. She had fashioned the yardsticks for measuring the universe. Astronomers would only need to find variable stars of the right kind, observe them for several consecutive days (or weeks), draw their light curves in order to measure their periods, and finally apply the relationship between period and luminosity discovered by Leavitt to determine the amount of light emitted by the star — the true absolute brightness. By comparing this to their apparent brightness, they could accurately estimate the star's distance from Earth.

On 12 December 1921, Henrietta Leavitt died of cancer. A few weeks earlier, Shapley "was to call on her on her death bed" as he explains in his memoirs.[5] Dava Sobel[6] points out that Shapley valued her as "one of the most important women ever to touch astronomy", since Leavitt had greatly contributed to both humanity's understanding of the universe and his own professional success. There was no doubt that Shapley's ability to apply Leavitt's discovery — the relationship between period and luminosity — had been a crucial element in the discovery of the real position of the Sun in our Galaxy. He had made the discovery that bestowed on him the reputation that led him to the directorship of the Harvard Observatory a couple of years before while living in Pasadena and working at the Mount Wilson

Figure 15.2 Henrietta Leavitt in her desk at the Harvard Observatory (c. 1916). Illustration by Isabel Gálvez (www.isabelgalvez.eu) inspired by a photograph in the Digital Collections of the Harvard Library.

Observatory. Harlow had harbored a great deal of doubt about moving to the East Coast. He had the feeling that the observatory that he was about to leave would have greater resources in the future (as turned out to be the case).

In 1919, the director of Mount Wilson Observatory George Ellery Hale (1868–1938) hired a new young astronomer for the staff, Edwin Powell Hubble (1859–1953), who like Shapley had been born in Missouri. Hubble arrived at the Observatory in the same year that the 100-inch Hooker telescope — the largest in the world at the time — was put into operation. Shapley was the son of a farmer and never made any effort to lose his Missouri accent. He was against American participation in the war in Europe, and he had aligned himself with political positions close to those of the Democrats. He couldn't stand Hubble's conservatism, or his affected way of dressing in the outfits from London, and "found his carefully cultivated Oxford accent a little off-putting."[7]

Hubble had hurriedly finished his thesis so that he could enlist as a volunteer in the United States Army. He joined the 86th Infantry Division, which was involved in the First World War but did not engage in combat. He did not immediately return to the United States after the Armistice was signed in November 1918 but instead spent a year at the University of Cambridge. Hubble thought that Shapley, by arriving at Mount Wilson immediately instead of going off to Army service, had pre-empted some of the astronomy he wanted to do. Harlow and Edwin never got on well. But both of them based a great deal of their work and professional success on the contribution quietly made by Henrietta Leavitt a decade earlier. She bound them deeply and was no doubt indirectly responsible for one of Hubble's greatest discoveries, which also literally ended the model of the universe that Harlow Shapley had so zealously defended in the Great Debate in Washington in April 1920.

15.3   *VAR!* Written on a Photographic Plate

On the night between 5 and 6 October 1923, Edwin Hubble had carried out a 40-minute-long exposure of the Andromeda Nebula with the Hooker telescope. Upon developing the photographic plate, he discovered three stars that had not previously been there and marked them directly on the back of the plate with an "N" for nova (see Figure 15.3). He had an extraordinary memory and was able to quickly recognize new objects if they appeared on photographic plates without needing to review previous ones. Novae are binary stars that experience a sudden and extraordinary increase in brightness. In a matter of days, their luminosity can increase more than 10,000 times. When reviewing that region of the sky against previous plates, he was pleasantly surprised to discover that one of these stars was not in fact a nova but rather a powerful variable star. He crossed out the "N" and wrote "VAR!" on the back of the plate. His work had begun. From that same night, he started to study this variable star so as to ascertain its period. During the following weeks, he completed the light curve and was finally able to determine when the cycle reached completion. The star had a period of

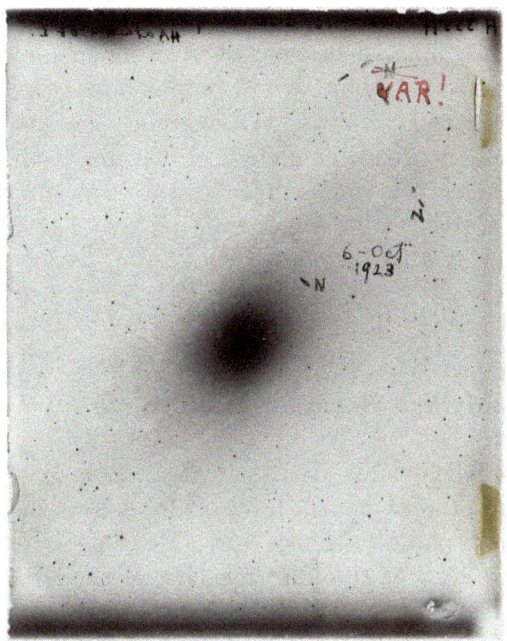

Figure 15.3 Hubble's "VAR!" plate of the Andromeda galaxy with a variable star taken on the night of 5–6 October 1923, with the Hooker 100-inch telescope of the Mount Wilson Observatory. Image courtesy of the Carnegie Observatories, Carnegie Institution of Washington.

31 days. By applying Henrietta Leavitt's period-luminosity relation, he worked out that the star ought to be located at a distance from Earth of nearly 1 million light-years. It was a surprising result. Not even the size attributed by Shapley to the Milky Way was so enormous. Shapley had argued during the debate that the diameter of our galaxy was 300,000 light-years in the face of his opponent Heber Curtis's position that it was a tenth of that size. If the variable star that Hubble had found in Andromeda was 1 million light-years away, it could not possibly belong to our galaxy.

Curtis was right. Andromeda was a different galaxy, an island universe. Over the following year, Hubble studied Cepheid variables in both Andromeda and other nearby galaxies. The results all pointed in the same direction. Their host nebulae were other galaxies like ours. The young Hubble wrote to Shapley, with whom he had never clicked: "You will be interested to hear," it began, before detailing the results. Shapley received the letter in February 1924. At the moment when he opened it, he was meeting in his office with the Englishwoman Cecilia Payne, who would soon receive a doctorate in astronomy earned at Harvard.[8] After giving the letter a somewhat nervous and cursory read, he gave it to Cecilia, saying to her as he did so, "Here is the letter that has destroyed my universe." Hubble also wrote to Curtis at the same time he wrote to Shapley. Curtis was pleased but said he wasn't surprised.

15.4   The Universe Became a Lot Bigger

Hubble first published his results in *The New York Times* on 24 November 1924. Just a month later, Hubble sent his results in the form of a research paper to Henry Norris Russell so that he could read it on 1 January 1925 at the joint meeting of the American Association for the Advancement of Science and the American Astronomical Society. Russell had been Shapley's mentor and thesis adviser. The setting was particularly appropriate. While Hubble was four thousand kilometers away in Pasadena, smoking his pipe, Russell was reading Hubble's text in his absence. Harlow Shapley and Heber Curtis were present in the audience. The heavens had opened. Miss Leavitt had provided the key, and all Hubble had needed to do was put it in the lock and turn it. And with this opening of the heavens, the universe became a lot bigger. But this is not the whole story, and it is important to clarify that, although Miss Leavitt showed the correlation, this had to be calibrated to be useful for estimating distances, because no Cepheids in those days had a geometric distance measured by heliocentric parallax (this technique was explained in Chapter 5). Shapley did the calibration using statistical parallax, which uses proper motions, radial velocities and the assumption that motions are random in space (in 3D). Hubble used Shapley's calibration in his 1929 paper[9] in the *Proceedings of the National Academy of Sciences*, with the bad luck of mixing different kinds of Cepheids that provide him a too small a value for the distance to Andromeda, although more than three times larger than Shapley's scale for the Milky Way galaxy.

We could imagine Shapley sportingly congratulating Curtis, telling him with a big smile something like, "Don't think that you have lost. What your old colleague from California and fellow Missourian has presented here today actually completes the work that has brought you most prestige. Years ago, you showed that the Sun wasn't at the center of our galaxy, and now we know that our galaxy is only one among thousands, perhaps millions, that inhabit this vast universe. We have learned, following in the footsteps down the path that Copernicus began on centuries ago, that we don't have any special position in the universe."

15.5 Cosmic Expansion

Milton L. Humason (1891–1972) was the boy who guided the mules hauling the heavy telescope parts to the top of Mount Wilson in California, where (as we described earlier) the best telescopes in the world were located in the late 1920s. Humason, who had still not completed his secondary education, showed a special skill for handling astronomical instruments, and he soon became part of the team of night-time assistants who monitored and maintained the telescopes. Humason's ability to obtain the spectra of faint galaxies earned him a great reputation, although his background was not connected with the academic world at all. Humason and Hubble soon became friends and working partners. First, Hubble in 1929 and then both Hubble and Humason together in 1931 proved that the galaxies surrounding us are moving away at a speed related to their distance from the observer. The further away, the faster they were receding, with the recession velocity proportional to the distance. This observation is known as Hubble's law and is one of the greatest discoveries of the 20th century. Hubble's law means we can estimate the distance to the galaxies by measuring their recession velocity, provided that the value for the expansion rate of the universe, Hubble's constant, is known. The most recent estimations place this value at around 70 km/s/Mpc, which means that for every 3.26 million light-years further away we look, the recession velocity of the galaxies increases by 70 km/s (252,000 km/h), although there is still some uncertainty about its value. Cosmologists have coined the expression

"Hubble tension" to denote the discrepancy between early- and late-time measurements of the Hubble constant. By early-time measurements, we mean that the Hubble constant is a parameter determined within a cosmological model from the observations of the cosmic microwave background (CMB). Conversely, late-time measurements of the Hubble constant are performed by measuring distances to standard candles and the recession velocities of the galaxies that host them. Cosmological observations based on the CMB radiation made by the Planck satellite[10] suggest a value for the Hubble constant of 67.4±0.5 km/s/Mpc, while observations of Cepheids with the Hubble Space Telescope provide a value of 73.04 ± 1.04 km/s/Mpc.[11]

We have already explained how Hubble measured distances using Cepheid stars. So, how can we find the other quantity at work in Hubble's law, the radial velocity of a galaxy (the velocity along the line of sight of an observer)?

The recession velocity of galaxies is measured using basically the same technique employed by the police to measure the speed of cars on the roads. This technique is known as the Doppler effect. This effect describes the frequency change of sound waves, and also of electromagnetic radiation, as a result of the relative movement between transmitter and receiver. A curious experiment to demonstrate this effect was carried out in the middle of the Dutch countryside, near Utrecht, in the spring of 1845.

Amateur meteorologist Heinrich Dietrich Buijis-Ballot (1817–1890) put together an experiment with an open train carriage, in which a trumpeter played a note while other professional musicians listened from the platform. They had to record the note they heard, and Buijis-Ballot expected that the same note repeated over and over again by the trumpeter on the train would be perceived as being more high-pitched as the train approached and low-pitched as it moved away. This is what would be expected if the effect reported by Austrian physicist Johann Christian Doppler in 1842 was verified. The results were not very convincing on the first attempt, because the noise of the train made it difficult to pick out the notes being played. The experiment was repeated on subsequent days and the results coincided with Doppler's predictions.

Sound waves produced by a musician moving away from the listener are heard with lower frequency (longer wavelength). Electromagnetic radiation experiences a similar shift: light from a star or a galaxy moving away from the Earth at a certain speed is received on our planet with a lower frequency and longer wavelength than if we and the source were at rest relative to one another. The wavelength of electromagnetic radiation corresponds to the distance between two consecutive wave crests. In the case of radio waves, this distance is much greater than in the case of the waves that make up visible light. Hubble realized that the wavelength of light from the galaxies was, in the majority of cases, redshifted; that is, it was found to have longer wavelengths than the wavelength from a source at rest relative to the observer.

In addition, the redshift of distant galaxies was greater the further away from the Earth they were, and others interpreted this fact as a result of cosmic expansion: if space expands, the light transmitted by distant galaxies has its wavelength stretched by the expansion. Hubble never quite endorsed this implication.[12]

The expansion of the universe is an expansion of space itself, not the motion of galaxies in three-dimensional space. That is, space is

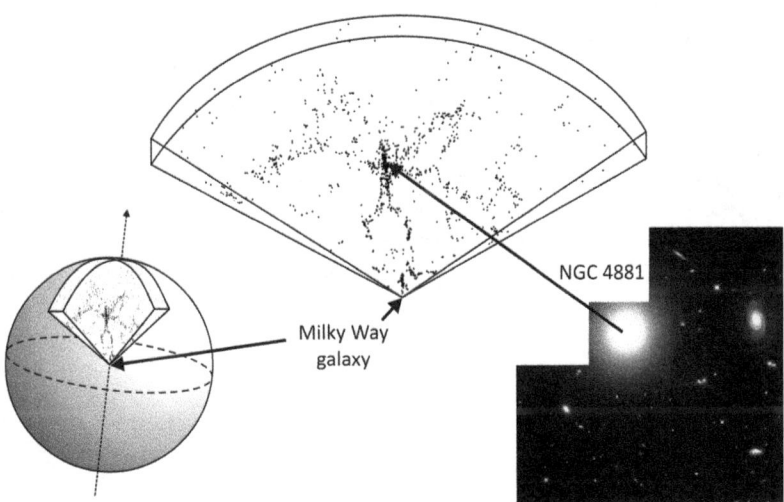

Figure 15.4 The slice of the Universe mapping the distribution of the galaxies obtained in 1986 by Valérie de Lapparent, Margaret Geller and John Huchra. It covers 6° in declination, 120° in right ascension, and a depth of about 700 million light-years, containing about 1,000 galaxies. The image of the galaxy NGC4881 in the Coma cluster comes from the Hubble Space Telescope. Credit: W.A. Baum and the WFPC tream, STScI. Montage by Fernando Ballesteros.

expanding and is dragging the galaxies with it. In the book *Blind Watchers of the Sky*, Rocky Kolb emphasizes this idea with his usual sense of humor, when he says that repeating three times per day the opening sentence of this paragraph is the key to understanding the expansion of the universe. Any observer, in any galaxy, anywhere in the universe would also see that all the other galaxies are moving away at speeds proportional to their distances. There is no center; there is no edge. The redshift produced by the expansion comes from general relativity and should not be confused with the redshift caused by the Doppler effect.[13]

Using Hubble's law, astronomers Valérie de Lapparent, Margaret Geller and John Huchra (1948–2010) embarked on a cosmic mapping exercise in 1986 showing the distribution of the galaxies (see Figure 15.4). To do this, the astronomers used a telescope based in Arizona and built up three-dimensional maps showing the cosmic macro-structure. In the maps it can be seen that the galaxies are not distributed randomly; they are concentrated in clusters and filaments that form a cosmic spider's web in which there are almost empty areas, like vast bubbles where there are very few galaxies. This view of the universe with structures on such an impressive scale surprised cosmologists a little, as they were expecting a rather more even spread, as suggested in the cosmological principle. But it is in fact a question of scale. In subsequent years, maps covering larger volumes have been produced showing that at scales more than 200 million light-years, the hierarchy of structures that give the universe this fractal nature does not continue, and the distribution of the galaxies becomes more even, validating the cosmological principle assumed by Einstein in 1917.

Chapter 16

What Is the World Made of?

O dark dark dark. They all go into the dark,
The vacant interstellar space,
the vacant into the vacant.

T. S. Eliot, East Coker, Four Quartets

Some of the ancient Greeks decided on four constituents Earth, air, fire, and water, while postulating a fifth element, or quintessence, for the planets and stars beyond the lunar orbit. Savants of the 17th and 18th centuries lit upon an increasing number of immutable elements, which, in the mid-19th century were organized into a periodic table by Dmitri Ivanovich Mendeleev (1834–1907) and shown also to be present in the Sun and stars by Robert Wilhem Bunsen and Gustav Robert Kirchhoff.

Ah! But how much of each of these 63 (known to Mendeleev, now 90) natural elements exists? The Earth provides a handy sample, and its rocky crust says that there is a good deal more of things like oxygen, silicon, magnesium, and iron than there is of gold, silver, and europium. Also quite sparse are the lightest element, hydrogen, and the noble gases that only barely form compounds, helium, neon, argon, krypton and xenon. In the first couple of decades of the 20th century, geologists compiled tables of the relative abundances of many elements on Earth and in the meteorites, and about half of them had been identified in the Sun from absorption lines recorded in laboratory experiments at the same wavelengths. A likely assortment was called "the Russell mixture," for Henry Norris Russell perhaps the most influential of American astronomers in the period. The dominant elements were those common on the Earth, found in the Sun and on an assumption called "the uniformity of nature" supposed to be the common ones in all the stars. Hydrogen and helium were present but very rare.

16.1   Cecilia Payne and Hydrogen

To Cambridge, Harvard University, came in 1923 Cecilia Helena Payne (1900–1979), fresh from the Cambridge (UK) influence of Arthur Stanley Eddington (a strong believer in uniformity) and from the labs of Ernest Rutherford. She carried with her an understanding of the beginning of quantum mechanics, in particular the idea of Meghnad N. Saha (1893–1956) that the fraction of atoms of a particular element that would have electrons in the right orbits to absorb particular wavelengths of light would be a (quite steep) function of the temperature of the gas. She also had just enough fellowship money to enable her to choose her own research topic, looking towards a doctorate. She chose to make use of some of the enormous number of spectrograms of stars of different colors, hence having different temperatures, sitting in the Harvard plate vaults.

There were blue stars whose spectra showed no lines except those of hydrogen and helium that had lost one of its electrons to high temperatures. At the other end were red stars whose spectra were dominated by molecules, and in the middle stars like the Sun with absorption features due to metals like sodium and iron, both with all their electrons and with one missing (ionized). Silicon was particularly useful, because in the array of spectra, she found it neutral, missing one, or two, or three electrons. Payne also supposed that moving along a temperature sequence, when a particular line just about disappeared, while some other line disappeared at a different temperature, the number of atoms absorbing a "just visible" line would be the same for any element plus electron configuration. Particularly informative were the K giants, relatively cool yet very bright stars whose spectra still showed hydrogen.

Now, what was she going to be able to say about hydrogen and helium? In their lowest energy ("ground") state, the wavelengths they can absorb are all ultraviolet "colors" not visible from the Earth's surface. The H and He lines in her stars' spectra had to come from atoms that already had one electron excited out of the ground state. This takes lots of energy. Yet, the lines were there at relatively cool temperatures (determined from those useful silicon levels).

How could this be? Only, she concluded, if there were lots and lots and LOTS of H and He atoms in the atmospheres of her stars (not including the Sun). As for relative amounts of everything in stars of different temperatures, these were nearly all the same, for nearly all her stars, and, for everything except H and He, and maybe oxygen, about the same as the Russell mixture. But she needed, literally, a million times as many H atoms as iron, and almost as many helium atoms, looking over the full range of temperatures and colors.

Payne wrote up her work and shared the draft with her advisor, Harlow Shapley, director of the Harvard College Observatory. A ha! He (and everyone else quickly) said, indeed all stars have the same chemical composition, and it is like the Sun and Earth, except, that dominance of helium and hydrogen. That cannot be right, he said. Shapley shared the draft with Russell, who was to be her external examiner. Yes, said, Russell, magnificent, indeed the best astronomy thesis I've seen except for Shapley's. But the dominance of H and He cannot be right. She should say it is somehow spurious.

The thesis was modified accordingly (and later she said she very much regretted having to do so, but it was politically the only possible choice). Thus, in 1927 H.N. Russell, R.D. Dugan, and J.Q. Stewart, in their classic text *Astronomy*[1] could write that "Miss Payne" (though her PhD had come in 1925) "had shown that the gas pressure was low in stellar atmospheres, that the hottest stars had surface temperatures up at least to 35,000 K", and that "the uniformity of composition of stellar atmospheres appears to be an established fact." But the behavior of hydrogen, helium, and to a lesser extent of oxygen was "puzzling." Staying on at Harvard, she became Shapley's employee badly underpaid compared to men doing the same jobs and required to devote more and more of her attention not to further analysis and interpretation of stellar spectra (where her heart lay) but to the establishment of brightness standards for stars recorded as images on photographic plates and to studying the variable brightness of Cepheids and other pulsating variables, novae and all.

Eddington, whom she had adored from afar beginning in 1919, when she returned to England post-PhD said, "not in the stars; on the stars," and indeed some chemical separation was part of the story, but the main plot line was and is that the stars, galaxies, interstellar and intergalactic stuff are made mostly of hydrogen and helium (left from the hot, dense early phase of the universe that we call the Big Bang).

Decades later, Otto Struve (Russian–American, 1897–1963) and last of a long familial line of astronomers, over the years President of the American Astronomical Society, the International Astronomical Union, and editor of the *Astrophysical Journal* while director of Yerkes Observatory declared that she had written the most important astronomical thesis of the century.

Payne married Russian refugee astronomer Sergei Illarionov Gaposchkin in 1934, and much of the variable star work was done with him (she is generally thought to have done most of the real work). Recognition came slow and late. She was finally appointed to a Harvard professorship in 1956 (but retired with a salary still considerably less than those of men with less seniority). The highest honor of the American Astronomical Society is the Russell lecture. He gave the first one in 1946, Shapley the 4th in 1950, Struve the 10th in 1957, and Payne-Gaposchkin the 29th in 1976, the first woman to do so, and with the first-ever female president of the American Astronomical Society, E. Margaret Burbidge to introduce her. Burbidge herself became the 37th Russell lecturer and second woman in 1984.

Born in Wendover, England, in 1900, Cecilia Helena Payne-Gaposchkin, a life-long heavy smoker, died of lung cancer in Massachusetts on 7 December 1979.

16.2   The Dark Side of the Universe

In modern cosmology, there are a number of invocations of invisible entities. There is evidence for two forms: "Dark Matter" and "Dark Energy," the nature of each of which is a complete mystery.

We are fortunate enough to be around at precisely the time in human history when we appear to have a credible understanding of the very early history and subsequent evolution of our universe. That understanding is formulated within the framework of Einstein's

Theory of General Relativity and direct empirical evidence that our universe was very hot and dense a finite time ago. The community of cosmologists has studied the consequences of this theory and confronted them with further observational tests, and we conclude, with confidence, that we have a viable solution to much of the mystery of our universe.

Of course, such a statement might equally have been made in regard to any of the models of the known Universe held by scholars at any time in the past. The outstanding examples in Western Europe were the Solar-centric model of Aristarchus, which was never adopted, and the geocentric epicycle model of Ptolemy which was so good at predicting the positions of the Sun, Moon and planets that it held sway, to the exclusion of all other ideas, for over 1,200 years. It was a great theory because it served its purpose — it worked. Yet it was based on what we now know to be a false premise because the planets are in fact orbiting about the Sun, and the Moon about the Earth. Could our consensus cosmology be in the same position — a great theory that is in fact wrong?

For us, and for some of our colleagues, that is an important question. It comes at a time when the cosmological paradigm in which we have such confidence tells us that some 95% of the contents of the universe are invisible and made up of some as yet unknown material or energy. Can we work our way around that? If the theory works, should we be worried?

Possibly the first astronomer to use the expression "dark matter" was the Dutchman Jacobus Kapteyn (1851–1922), in an article published in the *Astrophysical Journal* the year of his death[2] Kapteyn posited the existence of non-visible objects whose presence is necessary to explain the behavior of other visible objects that may be subject to the gravity of the invisible ones when he stated that "we have therefore the means of estimating the mass of dark matter in the universe." Also in 1922, the English astronomer James Jeans (1877–1946) claimed that "there must be about three dark stars in the universe for every bright star." For both these figures, the universe was made up of our Galaxy alone, and their calculations were conditioned by something that was not well understood back then: the significant interstellar light absorption produced by the abundant dust found around the galactic plane. In fact, when it was possible to detect less luminous stars, recognizable gas clouds, and remnants of stellar explosions, most of these dark objects moved into the bright category, with this first dark matter ceasing to be so conspicuous.

16.3   The Dance of the Galaxies Within a Cluster

Galaxies are not usually isolated in space; instead they tend to appear in concentrations that range from groups formed by a few dozen members through to rich clusters consisting of thousands. Inside each group or each cluster, galaxies move under the influence of the gravitational field provided by the entire mass of the group. So, for example, Andromeda, the biggest galaxy of the Local Group, is getting closer to ours instead of moving away from it. It has, therefore, a radial velocity that will actually make both galaxies (Andromeda and ours) collide in around 4 billion years' time.

The modern concept of dark matter that astronomers continue to discuss today appeared in 1933 with Fritz Zwicky. Using the 100-inch Mount Wilson telescope in California, Zwicky measured the peculiar velocities along the line of sight of eight individual galaxies in the Coma Cluster and obtained values with a range of about 1,000 km/s (which is technically known as velocity dispersion). At such velocities, galaxies ought to escape the gravitational attraction produced by the cluster's total mass, if that mass corresponded to the sum of the individual masses of the galaxies in the cluster. If we observe the galaxies in a cluster, we realize that they move, seemingly in an erratic fashion, around the cluster (like a swarm of bees). But they actually follow orbits that are perfectly determined by the cluster's gravity. Velocity dispersion is the statistical dispersion of the individual velocities around the mean. What Zwicky in fact did was to measure radial velocities — that is, in the direction of the line of sight — of some of the individual galaxies that make up the cluster. He did so through analyzing the redshift of the spectral lines of the radiation that reaches us from each of the galaxies.

As a result of cosmic expansion, the whole Coma Cluster is moving away from the observer at a speed of about 7,000 km/s. The radial component of each galaxy's own peculiar movement modifies this value, increasing it if this component heads away from the observer, or decreasing it if the peculiar velocity relative to the cluster center is towards the observer. These variations provided Zwicky with the velocity dispersion. Then, a consequence of Newton's theory of gravitation, called the Virial theorem, says that if you know the size of the cluster and that velocity dispersion, you can calculate the total mass needed for the cluster to hold together (see Figure 16.1).

Zwicky came up against the discovery that, when he applied the Virial theorem, the mass needed to make the cluster of galaxies stable was several hundred times that obtained by adding the masses of the stars in the visible individual galaxies. This discovery made him assume that there must be non-visible matter. He postulated the existence of an invisible or unknown entity in order to justify and explain an astronomical observation within the framework of accepted physics theories. As he wrote in German, he called this entity "Dunkle Materie." In his article published in German in 1933 in the Swiss journal *Helvetica Physica Acta*, he tells us that "Should this turn out to be true, the surprising result would follow that dark matter is present in a much higher density than radiating matter."[3]

Zwicky showed that the total dynamic mass in the cluster would have to be 400 times larger than the mass associated with the galaxies in the form of luminous matter. Interestingly, this discovery went unnoticed by the scientific community during the 30 years that followed. In fact, there are almost no references in the scientific literature to Zwicky's work between 1933 and 1975. This fact could be due to different reasons: the journal and the language where the result was published, a scientific community not ready to accept the existence of unseen entities or, even, the unpopularity of Zwicky due to his irascible personality.

Zwicky's "Dunkle Materie" was, he said, faint galaxies and small galaxies, from which he expected to see excess sky brightness from between the known Coma galaxies. He looked for this and probably detected it. Later observations have confirmed this, though there is not enough to bind the clusters at a mass-to-light ratio characteristic of a stellar population. The American astronomer Sinclair Smith (1899–1938) wrote less specifically of "a great mass of internebular material within the cluster," (meaning Virgo). At least, two more pre-war observers — Erik Holmberg (1908–2000) in 1937 on binary galaxies and Horace W. Babcock (1912–2003) in 1939 on the rotation

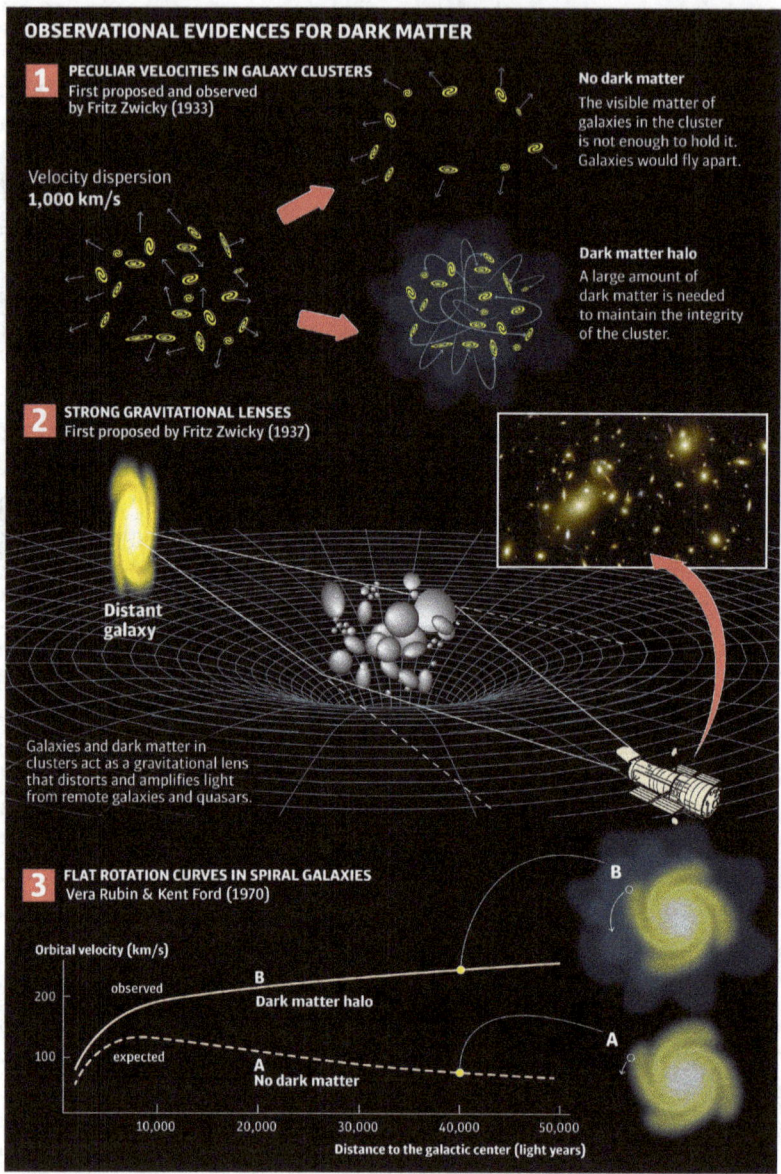

Figure 16.1 We illustrate in this diagram three of the main kinds of observational evidence of the need for dark matter. Credit: Infographic sketched by the authors and elaborated by Javier Pérez Belmonte.

of the outer parts of M31 — acquired relevant data but suggested no specific entities to account for the larger masses implied.

And then, there was a war, after which a large fraction of astrophysicists focused on stellar structure and evolution. From time to time, however, someone noticed that there seemed to be a good deal more than one solar mass per one solar luminosity someplace. In giant elliptical galaxies, the German–American astrophysicist Martin Schwarzschild (1912–1997), son of Karl Schwarzschild, said that there must be a great many white dwarfs left from early, vigorous star formation. It would be unreasonable to expect him to have mentioned neutron stars let alone further-collapsed entities. The most notable thing about this paper[4] is the graph he did not draw of mass to light ratio vs. length scale (although a table is included). The Dutch astronomer Jan Oort (1900–1992) analyzed the rotation curve[5] of the galaxy NGC 3115. He stated that "the distribution of mass in this object appears to bear almost no resemblance to that of light." To account for the missing mass, he had preferred a mix of faint stars plus gas in the then-undetectable phases for the disk of the Milky Way, which he had examined in much the same way that James Jeans and Jacobus Kapteyn had done but with more data and more elaborate mathematical techniques.

The length scale for nearly modern unobserved entities expanded with F. Kahn and L. Woltjer. They cannot have been the first to notice that the 100 km/sec velocity of approach between the Milky Way and the Andromeda Galaxy (M31) in an expanding universe called for some explanation. But they were the first to publish the arithmetic and a proposed binding medium in the form of some 10^{12} solar masses of gas. Indeed, the warm/hot intergalactic medium is still with us, though not quite so abundantly.

In fact, Zwicky's article acquired popularity in the 1970s when two discoveries were made that highlighted the need for dark matter. The first is connected with galaxy clusters. The second is connected with the existence of dark matter inside the galaxies themselves.

16.4   X-rays from Ultra-Hot Environments

Inside a cluster of galaxies, the collisions among them heat the gas that surrounds them to extremely high temperatures (between 10 and 100 million degrees). This gas, mostly made up of hydrogen and helium, is ionized, meaning that its atoms have lost electrons and have gone from being neutral to being electrically charged.

In the early 1970s, the first satellites to detect X-radiation observed clusters of galaxies as huge clouds of hot gas that emitted a vast amount of energy in this part of the spectrum. In fact, individual galaxies in clusters may not be directly detected when being observed in X-rays. This means that two different images could be obtained of the same cluster of galaxies. One came from photographic plates at the focal plane of a large optical telescope, as done by George Abell in the 1960s. This showed individual galaxies. The second image came from a satellite carrying detectors sensitive to X-rays, in which the cluster appeared as one big cloud of gas. This second method is more reliable than the first for detecting galaxy clusters, as with an optical image observers run the risk of assigning galaxies to the cluster that in fact do not belong there, as they are in the same line of sight but further away or closer.

The X-radiation of the gas in the clusters is thermal in nature, meaning that it is associated with the temperature of the emitting material. It is called braking radiation (or Bremsstrahlung in German) and consists of photons emitted by free electrons that alter their trajectories and speeds when they come close to ions but without being captured. This acceleration causes the emission of an X-ray photon. As the electron was free before this encounter and continues to be free after its interaction with the ionized atom, the radiation is also termed free-free.

X-ray observations allow the amount, temperature, and density of the gas inside the cluster to be measured. This hot gas would escape if it were not for the gravitational pull that is exercised by the entire mass of the cluster. The outward pressure of the hot gas is counteracted by the tendency towards gravitational collapse induced by the mass of the gas, the mass of all the galaxies and the mass of the possible dark matter contained in the system. All this sets up a hydrostatic equilibrium similar to the one that maintains the stability of stars. Analyzing X-rays from a galaxy cluster enables us to determine how much matter would be required to create the gravitational field capable of counteracting the pressure of the gas emitting that radiation (a higher temperature would mean more matter is required). The results of observations show that the mass of the galaxies in the cluster provides only about 10% of the total mass, whereas the hot gas accounts for around 20%. The other 70% is dark matter, the nature of which is unknown.

16.5   Flat Rotation Curves

Non-technical discussions of dark matter often begin here, with the idea that many galaxies are rotating so fast that stars in the outer parts of their disks would fly away if not held in by more gravitational force than can be provided by the visible stars and gas. Credit for persuading the astronomical community that this is so goes, rightly, to Vera Florence Cooper Rubin for work done in the 1970s and 80s (see Figure 16.2). The Large Synoptic Survey Telescope has been renamed in her honor (the Vera C. Rubin Telescope) for this major contribution to our understanding of the universe. As is nearly always the case in such matters, she was not the only astronomer, nor the first, to find such evidence.

Our Sun has a mass to light ratio of one solar mass to one solar luminosity: $M/L = 1$. A typical population of stars weighs in at M/L about 3, because faint stars are commoner than bright ones. Astronomers in the 1930s had reported ratios of 10 and 20 and 30 for M/L. Horace W. Babcock studied the Andromeda Nebula for his PhD thesis and found rapid rotation in its outer regions. He then turned to solar astronomy and never again looked at rotating (spiral) galaxies except to appreciate their beauty.[6] Particularly noteworthy is a 1930 paper by Knut Lundmark (1989–1958). He wrote in German[7] and reported what he called "Verhältnis: Leuchtende + Dunkle Materie/Leuchtende Materie," that is, the ratio of luminous + dark matter to luminous matter for six galaxies, including 20:1 for Andromeda (M31).

Numbers for a few other galaxies drifted along in the 1950s.[8] In the 1960s, Margaret and Geoffry Burbidge reported "On the rotation and mass" of a number of galaxies. They were foiled, unable to record faint outer regions because of the limited light sensitivity of photographic plates.

Rubin, who worked with the Burbidges on one of those papers, had access to a photoelectric detector of much greater sensitivity,

Figure 16.2   Vera Rubin in 1972 measuring spectra at the Department of Terrestrial Magnetism of the Carnegie Institution (Washington D.C.). Illustration by Isabel Gálvez (www.isabelgalvez.eu) inspired by a photograph from the Carnegie Institution of Washington.

partly developed by her co-author, W. Kent Ford II.[9] She could, therefore, extend rotation curves (as in Figure 16.1) further into the outskirts of her galaxies, where light was faint, but mass density still significant. We would expect rotation speeds to drop as the distance away from the galaxy center increases, the same as happens with the speeds of the planets as they orbit the Sun, but the speeds remained constant up to long distances from the center. This fact can be explained if we accept the existence of a large quantity of invisible mass in the galaxies, which, although it does not shine, does exercise the gravitational pull that explains this result. Her work was the more persuasive because she kept at it for many years, studying many galaxies in a consistent way. It helped also that contemporary optical astronomers understood the technique and the data processing!

Another new technology, radio interferometry at the wavelength of the 21 cm emission line of hydrogen also permitted exploring the outer reaches of galaxies. Reports of sizable M/L ratios began trickling in from Morton S. Roberts, using the 300-foot antenna at the National Radio Observatory, staring in 1966.[10]

Rather remarkably, a graduate student, David H. Rogstad, pursuing his PhD studies at the Owens Valley Radio Observatory[11] also discovered that radio-emitting hydrogen extended further out than where stars could be seen. He looked at 79 spiral galaxies and concluded that an average spiral had a mass-to-light ratio of 11.3, with the more tightly wound (Sb) galaxies at M/L=13.3 a bit heftier than the less tightly wound (Sc) ones at M/L=11.0.

Why wasn't this the defining moment for dark matter in individual galaxies? First it was a one-shot affair, not the start of multiple investigations and publications, though most of the data appeared in the *Astrophysical Journal* the next year. Second, how to interpret data from radio interferometers typically baffled Rogstad's typical contemporaries. But his thesis came to the hesitant conclusion that even his most conservative analysis "still puts a considerable fraction of the mass in a relatively invisible form."

We couldn't agree more!

16.6   Gravitational Lenses

Galaxy clusters and the enormous quantity of dark matter they contain are responsible for some of the most spectacular images taken by the Hubble Space Telescope. The Abell 1689 cluster is a powerful gravitational lens that distorts and increases the brightness of weaker objects located beyond the cluster which we would certainly not be able to see if it were not for the intensifying effect of the cluster acting as a lens. As we have already seen, the light emitted by a distant source bends when it passes close to a massive object (the so-called deflector) in its path towards the observer. Abell 1689, a rich and compact cluster situated at around 2 billion light-years away, also acts as a gravitational lens that enables us to observe galaxies that are between five and ten times further away.

These galaxies generate multiple, distorted images, similar to the ones that can be seen when looking at a distant light through an irregular piece of glass like the bottom of a wine bottle. In addition to the spiral and elliptical galaxies making up the cluster, a whole set of arc-shaped structures can be seen, which are the distorted images of very distant galaxies, so far away that they would not be seen if it were not for the brightness amplifying effect induced by the gravitational lens. In fact, these galaxies already existed when the universe was only 3 billion years old. To interpret the jumble of arcs seen in the photograph, we need to model the gravitational effects of the galaxies in the foreground cluster and well as the effect of the cluster as a whole. Detailed examination of these cosmic mirages enables us to draw important conclusions about the quantity and distribution of dark matter at the center of the cluster (see Figure 16.1). This means that gravitational lenses, predicted by Einstein's theory of general relativity, play a fundamental role in weighing the universe.

As we have seen, there are various clues that point to the existence of dark matter in the universe. It is exciting that different methods like the analysis of peculiar velocities, X-ray emissions and gravitational lenses all come up with basically the same amounts of dark matter in galaxy clusters. It should be emphasized that the measurements taken by astronomers of all visible and invisible matter in the universe indicate that in its entirety it makes up no more than 30% of the matter and energy that would be needed to explain the flat universe revealed by analysis of cosmic background radiation. So where is the rest? Only a short time ago, the analysis of very distant supernovae has suggested an answer. Perhaps what prevails in the universe is not just dark matter but also an exotic kind of energy that would cause cosmic expansion to be accelerating.

The "tipping point" in favor of dark matter (later to be narrowed to non-baryonic dark matter) was a pair of papers. The Estonian astronomers Jaan Einasto, Ants Kaasik and Enn Saar and the North American astronomers Jeremiah Ostriker, Jim Peebles and Amos Yahil both published in 1974, with each paper summarizing data from many observers and on many length scales which was organized to show monotonically rising M/L ratios when you look at larger systems. This could have been done by Schwarzschild in 1954 or, indeed, by anyone who brought together the clusters of Zwicky and Smith, the binary galaxies of Holmberg, the 1939 rotation curve of Babcock, and Hubble-type numbers for individual galaxies in the year or so just before German armies marched into Poland.

16.7 Alternatives to Dark Matter

The need for something to stabilize clusters of galaxies hit mainstream astronomy with a pair of conferences in 1961. At this point, some astronomers were prepared to abandon Newton (or Einstein) in favor of non-gravitational forces or matter streaming into the universe from some other dimensions (that was Ambarsumian). Either most out of date or most prescient was Abbe George Lemaître (1894–1966)who was still defending the universe of his youth with a cosmic egg or atom and expansion dominated by Einstein's infamous cosmological constant, Λ. His clusters were not actually bound but had temporary populations of galaxies being exchanged with the field. The first truly non-Newtonian gravity arrived soon after with the Italian astronomer Arrigo Finzi (1917–2012). Finzi's hypothesis was a modification of Newton's Law of Gravitation in such a way that the actual attraction at long distances would be stronger than the value predicted by Newton's Law, but probably the optimal version of this has not yet been put forward. Two decades later, Mordehai Milgrom proposed a different alternative to dark matter based on MOdified Newtonian Dynamics (or MOND for short). In this hypothesis, Newton's second law of dynamics is modified in such a way that when accelerations experienced by objects are smaller than a certain value, the force of gravity is inversely proportional to the distance, instead of to the distance squared. This modification works rather well for the flat rotation curves of spiral galaxies if the dynamics are a consequence of the luminous baryonic matter alone, with no need to postulate dark matter. Although MOND has successfully explained other cosmological observations, it does not work so well when reproducing the dynamics of clusters of galaxies and the observations of weak and strong lensing and the CMB.

The only possible conclusions are either that gravity becomes monotonically stronger on large scales or that the ratio of non-luminous to luminous matter increases with length scale. The latter is by far the majority view in the astronomical community and centers around something like 23 percent of the total density being in nonluminous, nonbaryonic dark matter.

16.8   Rescuing the Cosmological Constant

For many decades, cosmologists have been trying to quantify by how much the expansion of the universe was slowing down due to gravity. However, in 1998, two independent teams, one led by the astronomers Adam Riess and Brian Schmidt and the other by Saul Perlmutter, presented convincing evidence for just the opposite: an accelerated expansion. For this discovery, these astronomers were awarded the Nobel Prize in Physics in 2011. They used high-redshift Type Ia supernovae as standard candles. After ruling out possible systemic obscuration by dust or evolutionary effects, they interpreted the unexpected faintness of very distant supernovae as a consequence of being farther away, thus implying an acceleration in the expansion.

The reason for this acceleration is not clear. Cosmologists talk about dark energy without really knowing what it is. It is not matter, and so it cannot be detected by its gravitational pull, neither does it emit radiation. It is an energy associated with empty space, which acts as a repulsive gravity with negative pressure and which would be responsible for cosmic acceleration. Interestingly, this is the effect in models of the cosmological constant. Many people have rushed to retrieve this from Einstein's waste-paper bin and, although it may not be the only form of dark energy (there are other alternatives with intriguing names like quintessence), the cosmological constant is able to explain the acceleration in cosmic expansion shown by distant supernovae.

The story of the rehabilitation of Λ and its repurposing as dark energy, quintessence, or whatever is comparably long and complex. Do we insist that dark matter and dark energy are the only way to understand the current set of observations? Not necessarily. But whatever you put in their place must deal with individual galaxies and clusters (their dynamics and lensing on all scales) to replace dark matter and a range of details of the microwave background radiation, brightnesses of distant supernovae, and formation of galaxies to replace Λ. By the same token, Ptolemy could have said to Copernicus, "Yes, all right, but you will need almost as many epicycles, equants, deferents, and all to get precession, changes in orbital inclination, variations in orbit speed, and so forth right as I did." Indeed, that is more or less what part II of *De Revolutionibus* was about.

Should we regard the "discovery" of dark energy and the acceleration of the expansion in the universe as a scientific success? Certainly, in 1998, *Science* magazine considered this discovery as the breakthrough of the year and we agree with that decision. This discovery put together many astrophysicists, cosmologists and high-energy physicists in a common effort trying to understand the nature of dark energy.

But other physicists are more skeptical. For example, Lee Smolin, in his controversial book published in 2006,[12] argues "The discovery of the dark energy cannot be counted as success, for it suggests that there is a major fact that we are all missing." Of course, this statement does not reduce the merits of the Supernova Cosmology Project led by Saul Perlmutter and the High-z Supernova Search Team led by Brian Schmidt and Adam Riess and other observations supporting the accelerating world models. What it means is that the presence of a nonzero vacuum energy is a problem that has to be explained in much the same way the existence of the ether was a problem that had to be explained by the physicists prior to the Michelson–Morley experiment. In that case, the experiment acted to deny the existence of the ether, and that was the solution.

It seems clear that many of the pre-Copernican astronomers who made Earth-centered models gradually more complex to match observations better thought — according to historians, anyhow — that they were describing the phenomena, not explaining them. But as Rocky Kolb once remarked to one of the authors of this book, "Our goal must not be a cosmological model that just explains the observations, the ingredients of the cosmological model must be deeply rooted in fundamental physics. Dark matter, dark energy, modified gravity, mysterious new forces and particles, etc., unless part of an overarching model of nature, should not be part of a cosmological model. We may propose new ideas, but they must wither unless nourished by fundamental physics." In the final chapter of this book, we discuss many of these questions.

16.9   The Neutrino and the Higgs Boson

When, around 1930, beta decay[13] appeared to violate conservation of energy, momentum and spin, Wolfrang Pauli (1900–1958) postulated the chargeless, (nearly) massless, (nearly) undetectable neutrino to make everything come out even. And since this is being typed in a building named Frederick Reines Hall, you can be sure he was right. The neutrino at its birth was, nevertheless, an unobserved, and perhaps unobservable, entity postulated to preserve the cosmos in which energy (or mass–energy) is a conserved quantity! In fact, when something unknown needs to be postulated, there are other scientists that try to modify the accepted theories to accommodate the experiment. That was the case of Niels Bohr (1885–1962) with his statistical version of the conservation laws.

You may have played with a magnet sometime.[14] If you bring it close to some paperclips scattered on the table, you can see how it attracts them. If you bring it close to an assortment of iron filings, you can spot an interesting pattern: It is caused by magnetic force. You can't see it but it is there.

In the 17th century, Isaac Newton taught us that gravitational force is responsible both for an apple falling from a tree and for the planets going around the Sun. Another English scientist, Michael Faraday, introduced the concept of electric fields in the 19th century. We have all experienced how rubbing a balloon to electrically charge it causes it to attract small scraps of paper from a distance. Modern physics understands these interactions through the concept of a field. In the late 19th century, James C. Maxwell pointed out that two fields, the electric and the magnetic, are two sides of the same phenomenon: electromagnetism. The unification of these two force fields finally provided an explanation for one of the oldest mysteries of science: the nature of light.

Besides gravitational and electromagnetic forces, there are two other fundamental forces in nature. Their scope is limited to subatomic distances and they have no noticeable effect on our daily lives. They are the strong force and the weak force. These forces act at the level of atomic nuclei. We know that the matter around us and of which we are made is composed of atoms. An atom, in turn, consists of an electron cloud surrounding a nucleus. The nucleus is composed of protons and neutrons, which in turn are composed of three particles called quarks. We will return to these concepts with more detail in Chapter 18.

The strong force holds quarks together to form neutrons and protons in the nucleus of the atom. The electroweak force is the main responsible for the Sun shining and the natural disintegration of many nuclei. This interaction is behind the Higgs boson theory.

The modern view of forces is that they propagate through space because there is an exchange of small particles: two magnets repel or attract each other because there is a kind of flow of virtual elementary particles (photons) between them transmitting force. As for magnets, and electromagnetic interaction in general, you may not be surprised to learn that these mediating particles are photons, the same particles that make up light.

In 1967, the Nobel Prize winners Abdus Salam (1926–1996), Sheldon L. Glashow and Steven Weinberg (1933–2021) managed to find a mathematical model describing the operation of the electromagnetic force and the weak force simultaneously. And in doing so, they unified the two forces into one: the electroweak force. But there was a hitch: to avoid any mathematical problems or internal inconsistencies in the theory, the four mediating particles of the electroweak field must have no mass. But three of them do have mass and it has been measured. Who is right, then? A number of physicists independently came up with a mathematical trick to solve the problem. All at about the same time, Robert Brout (1928–2011) and François Englert in Belgium, Gerald Guralnik (1936–2014) and Carl R. Hagen in the United States, and Tom Kibble (1932–2016) and Peter Higgs in England had approximately the same idea. Suppose there is an additional field in nature, which interacts (mixes) with the previously unmixed and massless electroweak mediators and thereby confers mass upon them (except the photon, which continues massless and moving at the speed of light). This would also confer mass upon other elementary particles (quarks and leptons). The W and Z bosons are slowed down and given mass at the expense of the original fields of the corresponding sector, the name Higgs having won out for the field, sector, and the corresponding particle (perhaps because it was the easiest to spell?). The Higgs particle was discovered in experiments at the Large Hadron Collider at CERN, in 2012, and Englert and Higgs shared the 2013 physics Nobel Prize for their ideas. One of your authors claims a tiny bit of the credit, having nominated Higgs and Kibble that year.

Albert Einstein showed us that massless particles (like photons) can only travel at the speed of light; they cannot travel slower. Without the Higgs field, electrons and quarks, of which matter is composed, would have no mass and would travel to the speed of light, and in that case, it would be impossible for quarks to bind in groups of three to form protons and neutrons. There would be no atoms or molecules, planets or stars. All the basic components of matter would be running around, through the universe, at top speed. The Universe is as it is today because the Higgs field exists. Moreover, the Higgs field may be behind the origin of the universe itself.

In 2012, the Large Hadron Collider at CERN discovered the existence of a new particle, which could correspond to the Higgs field. With this amazing discovery, whose importance for physics can be perhaps comparable to that of DNA for biology, the Higgs model, which was just a bright idea, seems to have become a reality confirmed by experiments. Its discovery came about by making protons traveling at nearly the speed of light collide head-on. In this collision, the Higgs boson, i.e. the particle mediating the Higgs field, can be produced. This boson quickly disintegrates into other particles, which were detected at the LHC. And, as they say, where there's smoke there's a fire. CERN later checked that the particle which behaves like the Higgs was not an impostor but the real Higgs boson. However, it is still not certain that the Higgs particle discovered at the Large Hadron Collider is an exact match to the predicted, Standard Model Higgs boson. Further checks are needed and will be carried out.

Meanwhile, perhaps the most exciting possibility (also amenable to experimental verification) is that there is more than one Higgs, as predicted in some extensions of the Standard Model, like supersymmetric theories.

Part VI

The Dragon Lords

The Whole Shebang

Chapter 17

Our Cosmic Bubble

> **Professor Bondi:** *The universe is the way it is because it was the way it was.*
> **Student:** *But why was it the way it was?*

Hermann Bondi, Lectures King's College
London, 1960s

Cosmology, the study of the Universe, is a particular branch of science in two respects: we have only one instance to study and we cannot perform experiments on it. All we can do is observe the Universe, gather as much data as possible and interpret the data: a process that has been going on for thousands of years.[1]

In modern times, the interpretation of the data is done within the framework of mathematical models of the data that are based on known and tested laws of physics. The outcome of fitting a model is a set of numbers that describe the properties of our universe: its ages, density, rate of expansion, and so on. In constructing those models, we can also draw on knowledge from other fields of science. In that way, we are able, for example, to discuss the possible cosmic origin of the chemical elements in terms of well-understood science drawn from experimental and theoretical nuclear physics. Armed with that we can estimate the amount of the element helium that should be observed and set astronomers on the path towards measuring it.[2]

There is little or no room within this framework for the kind of philosophical introspection that was so common in the Late Middle Ages and before.[3] Cosmology today is largely, but not entirely, a data-driven science. Acquiring relevant data is technically difficult and measurements of cosmological phenomena are generally refined by successive generations of experiments. The best situation arises when a cosmological descriptor can be estimated in two different ways: we demand statistical consistency of the results. There are few things in cosmology that are as exciting as when two such estimates disagree at a significant level because it signals the possibility that we may be on the verge of a significant discovery.

17.1   Modeling the Universe

Modeling the known universe did not start with the advent of digital computers. Models predicting the motions of the planets and Solar eclipses have famously existed since ancient times.[4] They were surprisingly successful in allowing soothsayers to know the future location of the planets in the sky and thus make astrological predictions. The accuracy of the astronomical prediction was, however, not a guarantee of the accuracy of the astrological prediction.

Of course, questions are to be asked about the ability of a simple model to account for all cosmological data. It makes sense, on the grounds of making minimal assumptions, to start with the simplest possible model and modify or add features as new data become available. In areas where there are no data available, there is of course free-for-all speculation. There is nevertheless the constraint that, whatever are the wider consequences of such speculation, they should be consistent with the data that are available, or at least it could be tested through some plausible future experiment.[5]

But what do we do when faced with new data that are incompatible with our best models? That the existing model is based on data and the known laws of physics makes us reluctant to throw the model away. One way of handling the situation is to invoke some hypothetical entity that solves the problem without disturbing the successes achieved within the framework of the earlier versions of the model.[6]

Invoking some unobservable entity in order to provide an explanation for some physical phenomenon does not seem like a convincing form of argument. Nonetheless, there are countless examples of the success of this approach throughout scientific history where the only limitation has been the lack of instrumentation to make the key observation. Validation may not be instant or cheap: the LIGO experiment, built partly to detect gravitational waves, and the Large hadron Collider, built to detect the Higgs particle, took many decades in their building with no prior proof that they would in fact discover the phenomena they were built to find. Those projects were costed in tens of billions of dollars. When they achieved their goal, their principal investigators were awarded with the Nobel Prize.

17.2   The Standard Model of Cosmology

The discovery in 1965 of an isotropic background radiation field by Arno Penzias and Robert Wilson and its interpretation as the left-over radiation from the Big Bang by the Princeton group of Robert Dicke, James Peebles, Peter Roll and David Wilkinson (1925–2002) caused a revolution in our view of the universe. It established the view that the universe started with a hot Big Bang about 14 billion years ago. Cosmology became a branch of physics working in the framework of the theory of General Relativity combined with the dynamics, thermodynamics and physical properties of the constituent components. The race was on to explore this new universe in more detail by university research groups all over the world. Two groups were of particular note in the development of "Physical Cosmology"[7]: the group around P.J.E. ("Jim") Peebles in Princeton, who went on to be awarded the 2019 Nobel Prize for his work in cosmology, and the group of Yakov Zeldovich (1914–1987), who had he lived to 105 years of age, might have shared that prize.

The picture we have of our universe is that it is expanding, it is fundamentally homogeneous and isotropic, by which we mean that, in the large, it looks the same in all directions and around any place.[8] The detailed structure we see, planets, stars, galaxies, clusters of galaxies and larger more tenuous structures, has evolved through the action of gravity acting on almost exceedingly small primordial inhomogeneities. By "exceedingly small", we mean that the deviations from the perfect homogeneous state are to be measured in parts per million of the ambient homogeneous density. By "primordial", we mean that they existed at or near the start of the cosmic expansion: they were a consequence of the Big Bang and were the seeds that grew to form the grand cosmic structures. Throughout the cosmic expansion, this primordial structure is amplified by gravity and becomes an ever more prominent wrinkle on the otherwise smooth and homogeneous background fabric of space–time.

In order to put that in context, we should go back a further 50 years to the time when Albert Einstein was finishing his *Theory of General Relativity* which provided the framework on which to build mathematical models of the Universe.

In 1917, the First World War was raging on all fronts. The German army had intensified its military action as a result of which the United States withdrew its ambassador from Berlin and entered the war on the Allied side. In December, following revolutions in February and October, Russia withdrew from the war with a multinational armistice. The Dutch were neutral, but being surrounded by warring nations they suffered serious deprivations with a high influx of refugees from Belgium and wartime blockages on shipping. At this time, Einstein was in Berlin and Dutch mathematician Willem de Sitter was at the Leiden Observatory located southwest of Amsterdam.

The year before, Einstein had just published his seminal papers describing his theory of gravitation: the General Theory of Relativity.[9] These equations were formidably difficult to solve except in very special cases.[10] Karl Schwarzschild in Potsdam and Johannes Droste (1886–1963) in Leiden managed to find the first solution, now known as the Schwarzschild solution, which revealed the possibility of "Black Holes" in Einstein's gravity theory, point masses defining a spherical volume of space from which light could not escape.[11]

Two more solutions of the equations of General Relativity came in 1917, one from Einstein and the other from de Sitter. Both were described as "cosmological solutions" even though almost nothing was known about the universe beyond our own Galaxy. Einstein's solution was for a nonexpanding never-changing universe filled with matter, with a delicate balancing act between the pull of gravity and the pressure exerted by his infamous "cosmological constant".[12]

De Sitter's solution was empty of matter but expanding owing to the pressure of the cosmological term. For a variety of technical reasons, Einstein was not happy with the interpretation of de Sitter's solution.

The two solutions were opposite extremes. One, Einstein's, had matter and described a never-changing, static universe. The other, de Sitter's, was empty of matter but always expanding. The "compromise" solution describing a homogenous and isotropic distribution of expanding matter came from the Russian mathematician Alexander Friedmann (1988–1925) in two important mathematical papers published in 1922 and 1924, but Einstein was not happy with these either. It seems that Einstein was rejecting the notion that the universe might be expanding rather than pointing to any manifest errors in the interpretation of the work of either de Sitter or Friedmann. Friedmann died in 1925 and so there was no further discussion of the matter: his work did not get the attention it deserved. During the 1920s, Einstein was at the height of his fame, traveling around the world and being acknowledged by the public as the most intelligent person on the planet.[13] The risk of being criticized by Einstein almost certainly would have discouraged others from spending time studying these complex papers.

On the astronomical front, 1917 was the year in which Vesto Slipher (1875–1969) of the Lowell Observatory published the first list of velocities of 25 "nebulae" (later called "galaxies").[14] The velocities he reported were on average 30 times higher than the velocities measured for the stars in the Galaxy. From 1898 until his death in 1908, James Keeler (1857–1900) had used the Lick Observatory Crossley Telescope to take deep photographs of the sky and had noted numerous images of spiral nebulae on his plates – so many in fact that he estimated that there would have been over 120,000 of them if the whole sky had been photographed.

What is perhaps surprising is that nobody at the time put the two new cosmological solutions together with Slipher's hint that the 120,000 spiral nebulae were a system external to the Galaxy and flying apart faster than anything ever seen before. An important factor would have been the lack of communication between the relativity community of mathematicians and mathematical physicists on the one hand and the astronomers on the other. The mathematicians were keen to come to a deeper understanding of the new theory and to find more solutions to the Einstein equations. Most of the astronomers were observers who gathered data using the ever-more-powerful telescopes and instruments that were being built, but who were not well-versed in the deep mysteries of Einstein's theory. At the time, there were only a few who approached that almost ideological boundary, notably Willem de Sitter in Leiden and the eminent Arthur Eddington in Cambridge.[15]

Slipher continued gathering velocities for the spirals, increasing the number from 25 published in 1917 to 41 in 1922, which he sent to Eddington for publication in his forthcoming book, *The Mathematical Theory of Relativity* (CUP 1923). In the book, Eddington[16] remarks that "The great preponderance of positive (receding) velocities is very striking, but the lack of observations of southern nebulae is unfortunate and forbids a final conclusion". Quite true, of course, but hardly adventurous. Eddington was also understandably put off by the few negative (approaching) velocities in the list saying that this was difficult to understand within the framework of de Sitter's model. His reservation was that the most accurate velocity determination of the sample was the one for the Andromeda nebula ("M31" — our nearest great spiral galaxy) which showed that the nebula was coming towards us at the then-astonishing speed of 300 kilometers per second. Eddington had seen the relativistic connection between de Sitter and Slipher but withdrew from taking the essential final step.

It was not long before the connection between relativistic models and the velocity data of Slipher was made. In 1927, Georges Lemaître independently obtained the same solution of the Einstein equation that Friedmann had a few years before. His paper deriving this, written in French, was arguably one of the most important papers in 20th century cosmology: it was the first paper that brought a viable relativistic model and the data together to discuss the expansion of the universe.[17] Crucially, it also provided an estimate for the expansion rate of the universe 2 years prior to the American astronomer Edwin Hubble. However, the discovery that the universe is expanding and the first estimate of the expansion rate are attributed to Hubble: the "Hubble expansion" and the "Hubble constant". So what about Lemaître? Should he not get some credit?

The Royal Astronomical Society recognized the importance of Lemaître's 1927 paper and, following a talk given by de Sitter in 1930, had invited Lemaître to submit his paper to their widely read

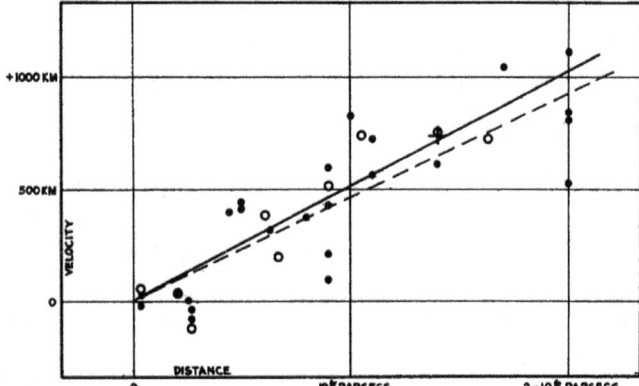

Figure 17.1 The 1929 Hubble diagram showing that the more distant galaxies are moving away faster than the nearer ones. The relationship is described by a straight line and is referred to as the "Hubble–Lemaître expansion law". The velocities were from Slipher's list. Hubble estimated the distances using the brightest stars he could see using the 100-inch telescope on Mount Wilson. His calibration for the intrinsic brightness of those stars was underestimated by a factor of about 10. From Hubble (1929), *Proc. Nat. Acad. Sci.*, 15, 168–173.

journal, the *Monthly Notices*. Before that Lemaître had worked for a year with Eddington in Cambridge on problems in General Relativity, following which he made a tour of the USA, starting at Harvard with Shapley, then on to Mt. Wilson Observatory where he met Hubble followed by a visit to Slipher at the Lowell Observatory. After he headed back to Cambridge and thence to the 5th Solvay conference in Brussels where he took the opportunity to have a talk with Einstein. Not surprisingly, perhaps, Einstein rejected Lemaître's thesis that the universe was expanding.

By 1927, Lemaître was well known and well respected. Lemaître had translated his 1927 paper into English, omitting a few key paragraphs on the expansion, and the edited 1927 paper was duly published in 1931, 2 years after Hubble's seminal paper showing the evidence for the expansion (see Figure 17.1). It seems that Lemaître himself had removed the sensational part of his 1927 French language version: the first calculation of the expansion rate of the Universe. Why he did that we do not know and there have been many speculations on the matter. The simplest explanation for this important edit is that in 1929 Hubble had written a paper finding the slope of the velocity-distance relationship that superseded Lemaître's 1927 work, removing the need to republish the earlier, less good version. The key difference between the work of Lemaître and that of Hubble was that Lemaître could only estimate the distances to the nebulae from their observed diameters: big ones are nearby, small ones are distant. An earlier paper by Hubble had provided the measurements of the diameters used by Lemaître. Hubble, with access to the world's largest telescopes, was able to use Cepheid variable stars and the brightest stars in nebulae as better distance indicators that Lemaître had access to, though both used Slipher's redshifts. But nevertheless, Lemaître's 1927 paper stands there as the testament to the first determination of the cosmic expansion rate.

After the Second World War, a new champion for the Friedmann model emerged: George Gamow. Gamow was born in Odessa in 1904, when the city was part of the Russian Empire, and had studied in Leningrad with Alexander Friedmann. Friedmann died in 1925, before Gamow was able to present a dissertation, and so he switched his field of research to the emerging subject of quantum physics and its application to nuclear processes. He made significant contributions to the field and got to visit the great European centers of quantum physics research. He spent the period 1928–1931 in Copenhagen working at the institute created by the 1922 Nobel Prize winner, Niels Bohr, after which he returned to Leningrad. In 1933, Gamow and his wife attended the 7th Solvay Conference in Belgium when, with the help of Marie Curie, they defected. In 1934, he accepted a professorship at George Washington University in Washington D.C. His work in nuclear physics continued, but he maintained a life-long interest in promoting the more physical aspects of what became known as the Friedmann–Lemaître cosmology and later labeled "the Big Bang theory" by Fred Hoyle.

George Gamow recognized that the early phase of that universe might be ultra-hot, like the interior of the Sun where nuclear reactions were taking place. Gamow, with his experience of nuclear physics gained prior to and during the war, saw the early Friedmann universe as a place which might have been hot enough that the chemical elements might be synthesized from a primordial soup of protons, neutrons, electrons, and photons. From that simple notion he and his collaborators, Ralph Alpher (1921–2007) and Robert Hermann (1914–1997) were able to predict that the universe would have 25%

of its mass in the form of the element helium, the rest being hydrogen with important traces of deuterium, tritium and lithium. Gamow went one step further and traced the history to the present day and showed that the universe would have cooled down to around 5° above absolute zero. This afterglow of the primordial explosion might still be measurable.

Some of the cosmological community were reluctant to accept a universe of finite age. Even Eddington, Einstein and Hubble expressed serious doubts. The doubts increased in 1948 following the publications by Hermann Bondi, Thomas Gold and Fred Hoyle of a "Steady State theory" of the universe. In that theory, the void left by cosmic expansion was filled with newly created material which would later form new galaxies. This was a theory which was consistent with the expansion data and, importantly, had no initial singularity a finite time in our past when the universe was infinitely hot and infinitely dense. In the Big Bang theory, all the material in the universe was created at a single instant, while in the steady state model, the process was continuous. This was achieved by postulating a "creation field" whose energy was tapped to make up for the deficit resulting from the expansion. In effect, this creation field had a negative mass that exactly balanced the positive mass that we see. The mean density of the steady-state universe was zero as in de Sitter Universe.

No mechanism falling within the scope of known physics was postulated for this continuous creation process. Nor for that matter was there any proposal for the sudden creation of the primeval atom at a single instant at a finite time in our past. The argument came down to whether one thought that instantaneous creation was any more or less plausible than continuous creation. In 1948, there was only a distant hope of using the Hubble diagram to discriminate between the two theories. There was little else to distinguish the two theories: the steady-state theory suggested that there would be galaxies of all ages in the visible universe: some old and some still in the process of forming. There would be galaxies of greater age than the age of the universe derived from the Big Bang theory. Moreover, the expansion should look more like a de Sitter expansion than Lemaître's Big Bang model. There is an irony in this latter prediction: at the very end of the 20th century, mapping the cosmic expansion using very distant supernovae revealed a de Sitter-like expansion! It was natural to interpret this as being simply a dominant manifestation of Einstein's cosmological constant. The established model now looks like some of the models proposed by Friedmann and Lemaître.

17.3   The Renaissance in Cosmology

A revolution in cosmology took place in 1965 when Arno Penzias and Robert Wilson announced the discovery of what they had been informed by the Princeton group of Robert Dicke, James Peebles, Peter Roll and David Wilkinson was almost certainly the left-over radiation from the hot explosive origin of our universe: the Cosmic Microwave Background radiation or "CMB" for short.

The discovery of this relic radiation was not entirely unexpected. Dicke had been developing radiometers that might be able to detect this radiation since the mid-1950s. By 1964, Dicke's group was running an experiment on the roof of the Princeton Joseph Henry Laboratories gathering data. Their experiment looked very promising and was showing some evidence for the sought-after signal when Penzias and Wilson announced their discovery in 1965. It was the time of the Cold War when communication between East and West were minimal and mutual hostility was at a maximum. At that time, the Moscow group of the great scientific polymath Yakov Zeldovich was focusing on the production of the element helium and searching past records for unrecognized detections of the radiation using radio telescopes working at microwave and lower frequencies.[18] Over the following decades, the Zeldovich group became an important force in driving the development of physical cosmology.

Penzias and Wilson worked with the Holmdel horn antenna at the Bell Laboratories in New Jersey, a satellite telecommunications antenna that they were using to measure the sky background as a source of noise over which the satellite communication signals would have to be detectable. During their experiments, they pointed the antenna towards the remains of a supernova in the constellation of Cassiopeia and they found an excess of antenna temperature which they thought at first might be due to something in the horn or amplifiers. They carried out various tests and the excess was still there, but the most surprising thing was that it did not matter which direction the antenna was pointing. Like several others before them they had no basis for interpreting the radiation as coming from the Big Bang[19] and most probably would not have done so had they not contacted the Princeton group.

Partly by coincidence, the two New Jersey groups came into contact with each other. Penzias mentioned the discovery of the radiation while on the telephone to a Washington radio astronomer, Bernard Burke (1928–2018). A few weeks earlier, a colleague from Burke's department, Ken Turner, had attended a seminar led by the young Jim Peebles in which he explained his theoretical calculations of the radiation that should be detected according to the Big Bang model. Turner put both groups in touch with each other. Penzias and Wilson realized the importance of their discovery. They came to an agreement and published the results in *The Astrophysical Journal Letters* in 1965. The article was preceded in the same issue of the journal, by another from the Princeton group in which they explained the cosmological origin of this radiation. The discovery was one of the clearest observational indications in favor of the Big Bang model and the third pillar (together with cosmic expansion and primordial nucleosynthesis) on which the model is based. It made front page news in *The New York Times* on 21 May 1965, and since then, it has been one of the main objects of study for observational cosmology.

The Princeton group had been scooped and Penzias and Wilson went on to get the Nobel Prize in 1978. Nonetheless, Peebles continued to define modern cosmology in terms of the physical processes taking place in the cosmic medium, building a theory for galaxy and galaxy cluster formation[20] and inspiring many wannabe cosmologists. He is regarded by many as the "father of modern cosmology". In 2019, he was awarded the 2019 Physics Nobel Prize[21] for "theoretical discoveries in physical cosmology".

So what is the direct source of the cosmic background radiation? 380,000 years after the Big Bang, the temperature of the universe was around 3,000 K, and the size of what is now the observable universe was about 1,100 times smaller than at present. At that moment, a phenomenon known as "recombination" occurred, when the first stable atoms of hydrogen were formed from pre-existing protons and electrons. Prior to that, during the first 380,000 years, the density and temperature of the universe were so high that energetic photons prevented the capture of electrons by protons. After that period, stable atoms of hydrogen gas were able to form and photons could

travel freely. The universe became transparent. We say that the radiation from this period came from the "surface of last scattering".
The universe continued to expand and cool, and the frequency of
this radiation fell.[22] By the time the universe reached the present
epoch, the radiation frequency had fallen to the band of centimeter
and millimeter wavelengths. This was the radiation that Penzias and
Wilson detected at 7.3 cm (4.08 GHz) with the Holmdel antenna.

The physics of this early period of expansion predicts a very specific shape for the energy distribution of cosmic radiation. The distribution of energy is described by a curve (called "the spectrum")
which displays the intensity of the radiation according to the wavelength. This is somewhat analogous to the sound generated by a
singing voice or musical instrument: a given note has a pitch (its
frequency) and the mixture of frequencies present in the sound (the
spectrum) tells us which musical instrument generated the sound or
even allows us to identify the singer.[23]

An important prediction, made even before the cosmic background radiation was discovered, was that the spectrum of the radiation should be precisely what we call a "black body".[24] Continuing
with our audio analogy, this is the unique signature of the instrument
that generated the spectrum.

In order to firmly establish that we were indeed looking at the
radiation left over from the Big Bang, it was essential to measure this spectrum: the short-wavelength side of the thermal spectrum could only be measured from space because of atmospheric
absorption. NASA's COBE satellite (Cosmic Background Explorer),
was launched in 1989. Measuring the spectrum was accomplished
with a specially designed instrument called Far Infrared Absolute
Spectrophotometer (FIRAS) on board the COBE satellite. FIRAS
showed, with unprecedented accuracy, that the spectrum of the
microwave background radiation matches that of a black body at
a temperature of 2.728 K.[25] The measurement error on this value
was 4 milli-kelvins. This is an absolute measurement that established once and for all that the origin of this radiation is the early
universe.

The COBE satellite carried another instrument to map the small
differences in temperature between two points in the sky with high
precision: the DMR, the Differential Microwave Radiometers. Differences between the temperatures at two places are comparative and
can be made with a precision of micro-kelvins. The DMR took four

years of scanning during the period 1989–1993 to produced the first full-sky map of the temperature irregularities. Analysis of the map provide the first evidence for variations in temperature at a level of 30 micro-kelvins. This was the first direct observational evidence for the existence of the hypothesized inhomogeneities that would lead to structure formation.[26] John Mather and George Smoot were awarded the 2006 Physics Nobel Prize for these discoveries.[27]

Owing to instrumental limitations, the COBE sky map was blurred on scales less than 6° on the sky.[28] Higher angular resolution maps of small patches of sky were generated from almost a dozen, mostly balloon-borne, experiments. The BOOMERanG[29] experiment of 2,000 revealed a fine-scale 1° feature in the rippled structure of the CMB sky fluctuations. This discovery was widely hailed as evidence supporting the need for an inflationary model of the earliest moment of the big bang. This[30] and the need for an inflationary model of our universe were made clearer by two subsequent space-based CMB missions: "WMAP", launched in 2001, and "Planck", launched in 2009.

A decade of technological advances in microwave imaging makes a huge difference. COBE revealed structure on an angular scale of 7° on the sky (14 moon-diameters), while WMAP could show structural details that were 20 times smaller. The Planck satellite, launched another decade later, had an angular resolution some one hundred times better than that of the COBE DMR with around 10 times the sensitivity to variations in temperature. The resulting temperature map showed these variations in remarkable detail. WMAP and Planck obtained data from multiple microwave wavelengths, which enabled the separation of contributions to the received signal from sources other than the primordial cosmic radiation: dust in our Galaxy and hot gas in galaxy clusters. They were also able to measure the polarization of the radiation.

Making the most of the COBE–WMAP–Planck data demanded sophisticated statistical analysis, it was a formidable task. Now, more than 50 years beyond that discovery, we have full-sky maps of cosmic radiation that have given us unforeseen insight into the nature of the universe when it was 380,000 years old. Not only have we confirmed the homogeneity and isotropy of the universe and seen the low-level ripples that are the seeds for the formation of structure, but we have a detailed menu of the constituents that made up the universe at that time and today.

Supernova studies that had started in the 1990s led to the 2011 Nobel Prize for Saul Perlmutter, Adam Riess and Brian Schmidt for "the discovery of the accelerating expansion of the Universe through observations of distant supernovae". The supernovae provided a radically different method for determining how the universe has evolved and what it is made of. The results from both the supernovae and from the microwave background mapping are in remarkable and satisfying agreement. Taken together the observational data from these experiments strongly support the Big Bang model of the universe as proposed by Lemaître, Gamow and Peebles.

However, it turns out that the stars and galaxies we see with our optical telescopes are but a fraction, less than 1%, of what is out there!

17.4   Dark Things We Do Not See

It is somewhat extraordinary that the analysis of maps of the cosmic background sky can tell us so much about what the universe is made of, how old it was when the radiation was emitted, and the age and expansion rate at the present universe. The reason we can do this is that the physical state of the universe at this special time is known with high precision from the CMB data: there is no speculative physics involved. Nevertheless, it does seem as if we don't even have to peer through a telescope to see what is out there![31]

The CMB modeling, together with other cosmological observations, determines the different components of the matter and energy budget in the universe, as depicted in Figure 17.2. Baryonic matter is made from neutrons, protons and all the atomic nuclei and is only 15.8% of the total *matter* (i.e. baryonic plus nonbaryonic matter) content of the universe.[32] However, not all of that 15.8% is visible: most of it, 90%, is nonluminous dark baryons. The material in the universe that we see in the form of galaxies and clusters of galaxies is only around 0.5% of everything that is out there! Fortunately, we can infer the presence of the dark matter and map its distribution through the weak gravitational lensing surveys of the large-scale cosmic structure.

The other 84.2% of matter is not baryonic; it must be made up of free electrons, photons and neutrinos, which we know about, and

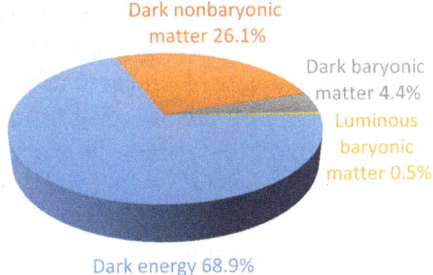

Dark nonbaryonic matter 26.1%

Dark baryonic matter 4.4%

Luminous baryonic matter 0.5%

Dark energy 68.9%

Figure 17.2 The contents of the universe. The universe is made up of baryonic matter (as are we) and nonbaryonic matter. But there is a dominant contribution from the so-called "dark energy", a substance we know absolutely nothing about. This conclusion uses CMB data and is amply confirmed by other methods of determining what is out there.

other hypothetical stuff we know little or nothing about like axions and weakly interacting massive particles (WIMPs). The kinds of nonbaryonic material we know about only account for a very small fraction of all the nonbaryonic matter. That is the mystery of dark matter: it appears that most of it it is not made from any form of matter that we currently know about: it is simply "dark matter".

Figure 17.2 also shows that there is a considerable amount of stuff, 68.9% of the total content of the universe, that is not even any form of matter as we know it! This is the so-called "dark energy" that is commonly attributed to the cosmological constant that Einstein introduced into his equations. As explained in Chapter 16, the dark energy component is invoked to account for the observed accelerated expansion of the universe.[33] The expansion of the universe is not controlled by the force of gravity alone: there is a "dark force" that has dominated the expansion during the past few billion years. We have no idea what this dark force is, but we assign it to the effects of Einstein's cosmological constant, the constant that was exploited by de Sitter, Friedmann and Lemaître.

17.5 Some Important Questions

One hundred years after Slipher first demonstrated that galaxies were receding from us and de Sitter and Einstein produced the first cosmological models, we can feel that we know a lot more about our universe. Most of that has been learned over the past 100 years. Moreover, that knowledge has been encapsulated in a physical model, largely expressed in the languages of mathematics and physics, that unifies everything we know about the universe from the point of view of data we have acquired over that time. This was not a matter of tinkering with a model to fit the data: the model we use today was set up during the first 50 years by Friedmann, Lemaître and Gamow. What we know now with reasonable confidence are things like the cosmic expansion rate and the other numbers that define the cosmological model, and the nature and quantity of the constituents of the universe.

Much of that is due to studies of the graininess of the structure seen in the CMB maps, most of which is corroborated by observations of distant supernova explosions. Observations of gravitational waves from merging black holes and neutron stars will provide a further check and refinement of that knowledge.

In particular, we have learned that Einstein was correct when introducing a cosmological constant into his equations. This is surprising, but what was more astonishing is the magnitude of that constant: it is far from insignificant and indeed dominates over all other cosmic constituents. Yet we have no idea what this "dark energy" is. We have also learned from the studies of galaxies and clusters of galaxies that their environments are dominated by some unseen "dark matter". We have no idea at this time what that dark matter is either, but we are confident that the answer will become known in the not-to-distant future. We cannot be so sure about the dark energy.

But even apart from the mysteries of dark matter and dark energy, there are other questions that niggle in the background. We start with some fundamental questions that have been asked many times:

1. What happened before "time zero"?
2. Why is the universe is expanding?
3. Why is the universe on large scales so homogeneous and isotropic?

The reason item 3 may be regarded as something worth questioning is displayed by the two points **A** and **B** in Figure 17.3, they are more than 90° apart on the sky. When the universe was 380,000 years old, light could have traveled at most a distance of 380,000 light-years. This defines a "causal horizon" about any point

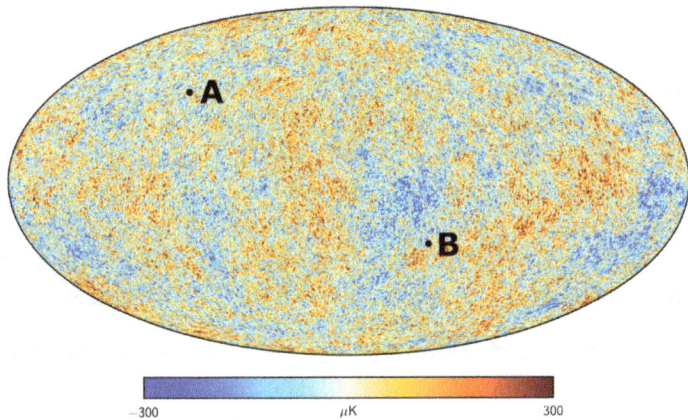

Figure 17.3 The Planck 2015 CMB sky temperature map. We are looking at the universe 380,000 years after the Big Bang. The color scale shows the temperature fluctuations which have been amplified for clarity, otherwise they would be invisible: they typically represent temperature deviations smaller than one part in 10,000. The character of the "graininess" of the image is what provides the scientific information. (Noncosmological influences have been removed.) Credit: ESA and Planck collaboration.

on the sky map.[34] Doing the calculation, that distance corresponds to an angle of about about 2° on the sky map, which is around 1% the height of the map. So, there was no way that the universe in region **B** could have physically synchronized with region **A**. As we shall see in the next chapter, this question of synchronization was a key issue in motivating the theory of the inflationary universe. Then, some more technical questions about the universe at large arise:

4. Why is the universe geometrically flat?
5. Why is the universe made of matter with almost no anti-matter?
6. Why are there no magnetic monopoles?

A geometrically flat expanding universe (item 4) is one that is just on the verge of future collapse without actually re-collapsing; it will expand forever. This requires a precise balance between the expansion rate of the universe and its density. What caused the universe to be so delicately balanced is another motivation for proposing the theory of the inflationary universe. Item 5 appears because under normal circumstances we would expect equal amounts of matter and anti-matter. We know of no way in which physical laws could permit favoring one above the other in these circumstances. The argument for magnetic monopoles (item 6) is similar, we would expect as many North Poles as South Poles.[35]

The final question results from detailed study of the graininess of the CMB maps produced by COBE, WMAP and Planck. This level of graininess determines the number of galaxies that form per cubic megaparsec of space. The nature of the graininess determines the relative number of galaxies and clusters of galaxies of different mass that can eventually form. This raises the following simple question:

7. What determines the kind of graininess that yields the currently observed large-scale structure of the universe? In other words, what determines the apparent texture of the map?

Of course, the answer to all of these question could be "The universe is what it is because it was what it was". However, those who study cosmology have fertile imaginations. There is no lack of putative answers, most of which fall within theories of an early inflationary stage of the cosmic expansion, an ultra-brief instant in which the universe expanded by an almost inconceivably large amount before settling down the universe as we know it. This is the subject of the following chapter.

Chapter 18

Towards Zero

It's extraordinary, the amount of misunderstandings there are even between two people who discuss a thing quite often — both of them assuming different things and neither of them discovering the discrepancy.

Agatha Christie, *Towards Zero (Ch. XI)*, Inspector Battle

When we look out into the universe, as deep as we can with our telescopes, we can look into the far past and see the universe since it was only 380,000 years old. To give some sense of scale to this, we can think of comparing a human embryo one day after conception when it is a fertilized egg that has not even divided[1] with an entire human lifetime of 100 years. On cosmic timescales, those 380,000 years are a short but important period of which we currently have no direct knowledge. Our only means of "looking" beyond that time-wall is to use the known laws of physics to run the expansion of the universe backwards in time, watching the universe get hotter and denser as it heads towards time zero: the "big bang".

The tools available to us are Einstein's theory of gravity, the laws of thermodynamics and an understanding of the nature of matter at ultra-high densities and temperatures. The nature of high temperature and density plasmas is the arena of High Energy Physics which tells us what kind of fundamental particles exist at different temperatures and what are their mutual interactions. This may lead to suggestions for cosmological experiments we can perform to validate the hypotheses on which our extrapolation into the past is based. We enter into the scientific discipline of "astroparticle physics".

This way of looking into our distant past is not a mere shot in the dark: understanding the state of the cosmic plasma at these early times will have consequences for the present universe that may well be visible. An example of this was Gamow's realization in the early 1950s that the element helium would be synthesized during the first minutes of the cosmic expansion if the early universe was hot. A decade later, this was supported by the first measurements of the cosmic abundance of helium. Here was evidence that the early universe was indeed a hot place which would cool off as it expanded towards the present time. Gamow and collaborators put the present temperature at around 6° above absolute zero. In one step, Gamow and his collaborators had taken us back to the first few minutes of the cosmic expansion.

In 1965, the prediction by the Gamow team was verified by the discovery of the radiation left over from the big bang by Arno Penzias and Robert Wilson. The Princeton Group of Robert Dicke, James ("Jim") Peebles, Peter Roll and David Wilkinson were themselves on the verge of announcing the discovery when they were scooped by Arno Penzias And Robert Wilson working in Bell Labs a few miles away. The Princeton group had fully understood what they were looking for. In a paper published in *The Astrophysical Journal Letters* that immediately preceded the Penzias and Wilson paper, they gave the physical interpretation of the discovery which they summarized with the graph shown in Figure 18.1.

The graph showed the temperature and density history of the universe back to the first 1/10,000th of a second after "zero time" when the expansion started. At that point, the temperature of the universe was about 10^{12} kelvins, 1 trillion degrees and its density was about 100,000 tonnes per cubic centimeter.[2] One second after the big bang the temperature had cooled to 15 billion degrees. Around 10 seconds after the big bang, the temperature was a few billion degrees when nuclear reactions can start to make the lighter elements deuterium, helium, lithium, beryllium and boron, a process which is completed in the following three minutes.

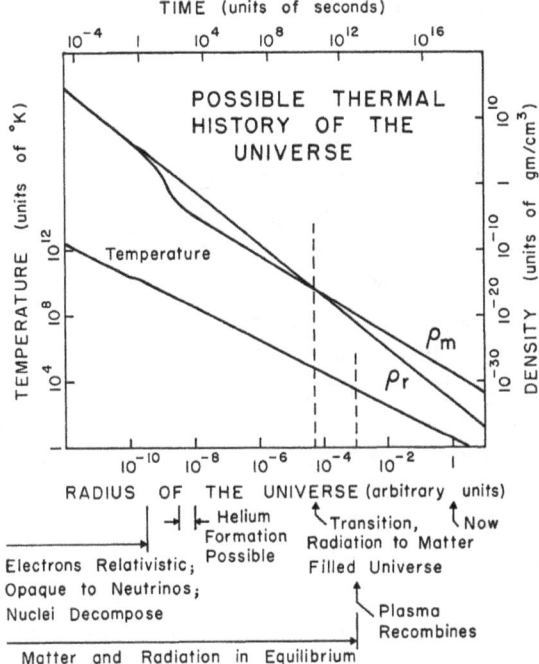

Figure 18.1 The history of the universe as presented by Dicke, Peebles, Roll and Wilkinson in July 1965. *Astrophysical Journal Letters*, 142, 414–419. This appeared as one of a pair of papers published together, the other being the Nobel Prize winning announcement of the discovery of the cosmic background radiation by Penzias and Wilson. In this one-graph-says-it-all picture, we see the history of the density and temperature of the universe right back to the first 1/10,000th of a second after time zero.

One might well ask "how can we possibly know that?" and "is there any way we can we test this?". Incredible as it may seem, because of our understanding of the properties of matter through the study of high-energy particle physics, we can know what the universe was like one-millionth of a second (10^{-6}sec) after time zero and perhaps even back to one-trillionth of a second (10^{-12}sec). Beyond that, it becomes truly speculative, in large part because of the uncertainties in knowing the physical properties of the cosmic plasma at that time. This is analogous to knowing that water is made from hydrogen and oxygen: but determining whether the state of the water is liquid, solid ice or gas and what the crystalline state of the ice might be from the atomic and molecular level is far from simple.[3]

18.1 Fundamental Particles

Everything in the world, past and present, is thought of as being made from a variety of particles that interact via the various forces of nature. In particle physics, the forces are carried by particles. Every time you have a force you have a particle associated with it.[4] There are four well-known forces: the gravitational force, the electromagnetic force, the weak force and the strong force. The gravitational interaction is by far the weakest of all forces, but also the one with the longest range of influence. The strong interaction is the strongest of these, but we shall see that it is merely a by-product of an even stronger force, the gluon interaction. The particles associated with these forces are "fundamental" in the sense that they themselves are not made up of any smaller units (think Lego bricks) but can combine to build other (nonfundamental) particles. The collection of indivisible matter particles and the particles of the fundamental forces is known as "elementary particles".

We will need to introduce one extra particle and its interaction in order to explain how these particles acquire the property of "mass". This is the famous "Higgs particle". Mass is a fundamental property in the theories of gravity as put forward by Newton and Einstein whereas in High Energy Physics "mass" is an endowed property that is explained by how an elementary particle interacts with the field associated to the Higgs particle. Our present understanding of the physics of these particles and their interactions is known as "the Standard Model of particle physics".

The 20th century was the century during which we came to terms with the submicroscopic world. It started with J.J. Thomson's discovery of the *electron* in 1897, for which he won the 1906 Nobel Prize. Further discoveries followed the development of quantum mechanics. 1932 saw the discovery of two new particles: the *neutron* by James Chadwick (1935 Nobel Prize) and the *positron* by Carl Anderson (1936 Nobel Prize). The neutron is a component of the atomic nucleus and the positron is identical to the electron except in that it has a positive electric charge. In 1935, Yukawa (Nobel Prize 1947) predicted the existence of a particle, the *pion*, which was found in 1947 in cosmic ray air showers by Cecil Powell.

By the end of the century, more than 200 subatomic particles had been discovered. All the discoveries had been made in cosmic ray observations and in particle accelerators where particles could be accelerated by electromagnetic fields to enormously high speeds resulting in high-energy collisions with targets. The collisions broke up the particles revealing their structure in the fragments.

There were two major directions of research in particle physics during the 1960s. One was the development of a framework within which to place and inter-relate the particles that were being discovered by successive generations of the accelerator. This effort was led by Murray Gell-Mann who found a way of classifying the particles using symmetry principles. This led to the predictions of new particles which were subsequently found. He was awarded the 1969 Nobel Prize.

Alongside this, there was an effort to bring the forces of nature into a single all-containing theory: a Unified Force theory. The fundamental idea was that the strengths of the forces could vary with energy, which in the cosmological context meant time. At high energies, there was only one force that would split into different forces at lower energy. The first major step in this direction was taken by Sheldon Glashow in 1961. His ideas were later taken up by Steven Weinberg and Abdus Salam who showed, in 1967, that the electromagnetic and weak forces were once a single force, the "electroweak force". The three shared the 1979 Nobel Prize. In cosmology, the splitting of the electroweak force into its components takes place one pico-second (10^{-12} seconds) after zero-time, at which time the temperature of the universe had fallen to 10^{15} K (a quadrillion, or million billion, degrees).

The manifest success of electro-weak unification stimulated the quest during the 1970s to go one step further: the Grand Unified Theory of the strong force and the electro-weak force. Grand Unification described a completely new state of matter at ultra-high density and temperature: it was clearly relevant to the problem of the pre-picosecond early universe. Working out the details produced a large number of highly technical papers. What is interesting is that although the Higgs particle had not yet been discovered, such was the confidence in its existence that all of these papers wholly embraced the particle and the important role it would play in endowing other particles with a mass, thereby influencing the state of cosmic matter. Andrei Linde, then at the Lebedev Institute in Moscow, himself an important contributor to this new field, wrote a review article in 1979 that became the basis for many of the developments of the early 1980s.[5]

In 1973, Glashow and Howard Georgi produced the first partially successful attempt at grand unification which became a template for all future attempts. This signaled the emergence of the "Standard Model" for particle physics which underwent several changes during the following decades. This was the collective effort of many people and several great experiments. The splitting of the GUT force took place somewhere between 10^{-36} and 10^{-34} when the temperature was in the range 10^{27} K to 10^{38} K, depending on which version of the GUT you use.

The Standard Model of particle physics is one of the most successful scientific theories ever conceived. It provides a quantitative description, with equations, of the world of elementary particles and their interactions. It has made predictions for previously unknown particles; it can be used to predict the masses of particles and the strengths of their interactions.

18.2   The Standard Model of Particle Physics

The Standard Model of particle physics emerged from the work of the 1960s. It describes 17 particles, 12 fermions that are "fundamental" and 5 bosons that are the carriers of the forces between them.[6] Figure 18.2 depicts the 12 fundamental particles which are divided into two families: the "quarks" and the "leptons". There are 6 quarks and 6 leptons. All matter in the universe is made from leptons and quarks.

The idea of quarks as fundamental building blocks originated in 1964 when Murray Gell-Mann and George Zweig were independently trying to understand the properties of protons and neutrons. Both supposed that protons and neutrons were built from three, as yet unknown, particles which Gell-Mann referred to as "quarks" and Zweig had called "aces". The idea was highly speculative, but in 1968, the first quarks were seen at the Stanford Linear Accelerator. These fundamental particles are not some kind of unseen entity of the kind that had been imagined since the times of Democritus, Epicurus and Gassendi.

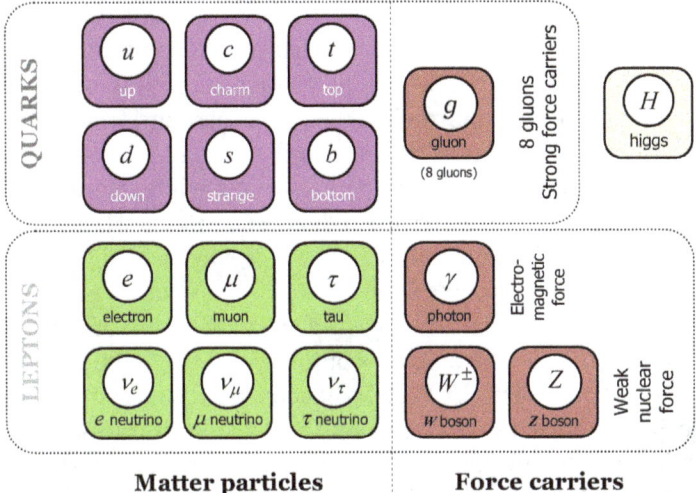

Figure 18.2 The Standard Model: fundamental particles and their force carriers. The diagram implicitly includes the anti-particles. Ordinary matter is built from particles and all anti-matter is made from the corresponding anti-particles.

The three lepton families consist of the electron, the muon and the tau particle, all of which have negative electric charge, together with their associated neutrinos simply referred to as the electron neutrino, the muon neutrino and the tau neutrino. These particles interact with each other via what is called the "electroweak force", a combination of the electric and magnetic forces and the weak nuclear force. The electromagnetic force is mediated by the photon and the weak force is mediated by what are called the *W and Z bosons*. Each of these particles has a corresponding "anti-particle", a particle having the same mass but having the opposite electric charge,[7] except for the force carriers that have no electric charge, which are their own anti-particle.[8]

The six quarks have names (referred to as "flavors"[9]): *up*, *down*, *charm*, *strange*, *top* and *bottom*. The particle mediating the force between the quarks, the "color force", is mediated by a group of eight particles called *gluons*.[10] The color force, the gluon, can bind quarks together to form other nonfundamental particles, notably the neutron and the proton that populate the centers of atoms. The proton is made from two up and one down quarks, while the neutron is made from one up and two downs. So, the fundamental constituents of matter that pervade our universe are the up and down quarks.[11] Other particles, the mesons, are made from combinations of one quark and its anti-particle. The force of interaction between these particles composed of quarks is called the "strong force", which is a residual force left over from the gluons that bind the quarks together.

In and around 1964, several people, among whom were the Belgian theoretical physicist François Englert and Peter Higgs from the University of Edinburgh, had suggested a way in which elementary particles could acquire their mass from a field which is now called the "Higgs field". This worked well within the framework of the Standard Model, according to which there would be an associated particle, now called the "Higgs particle". The subsequent discovery of the Higgs particle in experiments running at the LHC over the period 2011–2013 validated the mechanism by which particles acquired their mass and set the seal on the acceptance of the Standard Model. The story is told in Section 16.9 of Chapter 16. Englert and Higgs shared the 2013 Nobel Prize.[12]

Following the precepts of Quantum Field Theory, the all-pervading ultra-weak gravitational force should be mediated by a particle, referred to as the *graviton*. As of this time, the gravitational force has not been fully incorporated into the Standard model. The fact that the gravitational force between elementary particles is, in effect, negligible means that the Standard Model without the graviton works perfectly well at the accessible energy scales. However, the Standard Model has one important weakness: it is incompatible with the fact that the universe is not matter-antimatter symmetric. There is clearly something missing from the Standard model.

What does this tell us about the very early universe? The clue to understanding the very early universe is the fact that these fundamental particles and the elementary particles that build it, make their appearance in different energy ranges. Matter, like protons and neutrons that are made up of glued-together quarks, cannot exist at temperatures above 10^{13} K (10 trillion degrees). At higher temperatures, the gluons can no longer hold such particles together and the result is that we get a plasma that is made up of quarks and gluons: the "quark-gluon plasma". Conversely, at lower temperatures, the gluon will bind the quarks together to make "hadrons": neutrons, protons, mesons and so on, particles that makeup matter as we know it today and which interact through the strong force. When the universe was one-millionth of a second old there was a transition from a Quark-dominated universe to a Hadron-dominated universe.

This may feel somewhat like a fantasy. However, collisions in the Large Hadron Collider ("LHC") reach energies that correspond to temperatures a little more than 10^{17} K! The Higgs boson was first detected at the LHC at energies corresponding to a temperature of 1.5×10^{15} K. So, we are still in the land of known physics in this extreme regime. However, it must be borne in mind that these are the early days of exploring matter at these enormous energies and temperatures. There may be more to find that we do not yet know.

It must be emphasized that we are not indulging in some over-imaginative form of speculation. These are particles that we have "seen" in machines like the Large Hadron Collider: we understand their inter-relationship and interactions through the theory of the Standard Model of particle physics. They are not unobserved entities: we know about them through their interactions with other particles and we can determine the properties that serve to define them uniquely.

18.3   Ever Closer to the Big Bang

We can describe our universe back to the first "picosecond" (10^{-12} seconds) when the temperature of the universe was around 10^{15} K, a quadrillion (a million billion) degrees and its density was an inconceivable 10^{21} tonnes per cubic centimeter (think of the entire mass of the Earth squashed into a 1 centimeter cube). The ultra-brief period of time between when the universe was 10^{-12} seconds old and when it was 10^{-6} seconds old, a millionth of a second is referred to as the "Quark period" of the expansion. This is also the start of the era of the Higgs particle, which serves to endow mass to the particles it interacts with. Earlier, we mentioned that the Leptons interact through the electroweak forces which are carried by the photon and the W and Z bosons. W and Z acquire a mass because they do interact with the Higgs field.

Can we push back to times before that first pico-second? In asking that question we are pushing beyond the experimentally established

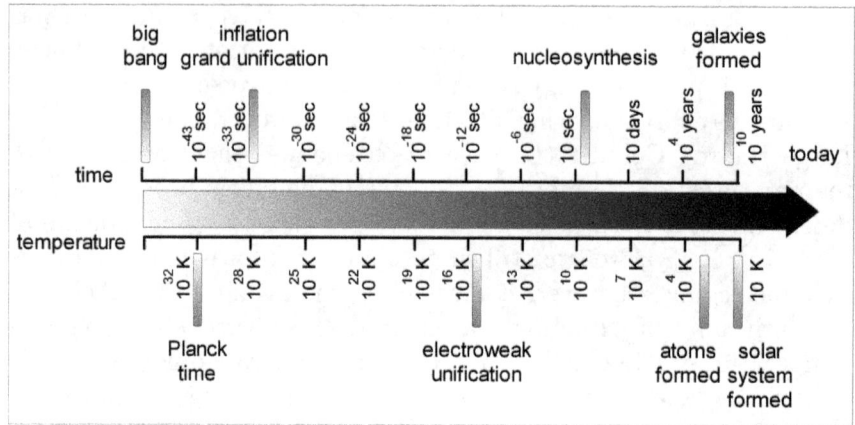

Figure 18.3 The thermal history of the universe showing the temperature of the cosmic plasma as a function of time since the Big Bang. Credit: Adapted and reproduced with permission from Brian Greene (2010). *The Elegant Universe,* W. W. Norton, New York, Figure 14.1 (p. 356).

laws of physics: we come into the universe of grand unification of the forces and quantum gravity. We have to enter into this regime in the spirit of exploration, we don't have much to go on. Within the framework of what we know today we can feel confident that the universe at this time was dominated by particles we have seen in experiments: the W and Z bosons and the Higgs boson. There may well have been other more massive particles present, but as yet, we have no direct evidence for their existence.

This pre-picosecond period is known as the "Electroweak Period". The start of the electroweak period is a matter of definition, some put it at 10^{-32} seconds after the zero time, while others push it further back to 10^{-36} seconds. To appreciate this in a slightly different perspective, the electro-weak period started when the temperature of the universe fell to a value between 10^{26} K and 10^{27} K degrees, also known as 100 to 1,000 yotta-degrees.[13] Where the start is put depends on the details of what you think happened before that. On the logarithmic timescale of Figure 18.3, the electroweak period looks very long: the universe will have expanded by a factor of 10^{12} or so, a million million times, during that period.

18.4   The Cosmic Inflation Paradigm

Prior to the electroweak period, it is thought that the universe underwent an important period of super-fast expansion. This period of inflation, those first 10^{-32} seconds or so of the existence of the universe, was the "Bang" in the Big Bang: it was the time when the entire future evolution of our universe was set by the laws of physics. As Hermann Bondi might have said in this case: we are what we are now because of what happened then.

The question has been asked as to how seriously should we take the extrapolation of our known science to this ultra-early inflationary period, those first 10^{-32} seconds, especially in view of the fact that we may never be able to test this picture. Is this just blind speculation? Or is it the birth of a new theory? The prevailing opinion is that early cosmic inflation is a hypothesis that is potentially testable. There is however a small but significant body of skeptics who do not buy into this and who prefer that we focus on alternative scenarios for those earliest moments of cosmic history. A third camp prefers to wait and see what evidence turns up before casting an opinion.

The phrase "cosmic inflation" simply refers to cosmological models that, for one reason or another, experience exponential expansion for a short period of time during the first microseconds of their existence. The term is borrowed from an analogy with the finance industry: exponential growth is a bit like "compound interest" on borrowed money — you pay interest on the sum borrowed plus the total interest that is due to date. As we know from experience with credit cards, this rapidly becomes radically different from paying the interest only on the original sum borrowed, "simple interest". The widely used analogy is of space that is the surface of an expanding balloon. As the balloon inflates points on the balloon recede from one another while the surface of the balloon becomes increasingly flatter. Pursuing this analogy a little further, we can appreciate how an ultra-large inflation factor causes the space to become flatter.

In mathematical terms, what generally drives this cosmological exponential expansion amounts to what in the Einstein equation looks like a negative pressure force. Not surprisingly, the very term "negative pressure" often causes some level of consternation since this is not a part of our normal experience of the way things happen.[14] In fact, it might be easier in the cosmological context to think of it as a negative mass: with gravity positive mass attracts while negative mass repels. However, one should avoid trying to take such analogies too literally!

In the strange world of quantum mechanics, strange things happen. Quantum theory tells us that the vacuum has zero density, but it is not empty! The vacuum is buzzing with invisible, "virtual", particles which come into visibility before disappearing again. Since the vacuum has no weight, this happens in such a way that its density is always zero. Yet, this strange vacuum has energy, the energy to drive inflation. This was first predicted by the Dutch physicist Hendrik Casimir in 1948 when investigating the properties of what are known as "colloidal solutions"[15] at the Phillips Physics laboratory, NatLab, in Eindhoven in Holland. If two mirrors are held in a vacuum with their surfaces parallel and very close to one another, there is a measurable attractive force between them.[16] The effect was first experimentally verified in 1958 by Marcus Spaarnay, also at Philips. This is now called the "Casimir effect". The strength of the force can be calculated with high precision using a theory known as Quantum Electrodynamics[17] and has been verified experimentally. The quantum vacuum is capable of driving inflation.

The cosmological solution of Einstein's equations discovered by de Sitter in 1917 describes a universe devoid of matter but expanding exponentially under the influence of Einstein's controversial cosmological term. There is no matter in this model universe, but it has a geometry. The force of the cosmological term causes the geometry of the universe to expand taking with it "test particles" (tracer particles) that are glued to the geometry. In this way, we talk of the expanding "fabric of space–time". The quantum nature of the vacuum mimics a cosmological term and can cause an exponential de Sitter-like expansion. In other words, de Sitter's solution describes an empty inflationary universe in which the cause of the inflation is the cosmological term, or if we go to an early enough quantum epoch, the quantum fluctuations that make up the vacuum.

The importance of the de Sitter model is that all parts of the de Sitter universe are in communication with one another during the expansion. This sounds like faster-than-light communication but it is not, it is probably best thought of as a trick of geometry.[18]

The existence of a very early de Sitter phase is a key element of the inflationary theory: it provides a plausible solution to the "horizon problem" that was described at the end of Chapter 17.

If we put the end of the de Sitter expansion far back enough in time, the entire volume of our universe will be causally connected during the accelerated expansion phase. This opens up a door to processes that can isotropize and homogenize the universe on a truly global scale, the scale of our present-day universe. Generally speaking, the time assigned to the start of inflation is the time of grand unification, around 10^{-36} seconds after time zero. It is perhaps the earliest moment when we are in the realm of almost-known physics. The furthest back we can go is the Planck time at 10^{-43} seconds. This is the time when the theories of General Relativity and Quantum Mechanics merge: the era of Quantum Gravity. At that time, the entirety of our presently visible universe is contained in a sphere $\frac{1}{100}$th of a millimeter across![19]

It must be stressed that simply postulating the existence of such an inflationary period is but a part of the solution to the horizon problem. We still have to figure out what stops the inflation. We also have to figure out how the universe gets reheated after the enormous expansion factor. Each version of the inflation theory makes its own assertions about this period and these become the basis of the experimental process of accepting or rejecting specific models.

18.5 Early Proponents of Cosmic Inflation

The period 1979–1982 was an important period in early universe cosmology research. During this period, a number of quite independent researchers were coming to the same or highly similar conclusions with their own similar versions of a period of primordial cosmic inflation.

The Russian–American physicist Andrei Linde wrote an important 50-page review article in 1979 about the properties of superdense matter at high temperatures. This was written from the point of view of a high-energy particle physicist, but Linde recognized that this would be important for cosmology and went as far as to describe how such a universe might condense into regions having different properties during its expansion and cooling: a "phase transition".

However, Linde made no reference to the possibility of exponential expansion. Then, in 1980, three authors from the worlds of theoretical astrophysics and high-energy physics wrote papers citing Linde about different aspects of what became known as the "Theory of cosmological inflation" or simply "inflation". The authors were Katsuhiko Sato of the University of Kyoto, Demosthenes Kazanas of NASA's Goddard Flight Center and Alan Guth of Stanford University.

These three did not cite one another's papers; their research was done independently and at the same time, and it was unlikely that they even knew about one another. Alan Guth was a particle physicist and published in a leading physics journal, while Kazanas and Sato published in astronomy journals. Possibly because of this, the "theory of the inflationary universe" is attributed solely to Alan Guth, and not to Sato, Kazanas, or any of their predecessors.

Inflation was the province of high-energy particle physics, and most of the papers on inflation were published in journals read by particle physicists, who themselves rarely, if ever, read articles in astronomy. We should recall that at that time there was no Internet. Nor was there a preprint archive: preprints were circulated to colleagues in the same field. International communication took place mainly through writing letters, visiting other institutions or international conferences which focused on bringing experts in specific fields together.[20]

Within a few years, we had numerous variants of the original inflation idea that appeared as the number of people working in the field inflated exponentially, as did the number of published papers. Since the pioneering work at the end of the 1970s and in the early 1980s more than 30,000 papers on Cosmic Inflation[21] have been published by almost 10,000 distinct authors![22] Inflation is now widely accepted among the physics community as a part of what has become known as the Standard Model for Cosmology.

The notion of cosmological inflation became a paradigm of modern cosmology. Prestigious prizes galore were handed out. Alan Guth and Andrei Linde shared the 2004 Gruber Prize "for their roles in developing and refining the theory of cosmic inflation". Guth and Linde later shared the 2014 Kavli Prize with Alexei Starobinsky. Previous to that Starobinsky had shared the 2013 Gruber Prize with Viatcheslav Mukhanov.

18.6   Evidence for Inflation

It is thought that cosmological inflation, in one form or another, might provide answers to most, if not all, the questions posed at the end of Chapter 17. These questions had been asked before the development of inflation theories, and so cannot convincingly be used as evidence supporting any of the inflationary models. The fact that such a simple idea as early time inflationary expansion can successfully address these problems is encouraging, but there has to be a point of validation in which some consequence of the models which was not a part of their construction can be explored observationally. We need an independent experimental prediction that can be tested.[23]

Two such tests have emerged. The first is to detect the consequences of inflation in maps of the Cosmic Microwave Background Radiation. The sought-after imprint is the polarization of the CMB due to primordial gravitational-wave ripples in the geometric fabric of space–time at the time of inflation. This is already an ongoing task with several experiments to map the CMB using equipment based at the South Pole. Nonetheless, even without that, it has been possible to use the CMB maps to impose significant constraints on theories of inflation. The other, more futuristic, experiment is to detect the background of primordial gravitational waves that have come from the first instants of the Big Bang. We have only recently started to detect gravitational waves from stars being swallowed by black holes, or from the merging of two black holes. These are comparatively strong signals from cataclysmic events, the first of which was detected by LIGO in 2015.[24] There is a long way to go in terms of reaching the sensitivity required to directly detect this primordial gravitational wave background.

18.7   Polarization of the CMB

In relation to the first test, we need a small diversion on what polarization is. We commonly appreciate what the "polarization of light" means through wearing sunglasses that have "polarizing lenses": glare from reflections is reduced. Light can be understood as a wave phenomenon in which the vibrations are in a plane perpendicular to the direction of travel of the light.[25] Normally, a light beam may consist of rays that have planes of polarization that are at a mixture of angles: that is a nonpolarized beam. If the rays are all aligned and have the same plane of vibration, the beam is said to be polarized in that plane. Light reflected off smooth surfaces is partially polarized.

Polarizing lenses are coated with a special laminate which has a vertical pattern that blocks the horizontal light component.[26] When light hits a reflective surface, only the horizontal component of the light is reflected so sunglasses can block glare from sunlight reflected off a swimming pool. If you look at the blue sky through polarizing lenses and tilt your head sideways, the intensity of the light from the sky varies: that is because the light from the blue sky is scattered sunlight and is polarized by the scattering process.

The radiation that comes from the CMB is partly polarized by a variety of physical processes. The polarized component is very much weaker than the CMB temperature fluctuations and so was not detected until 2002 when the Degree Angular Scale Interferometer (DASI) experiment at the South Pole detected the CMB polarization. It took 271 days of observation to generate a small 5° square polarization map of part of the sky. The little map was published in 2005. Since then there has been an avalanche of polarization experiments, many of which have been located at the South Pole.[27] The first BICEP series of experiments started acquiring polarization data in early 2006. Both WMAP and Planck satellites had acquired multi-frequency data showing the polarization.

Both the WMAP and Planck satellites produced whole-sky polarization maps. The Planck polarization map is shown in Figure 18.4 where the little "rods" indicate the strength and direction of polarization at points on the sky. The polarization data provide an independent check on the conclusions drawn from the temperature anisotropy data. But for checking on inflation, we are interested in the pattern of polarization.

The CMB polarization pattern on the sky can, in principle, be separated into contributions from two distinct and independent components. In one component, the so-called "E-mode", the rods tend to diverge away from local temperature maxima,[28] while in the other, the so-called "B-mode", they tend to loop, or spiral, around those maxima. It is difficult to discern these patterns by eye, but sophisticated data analysis can easily make the split if both modes

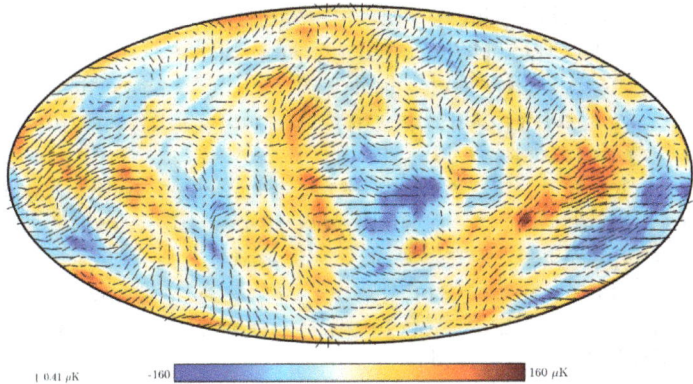

| 0.41 μK   -160 ████████████████ 160 μK

Figure 18.4 Planck 2018 polarization map. The small rods indicate the strength and direction of the polarization. The background image is the same as in Figure 17.3 smoothed to a 5° scale. Credit: ESA and Planck collaboration.

are present. The data shown in Figure 18.4 reveal the E-mode but provide only a limit as to how strong the B-mode could be without being detected.

These searches for the elusive B-mode are incredibly difficult. There are other sources of B-mode polarization that have to be removed before uncovering the sought-after signal. B-mode polarization also arises as a result of gravitational lensing between us and the CMB, and it is also generated by dust particles in our own Galaxy. These substantial effects have to be identified and removed without removing any of the CMB B-mode signals. The B-modes might be around 100–1,000 times weaker than the E-modes shown in the figure which are themselves 100 times weaker than the fluctuations shown in Figure 18.4.

We are looking for the proverbial needle in a haystack. Finding a B-mode pattern in the CMB would be strong evidence for primordial gravitational waves and, by implication, for a period of inflation and, indirectly, for quantum gravity. This is a Nobel Prize-winning challenge and it seems that this, the "Prize factor", was a key element of what happened next.[29]

18.8 The BICEP Fiasco

Undaunted by the challenge, a team from Caltech started work in 2002 on a new detector to measure the CMB polarization from the

South Pole: BICEP, Background Imaging of Cosmic Extragalactic Polarization. In 2003, Brian Keating was the lead author of a paper describing the BICEP experiment and BICEP was deployed in 2005 for a three-year data gathering stint. At that point, the collaboration grew and work was started on BICEP version 2. BICEP2 was a 26 cm refractor telescope with new detector technology and 10 times the number of sensors as BICEP. However, unlike BICEP which operated at three frequencies, BICEP2 only gathered data at one frequency: 150 GHz.[30] This was a crucial factor in what followed when in March 2014 they believed that they had incontrovertible evidence for the cosmological B-mode signal.

The discovery paper had been written and was about to be widely circulated but had not yet been subjected to peer review. Wanting to lay claim to being the first to discover the B-mode, the BICEP2 team gathered for a press conference at the Harvard Center for Astrophysics on 17 March 2014, St. Patrick's day, to announce the discovery of the B-mode polarization of the CMB. They claimed they had detected the elusive CMB B-mode polarization with a statistical significance of ">5σ" or 99.99994% and asserted, with even greater certainty, that they also had evidence for primordial gravitational waves.[31] Either of these assertions, if verified, would be Nobel Prize-winning science.

Selected press outlets had been given an embargoed Press Release the night before. The discovery would confirm the inflation model for cosmic expansion and the creation of the fluctuations that would lead to large-scale cosmic structure, and it would also be a signal that gravity, like the other forces of nature, was indeed quantized. *The New York Times* of 17 March ran a feature story[32] entitled "Space Ripples Reveal Big Bang's Smoking Gun", with a follow-up the following week.

The post-conference disaster unfolded quickly. In failing to seek peer review of their paper before their public announcement, the BICEP2 team failed to notice a fatal flaw in their experiment! Their use of only a single frequency to gather data meant that they could not adequately correct for Galactic dust.[33] This defect was quickly spotted by other cosmologists.

The obvious question being asked by many cosmologists worldwide was whether the polarization reported could simply be due to Galactic dust. On the same day as the press conference, Matt Strassler posted information on his *Of Particular Significance* blog about the press conference, and saying that "some people are worried about certain weird features of the data, while others seem less concerned about them". He expressed the sentiment, shared by many, that he was "disturbed that the media is declaring victory before the scientific community is ready to".[34] That was taken up in more detail two days later by Peter Coles, in his popular *In the Dark* blog: "BICEP2: Is the signal cosmological?" He expressed his concerns concluding that "the absolute minimum would be a detection of the signal in more than one frequency band" in order to be convinced. During the following days, others shared the concern.

The Planck team, which the BICEP2 team regarded as "the competition", announced their own findings in a draft paper posted on the important preprint archive, arXiv[35] on 5 May, "An overview of the polarized thermal emission from galactic dust", which presented detailed maps of the sky showing where there was substantial dust polarization. It was clear that the BICEP2 data was consistent with their having detected only the galactic dust. That re-ignited an avalanche of commentary. On 12 May, on his Résonaances site,[36] Adam Falkowski discussed what he thought to be the problem with the BICEP2 findings. Two days later, this was taken up once more by Peter Coles on his blog.

The BICEP2 paper was published on the 20 June 2014 *Physical Review Letters* with a "note added" at the end of the paper remarking that the first data release data from the Planck satellite showed that the BICEP2 modeling of their data had underestimated the contribution from dust in the interstellar medium. Serious doubt had been cast on the BICEP2 claim and with that, any chance of a Nobel Prize was gone. However, the BICEP2 team has always insisted that their data acquisition was not at fault, the map they produced was a correct B-mode map but that their only problem was in not having enough information about the non-CMB contributions: dust and gravitational lensing.

The BICEP2 group had made a number of strategic errors. Early on in the project, BICEP1 detected signals at three frequencies, yet they chose to have BICEP2 working at one frequency, 150 GHz. With that, it was inevitable that they would have to use someone else's data to have any chance of revealing the CMB signal. Their solution was to digitize Planck CMB polarization sky maps made at different frequencies that had been shown at Planck presentations. The maps were preliminary and the digitized versions would hardly have been accurate enough for the job.

The decisive factor in revealing their error was the Planck 353 GHz map and the analysis showing that dust was far more of a problem than had been expected. So why did they not collaborate and delay publication until they had everything they needed? It is clear from Keating's book[37] that there was intense competition between teams searching for B-mode polarization and it was feared that Planck was on the verge of scooping the honors and with that the Nobel Prize. The prize and the question of which three people would go to Stockholm is a central theme in Keating's book.

18.9 Is Cosmic Inflation a Theory?

The presence of the B-mode polarization component of the CMB is consistent with an inflationary period in the early evolution of the universe. However, the evidence to date is not sufficient to accurately define what kind of inflation took place or even to rule out other noninflationary causes. The Planck observations served to reject some models, but that still left a large number of more complex models which needed to be specified by more parameters than the simpler models. If this process of inventing a new variant of the inflation theory each time one is excluded continued what would have been proved?

Jo Dunkley, Professor of Astrophysics at Princeton University, explains in her book *Our Universe: An Astronomer's Guide*[38]: "if the polarization signals are discovered, it will be marvelous. We will have stepped closer to the limits of what we might ever know about our origins... It is also quite possible that the theory is wrong and we will not find any hint of these ripples."

In May 2013, Anna Ijjas, Paul Steinhardt and Abraham Loeb ("ISL") published a paper, "Inflationary paradigm in trouble after Planck2013", reviewing the status of the inflationary models in light of the 2013 Planck satellite mapping of the CMB.[39] They concluded that the current versions of inflation lead to a view of the ultra-early universe as consisting of differently evolving bubbles each of which would evolve into a different universe than our current one: a "multiverse". In that case, the one in which we live and its properties have emerged by chance: there is neither explanation nor possibility of testing any putative explanation for "why is our universe the way it is?" At the same time, ISL produced a new noninflationary theory which proposed that the universe originated in the bounce of a cyclic universe.[40] During the final stages of the collapse, all structures would be wiped out during a slow period of conflagration ("ekpyrotic"), which would then re-expand in an inflationary episode. This avoided the requirement that our universe evolved from some random bubble.[41]

ISL presented their views in an article published in the February 2017 *Scientific American*, entitled "Pop goes the universe" arguing that we should change our focus away from inflation and move towards contemplating other possibilities. That turned out to be highly controversial.

Scientific American is a far more widely read and public forum than the learned journals of scientific research. Any criticism of an accepted theory in the magazine would draw a significant amount of attention from both the public and from the national funding bodies who had supported the criticized research. It is not surprising that a strong rebuttal of the ISL thesis appeared in the July 2017 issue of *Scientific American*, entitled "A cosmic controversy". This was signed by 31 leading scientists from the cosmology and high energy physics communities, including four Nobel Prize winners and Stephen Hawking, and of course some of those who were the drivers of inflation theory, Alan Guth, Andrei Linde and Alex Starobinsky. The journal allowed ISL a small amount of space for a rebuttal. The process of theorizing about the very early universe and designing experiments to test these ideas continues unabated. We advance ideas, argue about them and perform experiments, rewarding the advances made with honors and prizes. This is the way science is done, although it is sad when the honors and prizes dominate sentiments.

In a final analysis, the issue is whether we think we are trying to explain the entire universe, even the bits we do not and cannot ever see, or just our local patch. The former is indeed unattainable while the latter seems like a reasonable goal for scientific inquiry. The debate is perhaps more relevant to the question of whether we should spend time contemplating issues such as the "true" dimensionality of our universe: a question that is almost certainly beyond the limits of our perception or our experimentation.

The fact of the matter is that scientists merely build models of reality (whatever "reality" may mean). By referring to well-defined, even abstract, models, we feel satisfied that we can come to terms with what we see in terms of the known laws of physics. The fewer assumptions that we have to make to get to that point of comfort, the better. Inflation does just that: it answers Alan Guth's fundamental question of why the universe is homogeneous and isotropic. Many scientists are "seekers after understanding", which demands consistency, while others are "seekers after truth", which demands a sense of belief. It is easier to satisfy the first lot and impossible to satisfy the others.

Part VII
The Dragon's Future
Science Reinvented

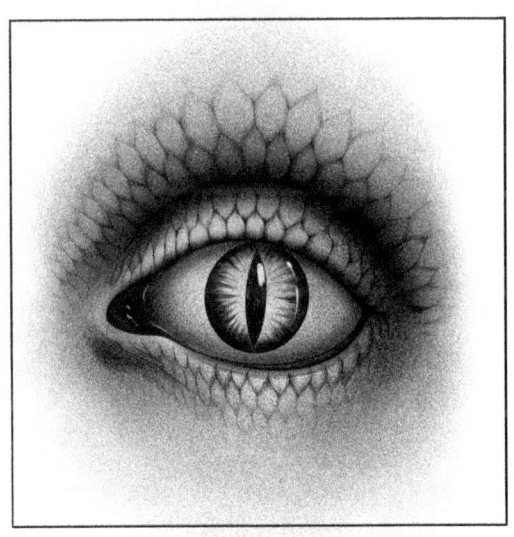

Chapter 19

Machines Reinventing Science

> *I propose to consider the question, "Can machines think?" ......We may hope that machines will eventually compete with men in all purely intellectual fields.*
>
> A. M. Turing, "Computing Machinery and Intelligence", 1950

> *Success in creating AI would be the biggest event in human history. Unfortunately, it might also be the last, unless we learn how to avoid the risks.*
>
> Stephen Hawking, Brief Answers to Big Questions, 2018

In 1840, the Cambridge University mathematician, Charles Babbage (1791–1871), had just retired from the prestigious position of Lucasian Professor, when he gave a series of lectures in Turin about his work developing a steam-powered machine that could be programmed to perform mathematical calculations: the mechanical Analytic Engine.[1] In the audience was a young Italian military engineer, Luigi Menabrea (1809–1896), a mathematician who would later become the 7th Prime minister of the new Italy.[2] Menabrea had been directed to take notes on Babbage's lectures which were published as an article in a French-language journal in 1842.[3] The following year an English language translation of the article was published with the addition of a long insightful section titled "Notes by the translator", referred to as the *Notes* in the following. The *Notes* consist of six subsections labeled A–G and are some three times longer than the original article!

The translator of Menabrea's paper and author of the *Notes* was Augusta Ada King, Countess of Lovelace (1815–1852), simply known today as Ada Lovelace.[4] Her text is simply signed "A.A.L.". Lovelace's name does not appear in the paper.[5] Ada Lovelace was justifiably concerned that her work might not be taken seriously if it were known that the author of the *Notes* was a woman.

The Countess of Lovelace was an active participant in the titled people's social scene[6] and so it was widely known who A.A.L. was. That probably contributed to the fact that the article was largely ignored for over a century, but the widespread criticism of Babbage's work and his difficult personality would also have been a factor: there was little popular or scientific enthusiasm for his project.[7] A century later, the article is recognized as one of the most insightful papers ever written on the subject of Scientific Computing, laying out what distinguished a calculator from a computer and presenting in great detail what was the first-ever computer program for executing a mathematical task.[8]

Lovelace's computer program was specifically written for Babbage's *Analytic Engine* which, as Lovelace points out, had all the essential attributes of what is now the modern computer. It had an input mechanism for both acquiring data and getting programming instructions from the outside world, it had a processing unit that was able to perform basic algebraic operations and which had internal storage to save important data. It also had a programming language which could deliver instructions to the processor, and it had an output mechanism whereby the results of the calculation could be printed. With those features, it could execute a task without user intervention.

The input of program code and data to the machine was based on the punched cards used in the Joseph Marie Jacquard (1752–1834) weaving loom, invented in 1804. A deck of "Jacquard cards" was used to present instructions for weaving complex patterns into fabric.

Each card defined one row of the design: a complete "weaving program" may require tens of thousands of cards.[9] By contrast, the Analytic Engine was designed so that the programming language included instructions that directed the machine to test conditions which could change the course of execution of the program. The machine could execute loops, i.e. execute the same set of instructions any specified number of times. It could also branch to another part of the program and it would stop when the task was completed. This made the Analytic Engine the first "Turing complete" computing machine.[10] In her *Notes*, Lovelace describes the process executed by the Analytic Engine: "We may say most aptly that the Analytical Engine weaves *algebraical patterns* just as the Jacquard loom weaves flowers and leaves."

Except for an early small-scale prototype built in 1832 with the help of the tool-maker Joseph Clement (1779–1844), Babbage's earlier Difference Engines were never built despite being well supported by grants from the British government. However, the Clement prototype inspired the Swedish father and son, Per Georg and Edvard Scheutz, to build their own version of the Difference Engine. Starting in 1837, they were able to produce a working version in 1843, on the basis of which they received funding to refine the design and build more for commercial purposes. By the 1860s, several others became involved in building difference engines. The culmination of this was the advent of the desktop calculating machines around the turn of the 20th century.[11] In 1991, the London Science Museum managed to build and put on a display of a version of Difference Engine No. 2. The *Analytic Engine* was never completed during Babbage's lifetime.

Lovelace remarks in the final Note G of her text that "The Analytical Engine has no pretensions whatever to *originate* anything. It can do whatever *we know how to order it* to perform. It can *follow* analysis; but it has no power of *anticipating* any analytical relations or truths" [her italics]. This has remained true of all programmable computers until the turn of the 21st century when we saw the emergence of "Machine Learning". Until then, digital computers were used in science for statistical analysis and simulation of data using implementations of the laws of physics in computer programs of the sort envisaged by Ada Lovelace. As Lovelace recognized in regard to the Analytic Engine, it could even be used outside the realm of mathematics to write and edit music.[12]

Machine learning, in all of its diverse forms, focuses on presenting a machine with a well-curated large data set from which the machine will "learn" about patterns in the data. The programmer will not try to explicitly specify the nature of the sought-after correlations, as would have been the 20th century approach, but instead construct a programmed framework which is capable of providing the sought-after correlations. Consider the problem of facial recognition in computer video sequences: the 20th century approach would have been to first find a face in an image and then to specify a number of metrics, like the distance between the eyes, with which to characterize a particular face. This would have been a formidable task of programming even without side effects, such as variations in facial pose and lighting. The 21st century approach is to train a facial detection engine to recognize a face by showing it thousands or even millions of images of faces until it can classify the faces. When the process is complete we do not know what the method of classification was nor the details of the recognition process, but it works incredibly well. The machine has, in effect, programmed itself to do the job.[13] The human programming skill comes in the building of a framework which is capable of analyzing image data containing a large number of examples. Once the process of learning is complete, the machine can find and recognize faces using no more computer power than is available on a mobile phone.

Applications of this new technology are found in image and speech recognition, product recommendations, medical diagnosis, language translation, fraud detection and stock market trading to name but a few. There are danger warnings: not all applications are benign. For example, there are systems that can write a spoof scientific paper based on a few scientific keywords. The resulting paper seems credible and a few have been accepted for publication by referees of journals, but the science is fake! As Ada Lovelace says in the article: "There are in all extensions of human power, or additions to human knowledge, various *collateral* influences, besides the main and primary object attained". We have been warned.

Can such a machine discover new laws of science or is it just some sophisticated data-fitting system? Given the positions of the planets in the sky over a period of time, and without any prior knowledge of the laws of physics, (a) could it have made future predictions of the positions in the sky, given the information possessed by Kepler prior to his great work, and (b) could it have come up with New-ton's laws of motion and the inverse square law for gravity, given the information Newton had prior to the publication of the *Principia*? Kepler, one of the greatest scientists of the past thousand or more years, managed to achieve (a) of course but did not get as far as (b), the inverse square law of gravitation. As to whether a machine could do this, the answer to the first, (a), is likely to be "yes". The answer to the second is "maybe".[14]

We start with a case where a Machine Learning program has already solved a problem that humans have not succeeded in solving: predicting the final shape of a string of amino acids. We then move on to address the second question of the previous paragraph, could a machine have discovered Newton's Laws knowing what Newton knew in, say, 1681, 5 years before the first edition of *Principia* was published? In other words, could a machine have been as creative as Newton given what he knew at the time?

19.1 Solving the Unsolved

At the end of 2021, a paper appeared in the journal *Nature* that shook the world of science[15]: a machine had made a huge step towards solving an important problem that had remained unsolved for over half a century.[16] The "machine" was a program known as AlphaFold 2 running on a huge computer.[17] The problem was the "protein folding problem", the attempt to calculate the final 3-dimensional shape adopted by a long chain of amino acids as the chain interacts with itself and tangles up to form a protein. Understanding this would provide a way of creating new proteins for treating various medical conditions. Proteins control how our bodies work and how vaccines and drugs can be made to defend against serious illnesses.

AlphaFold had not been given any equations to solve, it was sim-ply presented with a vast database of known protein structures.[18] In the 14th running of a competition between 146 research groups, AlphaFold scored more than the 2nd, 3rd and 4th placed groups put together, and, in terms of accuracy, scored even higher on the subset of proteins that were considered hardest. AlphaFold is an extreme example in the field of Machine Learning.

Proteins are necessary for almost every activity in the body. A protein is a string of molecules known as "amino acids": there are 20 amino acids found in the human body.[19] Proteins are strings of anything from a few dozen to several thousands of these amino acids hooked together end to end. The typical length is around 400 amino acids. The heart of the problem is that these strings of amino acids are far from straight: the interactions between the different parts of the string and the nature of the joins between the individual amino acids causes the string to twist and fold in on itself to produce a tangled three-dimensional structure the shape of which is always the same for a given string of amino acids.

The key to how a protein interacts with its environment is its shape, the shape of the tangled knot of amino acids. Proteins can combine with one another or with other molecules (like sugars, fats, DNA or RNA) provided their shape allows it — think of a key fitting in a lock, the key must have the right shape to work.[20] A protein that is misshapen will make us ill since it cannot perform its function.

A biological virus is simply some genes, strands of DNA or RNA, wrapped up in a protein coat. Its only "purpose" is to take over a cell so as to make copies of itself.[21] The SARS-CoV-2 virus which is the cause of the coronavirus 2019 disease (COVID-19), is wrapped in a sheath made from several proteins, notable among which is the *Spike Protein* which enables the virus to enter a cell and replicate itself: the Spike Protein is the entry key.[22]

There is considerable optimism that machine learning will help to better understand the causes and provide effective treatment regimes for medical conditions such as cystic fibrosis, Alzheimer's disease, Parkinson's disease and Type II diabetes where protein misfolding is thought to play a role.[23]

The machine learning revolution is not confined to biochemistry. We often read of the AI revolution in image science, natural language processing and the world of finance, none of which are governed by underlying equations expressing fundamental laws of nature. A similar revolution is occurring in the physical sciences where we now have the capability of generating huge experimental data sets that are difficult to visualize and even more difficult to analyze. With machine learning techniques we can discover patterns in vast datasets that would not readily be found by the methods of 20th century data analysis. Of course, recognizing that there is a pattern does not tell us the underlying physical cause of the phenomenon, which is really what the scientist wants to understand.

This is a situation where the data precedes the understanding of the underlying physics: "data-led science". Often, in the past, we saw theories emerge on the basis of existing studies driving new experiments that would serve to validate, or invalidate, a theory. Notable examples of this are Newton's theory of gravitation, Maxwell's theory of the electromagnetic field and Einstein's general theory of relativity. Early data stimulated multiple theories, conflicting schools of thought emerged as in the Ptolemaic and the Copernican/Keplerian view of the planetary system or the Big Bang and Steady State theories for cosmology. The first was resolved by Newton and Halley with Newton's fundamental theory of gravitation, while the second was resolved by a key discovery, the Cosmic Background Radiation field left over from the Big Bang. Of course, Newton's theory has been replaced by a more fundamental idea of gravity: Einstein's General Theory of Relativity, which embraces Newton's theory in the everyday limit of weak gravitational fields.

The new tool in scientific research is loosely described as "Artificial Intelligence" which has had resounding impact in a number of areas already. The question is whether this tool can ultimately be creative in the same way as Kepler, Newton, Maxwell and Einstein, and a few others thereafter.

19.2 The Shape of Things to Come?

The subject of Machine Learning, a branch of what is commonly called "Artificial Intelligence",[24] has been growing exponentially over the past two decades. During that time, it has achieved remarkable successes in tasks such as facial recognition, biological classification, natural language processing, database searching, analysis of MRI scans and control of fusion reactors. In all areas of astronomy and physics where large datasets are encountered, machine learning in one form or another is emerging as the way to go. Mathematicians are looking into the problem of verifying the proofs of mathematical theorems that extend over hundreds of pages and even looking further at the possibilities that machine learning could prove theorems. This is currently at the stage of AI's guiding human intuition by revealing hidden patterns in the problem.[25] Ultimately, it might be that the AI will make suggestions and conjectures about the subject under discussion.

It is commonly said that we have no idea how AlphaFold came to its conclusions except to say that it recognized patterns in a vast sample of proteins for which the structure was known in detail. These patterns lay buried in the data. Put in more formal language, the program could discern correlations in the data between the nature of the amino acid chain and its final protein structure. Once it had learned about these correlations, the program could then go on to classify and analyze amino acid chains it had never seen and build an accurate 3-D picture of the folded proteins. It had succeeded where previous attempts at solving equations of the relevant laws of physics had failed! It looked as though AlphaFold had rediscovered those laws of physics.

This is impressive, of course, but it was precisely what both the architecture of the program, a "neural network", and the special parallel processing capability of the computer were designed to do. An important factor was the existence of a carefully curated database of amino acids with their associated proteins.[26] The surprising aspect of AlphaFold was that it was able to generalize from the patterns it found in that database, patterns so complex that they could not have been discerned by humans.

Concern has long been raised about the use of what are loosely referred to as "algorithms that few understand" and that are used to classify, make recommendations and decisions, predict markets and so on.[27] It is hardly surprising that using an algorithm apparently designed by a machine whose inner workings are not fully understood might be perilous, though we should concede that in many areas, humans designed algorithms can fail. There is less concern in those physical sciences where experimental validation is feasible. We do not feel we need to understand the details of the internal combustion engine or tire manufacturing processes in order to feel confident in driving motor vehicles, but we will be wary of allowing an automatically piloted car on the roads.

It should be emphasized that learning and generalizing from data to achieve a goal is not at all new. A remarkable example of this is provided by the recently decrypted Babylonian cuneiform tablets at the British Museum in London. The tablets come from the period 350 BCE–50 BCE but most probably reflect many centuries of acquired knowledge and data. The tablets describe a method for calculating the path of the planet Jupiter in the sky.[28] The text on the tablets shows the first use of a sophisticated mathematical algorithm, "trapezoidal rule", for computing the distance (on the sky) that an object would move. This was the period of history during which it is believed astrology was first developed as a predictive "science". The court astrologers were perfectly happy to use their predictive method even though they did not have an underlying theory for the motion of the planet in the sky — they simply used the data on hand and some remarkably sophisticated mathematics. The trapezoidal rule was reinvented several times since then, notably for the same purpose by both Kepler and Newton who needed it to verify the laws of planetary motion that they had derived. It is frequently used in modern scientific computing.

19.3   Newton and the Motion of Comets

There has been much discussion of what is seen as a paradigm shift in the way we do science. The usual argument, using Newton's work as an example of past science, goes something like this: an apple fell on Newton's head, he thinks about it and comes up with a theory consisting of the famous equation "F = ma", his three laws of motion and his inverse square law for the gravitational force. With this, he could "derive" Kepler's three laws of planetary motion. Everything was consistent with what was known, but to prove that Newton's theory was more than a fitted model, he needed a prediction that could be verified. His friend, Edmond Halley used those laws to predict when the comet of 1682 would come back. The answer was "end of 1758", a time when both Newton and Halley were unlikely to be around. The discovery of the comet on Christmas day 1758 settled the matter — Newton really had found a fundamental theory underlying the movement of the planets and comets.

The story in fact starts with a spectacular comet, not an apple. This was the Great Comet of 1680,[29] the most brilliant comet of the 17th century, so bright that it could be seen close to the Sun in the daytime.[30] The comet was a rare "Sungrazer" passing within a quarter of a million kilometers of the Sun's surface[31] on 18 December.

The comet was seen again in Europe a couple of days later after which its tail was reported to extend from the horizon to the zenith.[32] At the time, it was widely believed that this was not the disappearance and reappearance of the comet as it passed around the Sun: this was seen as the coincidental appearance of two separate comets great comets in a spectacular cosmic display. The eminent Isaac Newton, Christopher Wren, Robert Hooke and John Wallis were among the many who, at the time, shared that belief.[33] All four developed methods for computing straight-line orbits.

The idea that comets moved on straight-line paths in space was discussed by Kepler in his book *De Cometis Libelli*, published in 1619. Kepler had made a detailed study of the path of the bright comet of 1607 (Halley's Comet, as was later recognized) and the three comets of 1618[34] and concluded from the observations of their trajectories in the sky that (a) the paths could only make sense if the Earth orbited the Sun and (b) their paths in space were straight lines. The orbits he calculated were a rather poor fit to a straight line, but it was the best he could do unless the orbits were vast circles about a distant center. In 1680, Kepler's opinion was highly influential.

1680 was an important year for Newton: the appearance of the great comet set him thinking about the paths of the comets in space. To that end, he made many accurate observations of the positions of the comet and was perhaps the last person to see it in 1681. There followed a period of almost 6 years during which he worked on determining the paths of comets. He finally settled on fitting a parabolic orbit around the Sun. The idea of such an orbit was first proposed by Hevelius in his 1668 *Cometographia*, his study of Tycho Brahe's data on the Great comet of 1577. Following the work of Hevelius, Georg Dörffel (1643–1648) started observing comets and derived a method for fitting Hevelius' parabola to his data on the great comet of 1680. Fitting a parabolic orbit to his early data of the positions on three nights in November 1680 allowed him to later confirm that this was the same Comet that appeared in the New year of 1681. His work was published in his 1681 *Astronomische Beobachtung des Grossen Cometen*. Newton was apparently unaware of this.

By 1684, Newton had found how to compute a parabolic orbit for a comet from three observations of its position in the sky. This did not involve the use of the inverse square law; it was simply a computational procedure fitting a parabola in space to observations from the Earth.[35] The inverse square law only came later as a consequence of his research on comet orbits. Newton then went on to apply his work to fit parabolic orbits for the motion of historical comets for which there was adequate data. Newton sought to look for similarities in the orbits in the hope of seeing which of them were the same comet observed on different occasions.[36] Newton drew up tables identifying many potential returning comets, notably identifying 12 possible returns of the comet of 1618 to which he assigned a period of 105 years.[37]

It was around August of 1684 when, in Cambridge, Halley and Newton discussed whether the orbits of comets could be explained by an inverse square law of gravitation.[38] Following this conversation, Newton sent Halley a nine-page document, *De Motu Corporum in Gyrum* (*On the Motion of Bodies in Orbit*) which Newton presented to the Royal Society in December 1684. This is where Newton discusses, for the first time, the force that causes the motion of the planets to deviate from a straight line. He calls that the "Centripetal Force" and shows that it must vary inversely as the square of the distance of the planet from the Sun and argues that orbits can be elliptical, parabolic or hyperbolic. *De Motu* was, in effect, the first draft of what in 1686 was presented to the Royal Society: his *Philosophiæ Naturalis Principia Mathematica*, generally referred to as his *Principia*.

In the mid-1690s, Halley was encouraged by Newton to revisit the question of cometary orbits using refinements of his method for computing orbits. Halley computed parabolic orbits for 24 comets observed since 1337. He was struck by the similarity of the orbits of the comets of 1682, 1607 and 1531, later adding the comet of 1456 to the list of reappearances. He judged that this was the only significant match in the entire sample. He predicted that it would return in 1758. Some 20 years later, Halley was able to fit elliptical orbits to those appearances and showed that the comet traveled far beyond the orbits of the known planets. This was the comet that now bears his name.

The comet was recovered on 25 December 1758 and passed closest to the Sun on 13 March 1759. The return of the comet as predicted was a posthumous triumph for both Halley and Newton, and outstanding support for the mechanistic theory of gravitation and its effect on the motion of heavenly bodies. Comets were subject to the laws of Nature, their paths across the sky could be predicted and they were not random portents of doom and disaster.

19.4   The Clockwork Universe

When Newton published his *Philosophiæ Naturalis Principia Mathematica*, written in Latin, on 5 July 1687, he gave his Laws of Motion and his theory of Gravitation to the world. The notion that a body would move in a straight line at constant velocity unless acted upon by a force was not new. Nor was the inverse square law for the gravitational force: the French astronomer Ismaël Boulliau had, some 40 years earlier, postulated that the force driving the planets was an inverse square law.[39] Kepler and Galileo, and before them the 13th century Franciscan friar, Roger Bacon, had all presented their own versions of the Principle of Inertia: that bodies not acted on by forces will continue to move in a straight line. Galileo's assertion that in a gravitational field all bodies suffer the same acceleration independently of mass or shape became part of Newton's Inverse Square Law of gravitation. While the story is that Galileo established this by dropping objects from the Tower of Pisa is probably a myth, almost 400 years later this was wonderfully demonstrated on the Moon in 1971 when Apollo 15 astronaut David Scott dropped a feather and a hammer at the same time to show that they hit the surface simultaneously.

Newton presented these ideas in a coherent and quantitative way, using mathematical techniques he himself had developed. He came up with one of the most important equations in all of science, "F = ma", force is mass times acceleration, with which we could determine the effect of a given force on the motion of a body. With this, he went on to calculate the orbits of the planets and derive Kepler's three laws of planetary motion. The motions of projectiles, falling rocks, and the motion of the planets could be predicted: they were deterministic. He tried to derive the Moon's orbit, which is an ellipse that wanders around due to the additional force exerted by the Sun, but with little success. But nevertheless, it was incredible stuff and the theory was rapidly accepted by his inner circle of scientific colleagues. However, and not surprisingly, there was a degree of skepticism voiced among the wider scientific community.

What had Newton contributed that was so important? In the first place, he gave a definition of "force" and asserted that bodies would move at uniform velocity in a straight line unless acted upon by a force. Application of a force would cause a body to accelerate. This was a radical departure from the Aristotelian view that had been the commonly held opinion for almost 2,000 years. In order to express this in a useful way, Newton had to derive a new form of mathematics, calculus, with which he could express his ideas mathematically and make calculations of the movement of bodies acted upon by forces. This was the first time since the Greeks had explained everything in geometric terms that mathematics had been involved in the laws of nature. Newton's mathematics described a mechanistic description of motion. There was danger here for Newton: he was asserting that the motions of the planets were governed entirely by physical laws that allowed any person to compute where they would be at any time in the future. There was no role for Divine Intervention: the world was a machine, like a clock, that would simply run forever by itself. This "Clockwork Universe" was denounced as a heresy by the great German polymath, Gottfried Willem Leibniz (1646–1716). Both Newton and Leibniz were religious believers: this was an argument about two world views. Newton believed in God's ability to act however and whenever He chose, while Leibniz's view was that God had the ability to see ahead of time exactly how every conceivable event would play itself out.[40]

So, why was the mathematical aspect of his theory so important? First, Newton's mathematics provided a precise framework, a language in fact, for discussing the laws of motion: there was no ambiguity. Second, using Newton's mathematical apparatus, anyone could compute the consequences of those laws in specific situations that could be tested by confrontation with data. That is what Halley did. The return of the comet was a posthumous triumph for both Halley and Newton, and outstanding support for the mechanistic theory of gravitation and its effect on the motion of heavenly bodies. This established that all bodies in our Solar System orbited about the Sun and that was because of the all-permeating gravitational field.[41]

What would have happened if the comet had not returned as predicted in 1758–1760? Would Newton's theory have been abandoned, and if so, what would have replaced it? The answer is "not much" except perhaps for a large number of skeptical people who would then argue that, as usual, those mathematicians didn't know what they were doing. It was recognized before the recovery of the comet that Halley's work should be repeated in order to make sure of the predicted date and to provide accurate predictions of where in the sky the comet should be found if it was indeed returning as predicted.

In 1757, a team was set up in Paris to do the calculation. According to Halley, the comet was expected to reappear during the following year and so the time for completing such a complex calculation was extremely short.[42] The team was Alexis Clairaut, Joseph-Jérôme de Lalande (1732–1807) and Nicole-Reine Étable de la Brière Lepaute (1723–1788: see Figure 19.1). Clairault was already famous for his work on the three-body problem and ambition. He had won the 1751 St Petersburg Prize for his essay *Théorie de la Lune* in which he had presented the theory and calculation of an accurate orbit for the Moon without modifying Newton's inverse square law.[43] d'Alembert and Euler were also considered candidates for the prize. Lalande was an up-and-coming equally ambitious young astronomer who knew Nicole-Reine Lepaute through his discussions about astronomical clocks with her and her husband, Jean André Lepaute, the clock-maker to the king. Lalande recognized Lepaute's exceptional skill with mathematics and saw that she had independently, as an enthusiastic amateur astronomer, constructed her own astronomical tables for planetary positions.

Due to the pressure of time, the three worked long days together, writing with quills dipped into ink and following the methods of calculation laid down by Clairault. If they had not worked in parallel on different parts of the calculation they could not have finished the calculation in time. Their prediction for the date of perihelion passage was only 19 days from the actual date, 13 March 1759, 77 years after its last apparition. After working on the Comet Halley project, Nicole-Reine Lepaute went on to calculate the exact time for the solar eclipse of 1764 and the ephemeris for the forthcoming Transit of Venus. She wrote a number of significant papers with Lalande. We had the first mechanistic theory for our Solar System — the "Clockwork Universe" as it was understood it at that time.

But that was not the end of the story. The difference between the predicted and actual time of perihelion passage of Comet Halley gave impetus to the widely held concern among many scientists of the time that mathematics was not the answer to everything in nature,

Figure 19.1 Nicole-Reine Étable de la Brière Lepaute (1723–1788). The youngest daughter of Philippe II d'Orléans, the regent of the kingdom of France during 1715–1723 while Louis XV was a child. She was widely recognized as a formidable mathematician and was the first woman to be elected as an honorary member of the Academy of Sciences of Béziers. Illustration by Isabel Gálvez (www.isabelgalvez.eu) inspired by a portrait by an unknown artist.

nor even the motions of the planets. The very idea that anyone should calculate the future using mathematics was even lampooned by the popular English political pamphleteer and satirist Jonathan Swift in his 1726 book *Gulliver's Travels*.[44] d'Alembert (1717–1783) attacked the work, arguing that the problem was evidently in the method of calculation, blaming Alexis Clairault for the "error". We might wonder whether this was personal animosity: Clairault had won the St. Petersburg prize.

There are several ironies to this story. The trio could not have got the right answer! We now know that the planets Uranus and Neptune, which were yet to be discovered, also disturbed the orbit of the comet in a way that was in part responsible for the difference between their calculation and what was observed.[45] Following the discovery of Uranus in 1781 by William Herschel, the next return of the comet in 1835 was predicted with an accuracy of a few days.

Uranus had been seen many times but interpreted as a star until William Herschel saw it moving over a period of nights during March 1781. Herschel at first described it as a comet, but by 1783 it was clear that this was a planet moving in a circular orbit beyond Saturn. The final irony is that the planet Neptune had been observed by Lalande's group in Paris on two nights, the 8 and 10 May 1795 when a "star" was observed at two slightly different positions on the two nights. Lalande was aware of this but the positions were marked as uncertain and he attributed the difference in reported positions to observational error. We now know that this was Neptune. Lalande was not alone: Galileo had unknowingly observed Neptune while viewing Jupiter on 28 December 1612, and the 27 January 1613. Neptune was only discovered in 1846.

Perhaps, the discovery of Neptune was the final test of Newton's theory. If so, it took over 150 years to validate Newton's theory of the laws of gravity. Newton had both given an explanation for the motions in the Solar System, and, importantly, provided a better way of computing them. Newton had provided equations that could be solved and independently verified. Not all scientists worried whether this was the "right" theory of gravitation as long as it worked. Ultimately, it was seen to fail with the computation of the orbit of the planet Mercury. The precession of the orbit could not be explained. Solving that required embracing Einstein's version of the gravitational theory in which the underlying phenomenon was the distorted geometry of space-time rather than some mysterious force which happened to vary inversely as the square of separation of the bodies. Einstein's theory of space–time and gravitation has passed all the tests it has been subjected to and it is certainly one of the triumphs of the human mind. Nonetheless, as already remarked, Newton's theory is the weak gravitational field limit of Einstein's theory, and so we can still use Newtonian theory in doing the calculations of everyday life, whether it be ballistics or the flight or an aeroplane or rocket going to the Moon or Mars. We have no need to abandon Newton.

19.5 Mathematics — The Language of Science

Isaac Newton was perhaps the first scientist to develop a mathematical theory for nature in a form that resembles the way we use mathematics in science today. We use symbols like v, x and t to represent the values of variables like velocity, position, and time. In providing his Laws of Motion and the Inverse Square Law of gravity, together with some worked examples to help, anyone else could, in principle.[46] This is what Edmund Halley went on to do when calculating the orbit of his comet and showing that Newton's Laws of motion and gravity extended far beyond the reach of the known Solar System. The tricky bit was Newton's invention of fluxions, his version of what we now refer to as the differential calculus, which allowed the computation of, among other things, velocities from positions and acceleration from velocities. The inverse process, now referred to as "integral calculus", provides for computing velocities from accelerations and then positions from velocities. Calculus in a variety of forms is an essential part of modern science. We use calculus to compute the behavior of atoms and molecules, the flight of aircraft, mudslides, the formation of galaxies and the science of the Big Bang.

Controversially, Leibniz and Newton had each independently come up with the same calculus, expressed in different notations. Leibniz published first. At the time, Leibniz accused Newton of simply rewriting his work in a different notation while Newton accused Leibniz of stealing from his unpublished notes. It was an acrimonious row that lasted many years until Leibniz's death in 1716. Nowadays, we use Leibniz's version with its simpler notation. But, as usual, most things do not just appear out of nowhere. The remarkable Dutch engineer Simon Stevin (1548–1629) had used what amounts to the calculus, albeit in a rather cumbersome form, in 1585 in his short book *De Thiende* (*The Tenth*),[47] where he introduced for the first time the decimal system and the controversial notion of *infinitesimals*. Both Kepler and the Babylonians knew about and used "quadrature", an essential part of integral calculus.

Mathematical notation is a very important aspect of making mathematics comprehensible.[48] Mathematics had evolved since the time of Babylonian astronomy and Euclid's *Elements*. It was almost 2,000 years later that symbols like x, y, a, ξ, α were used to represent something like the price of a sheep or a bag of potatoes or the mass of the Earth. Before the 16th century, mathematicians did not write equations but used words to describe the various quantities when expressing relationships, as in "the apparent loss of weight of a body placed in water is equal to the weight of the displaced water", which we all know as Archimedes Principle (c. 246 BCE) with Archimedes shouting the word "Eureka!".

Negative numbers were not unknown before the 12th century but only came into general use in the 16th century. If you asked a learned person what zero minus two was the likely answer was "zero": if I have no apples in my basket and you remove two you still have no apples in the basket! Negative weight or length was unthinkable. This was because people generally thought of numbers as quantifying something, you cannot have negative amounts of anything. Nor was calculating very easy until the Arabic notation for numbers was adopted: it was hard to imagine "VI apples at XIV sestertii each", far easier to say "six apples at fourteen sestertii each" and to multiply 14 by 6 rather than cope with XIV times VI. Even after the widespread adoption of Arabic numerals, multiplication was (and still is) a clumsy process, and division is worse still. It is hardly surprising that the people who could use an abacus were much in demand. With an abacus, it was possible to add, subtract, multiply and divide numbers but only if you knew how.

Geometry was pictorial and so the square on the hypotenuse of a right-angled triangle could be simply illustrated by drawing the triangle and the appropriate square. However, describing algebraic expression in words is not very easy as in the 10th century Persian scholar Muhammad ibn Musa al-Khwarizmi's (c. 780–850) description of the way he invented solving a quadratic equation:

> You divide ten into two parts: multiply the one by itself; it will be equal to the other taken eighty-one times.
> Computation: You say, ten less a thing, multiplied by itself, is a hundred plus a square less twenty things, and this is equal to eighty-one things ...

This is only the start of his explanation of the method for solving a simple quadratic equation. The "thing"referred to here is what we usually call the unknown variable denoted by "x".[49] He introduced the use of a symbol for zero, "0", and constructed tables of the trigonometric functions (sines, cosines and tangents) that are used to describe and calculate with triangles.

The "+" sign appeared in the mid-14th century, the "−" only in the late 15th century and the "=" sign in the mid 16th century. The use of symbols to represent quantities ("things" as al-Khwarizmi referred to them in the above quote) was summarized in the late mid-15th century work of Abu'l-Hasan ibn Ali al-Qalasadi (1412–1486) who was from Andalusia in southern Spain. It was only during the period spanning the century 1650–1750 that we saw algebra written in a form that is vaguely familiar to us today.[50]

Kepler, when doing his calculations of orbits of planets, needed to work with negative numbers, trigonometric functions and to be able to multiply and divide numbers. Otherwise, he would not have been be able to say where on the sky a planet would be observed in a few months time. Graphical methods had been developed for calculating circular and epicyclic motion, but Kepler was for the first time concerned with ellipses. He also needed to calculate areas swept out in different parts of an orbit and to calculate velocities. This was a task for what at the time was leading-edge mathematics expressed in algebraic equations.

One of the most tedious tasks for anyone calculating orbits of planets was the problem of multiplying decimal numbers. Kepler was aware of the work of the Scotsman John Napier (1550–1617) and his invention of "logarithms" which turn the process of multiplication into addition. Napier's book, *Mirifici Logarithmorum Canonis Descriptio* (*Description of the Wonderful Rule of Logarithms*), was published in 1614. Kepler read this in 1619 and adopted logarithms for the construction of his famous *Rudolphine Tables* of 1627. As a consequence of his own investigations into logarithms, Kepler published his own book on the subject in 1624, the *Chilias Logarithmoria*.[51]

We see the mathematicians of the 17th century from Descartes to Newton and Halley developing a fledgling form of algebraic notation to express inter-relationships between physical quantities. This was important because science was making a transition from philosophical introspection to quantitative expression of physical variables that could be measured and interrelated. There is no way other than via mathematical statements that we can write down and manipulate a general law governing physical quantities. Without algebra, we would have only lists and catalogs of numbers reporting what we had observed.

The mathematics used by Newton and Halley would seem as abstruse to many of their contemporaries as Einstein's equations did to his fellow scientists at the start of the 20th century. This makes theories difficult to understand and appreciate without a huge investment of time in learning these specialized mathematical techniques. We know from Einstein's *Zurich Notebook* of 1913 that even Einstein himself initially experienced great difficulties with manipulating tensors, the fundamental mathematical objects of his theory.[52] This aspect of modern science occasionally generates a sense of suspicion that mathematics, in its obscurity, is hiding something bad and leads to the conclusion that there must be something wrong with the theory. Advanced mathematics looks like window dressing and in that sense is seen as a sort of the modern equivalent of the epicycle.

Notwithstanding this mathematical complexity, Einstein's General Relativity has passed every test and has been used to solve some major problems in astronomy and astrophysics. Consequently, Einstein's theory is widely thought to provide sound a basis for describing our universe as is currently available, at least insofar as the quantum world does not interfere. The theory even plays a vital role in our everyday lives in maintaining the high accuracy of Global Positioning Systems (GPS).

People are likewise distrustful of an AI machine "doing science" simply by analyzing vast data sets — there is no obvious way of validating the processes that lead to the solution of a problem, even though that solution may appear to be manifestly acceptable. There are serious concerns about the curation of scientific data sets for this purpose, this is a place where unintentional biases may creep in and skew the results of the analysis. But, in fact, that has always been a problem in data analysis.

19.6   The 21st Century Science

The scheme of *theory $\rightarrow$ hypothesis $\rightarrow$ observation $\rightarrow$ confirmation* is widely advocated. It is the 20th century Gold Standard for assessing the validity of a theory.[53] However, science does not always work that way. A theory rarely pops up out of nowhere without reference to other information, in other words data, experienced by our senses or measured by our instruments. Moreover, "confirmation" is only a statistical matter. Even after analyzing the data, there may be a deviation between prediction and observation that could be real tension between data and theory, as happened in the case of the 1758 recovery of Halley's Comet or sometimes just the consequence of accumulated computational and observational errors. Historically it is not unusual that a perfectly workable theory is derived from observation, and, like epicycles, is able to make pretty good predictions while in fact being "wrong".

During the past half century, we appear to have been heading at an accelerating pace towards a period of data-driven science: *observation → pattern → hypothesis → theory.* That last step, building a theory, may be beyond our mathematical ability to construct[54] until a rare individual has a stroke of genius. A problem with this *observation → pattern → hypothesis → theory* way of doing things is that when something does not quite fit tensions arise and we look for possible anomalies in the data. If we don't find an anomaly we "fix" the problem by adding additional hypotheses so that the tension is resolved. That sounds once again like adding epicycles to establish and improve what might be a flawed theory.[55] As previously stated, we are, for a good reason, unwilling to abandon our fundamental ideas about, say, quantum mechanics which, though counter-intuitive, has been highly successful in explaining the submicroscopic world. The good news is that we have not had to abandon Newton because of Einstein's relativity, and nor have we abandoned Maxwell's theory because of quantum electrodynamics. The old theory becomes a limiting case of the new one.

There has always been a strong interplay between these two approaches to science. As the above discussion of how Newton's theory of gravity came about shows, the evolution of science is a combination of both, oscillating between the two on long timescales.

A further example of this is provided by the theory of electricity and magnetism which, during the first half of the 19th century, had been explored in a succession of many now-famous experiments. Among the scientists involved were Ampère, Coulomb, Faraday, Gauss, Henry, Ohm, Ørsted and Volta whose names now label many of our more familiar electromagnetic units. Nobody had been able to knit all of their experiments into a coherent whole when in 1857 Michael Faraday wrote a letter to the Scottish mathematician James Clerk Maxwell asking him to tackle this issue. Maxwell's great realization was that electricity and magnetism were aspects of a single underlying physical entity: the electromagnetic field. In 1862, Maxwell described his vision in a set of equations that now bear his name. The equations presented a unified view of electricity, magnetism and light. For Maxwell, light was an electromagnetic wave phenomenon. The sad thing was that hardly anybody took his work seriously, how could the world be filled with waves, and how could those waves propagate in a vacuum? It is also likely that few read his dauntingly mathematical articles. Eight years after Maxwell's death his waves were discovered in 1887 by the 30-year-old German physicist Heinrich Hertz, and from there came the dramatic social changes arising from Maxwell's vision.[56]

We are not about to give up on those famous equations of science that have proved invaluable in calculating engineering in our daily lives. We have already discussed Newton in this respect, but there are also other great equations like Maxwell's equation for the electromagnetic field and Schrödinger's equation that underlies all of quantum mechanics. There are also less fundamental though no less important equations like the Navier–Stokes equations describing fluid movement, whether the fluid be our atmosphere or the blood flowing through our arteries or gas flows in galactic nuclei.

In the future, machine learning will prove to be a vital tool in the analysis of complex situations that are not amenable to the 20th century technologies and mathematical techniques. We may find solutions to difficult real-world problems like turbulent flow, climate change or the mechanisms of the brain without there being a mathematical understanding of why the solution works. This has already started in those first achievements of the protein folding problem. We must simply await the next Einstein to make sense of it and the subsequent experimental verification.

Stephen Hawking famously caused uproar in the philosophy community when in 2011 he said[57] that "... philosophy is dead. Philosophers have not kept up with modern developments in science. Particularly physics. Scientists have become the bearers of the torch of discovery in our quest for knowledge." That same year, the Cardiff philosopher, Christopher Norris wrote an eloquent and robust rebuttal of this, following Kant in concluding that "philosophy of science without scientific input is empty, while science without philosophical guidance is blind".[58] Many scientists might simply feel that philosophers are verbally over-pedantic just as philosophers might feel that scientists are obsessive about abstruse mathematics. There is nonetheless a strong and widespread interest in the wider community in debating or arguing about what we think we know and what we probably cannot know. So, following Norris, we end with a simple philosophical statement that is probably itself controversial.

Ultimately, mathematical equations are always just models of what we think is the "real world". Reality hides behind the shield of our senses. By the same token, other calculating tools, such as AI programs and advanced instrumentation, are just providing models too. Reality always remains elusively beyond the edge of certainty, so science is always open to reinvention.

Notes

Chapter 1

1. Letter to J. W. Strutt (Lord Rayleigh), 6 March 1887, in D. M. Livingston. (1973). *The Master of Light*. New York: Scribner's, pp. 117, 123.
2. A. E. Moyer. (1987). Michelson in 1887. *Physics Today*, 40, 50.
3. In China, the basic elements were Fire, Earth, Metal, Water and Wood, a five-fold system with complex inter-relations between the elements, https://en.wikipedia.org/wiki/Wuxing_(Chinese_philosophy). Earth, Air, Fire and Water appear as fundamental constituents of the world in several ancient philosophical/religious belief systems: Hinduism, Buddhism, Zoroastrianism and Babylonian mythology.
4. Although the authors of this book are inclined to think that sooner or later the nature of dark matter will be ascertained and its existence proven, this may not happen. If it does not, there would be a need for rethinking cosmology so we are able to explain current astronomical observations without the need to resort to dark matter. There are proposals of this kind; we will see them in Chapter 16. In any case, and given that the majority of cosmologists today evidently do not doubt the existence of dark matter, it is fitting to recall a remark attributed to Russian physicist Lev Landau (1908–1968): "Cosmologists are often in error, but never in doubt."
5. E. Carbonell and R. Sala. (2002). *Encara no som Humans*. Ed. Empúries. Barcelona.

6. The Bible, New International Version, *Genesis,* 1, 4–9.
7. "Our ability to understand the universe and our position in it is one of the glories of the human species. Our ability to link mind to mind by language, and especially to transmit our thoughts across the centuries is another. Science and literature, then, are the two achievements of *Homo sapiens* that most convincingly justify the specific name," Richard Dawkins from his commentary to the anthology *The Oxford Book of Modern Science Writing* (2008). Oxford University Press.
8. M. S. Longair. (2001). The technology of cosmology, in V. J. Martínez, V. Trimble and M. J. Pons-Bordería (eds.), *Historical Development of Modern Cosmology.* ASP Conference Series, Vol. 252, p. 55.
9. D. de Solla Price. (1974). Gears from the Greeks. The Antikythera mechanism: A calendar computer from ca. 80 B.C. *Transactions of the American Philosophical Society,* 64(7), 1–70.
10. T. Freeth, D. Higgon, A. Dacanalis, *et al.* (2021). A model of the cosmos in the ancient Greek Antikythera mechanism. *Scientific Reports,* 11, 5821.
11. F. Charette. (2006). High tech from ancient Greece. *Nature,* 444, 551–552.
12. The paradox refers to a race between Achilles and a slowly moving tortoise. They both start moving at the same time, but if the tortoise starts a few meters ahead of Achilles and continues to move forward, Achilles can run at any speed and will never catch it up, because Achilles must first get to the point where the tortoise began, by then the tortoise will have advanced, even a small distance, to another point; by the time Achilles covers the distance to this last point, the tortoise will have moved ahead further, and so on. A little calculus goes a long way here!
13. As explained by David Lindberg (2007) in *The Beginnings of Western Science,* The University of Chicago Press: "In Aristotle's sublunar realm, natural motion ceases when the moving object reaches its natural place, and violent motion comes to an end when the external force no longer acts."
14. Aristotle distinguishes between the heavenly region where movement is perfect, eternal and unchanging and the region under the Moon (sublunary) where there are changes and imperfect movements. The heavenly region, where the planets and stars

are located, is made of the fifth (hence "quintessence") element: the ether.

15. In modern times, an "Epicurean" is widely seen as a *bon viveur* who is overweight, enjoys excess eating, drinking, brawling and sex and suffers from gout. However, Epicurus did not advocate that life should be one bacchanalian party. For Epicurus, the goal was to achieve a sense of inner tranquillity through abstinence from unnecessary desires and contentment with simple things, and through philosophical introspection rather than sex, food and drink. Lucretius, as a disciple of Epicurus, advocates this in *De Rerum Natura* as a path through your one life.

16. The actual number depends on which published version of the poem is used. The poem is written in "heroic" form consisting of pairs of lines having a particular rhythm. According to Gregory Curtis, this is the same sort of verse structure as used in rap and hip-hop music: https://gregorydcurtis.com/on-lucretius/ (Part One in a series of four articles on Lucretius' poem).

17. The history of the poem is authoritatively presented in the formidable work of G. Passannante. (2011). *The Lucretian Renaissance: Philology and the Afterlife of Tradition.* University Chicago Press, where he traces the influence of the poem from Roman times through early modern Europe and as far as the 20th century when Einstein praised it as a work of "magic." See also the collection of articles in S. Gillespie and P. Hardie (eds.) (2007). *The Cambridge Companion to Lucretius.* CUP.

Chapter 2

1. *De Rerum Natura* was known in the Carolingian court circles in the early 9th century (Lindberg, *loc. cit.*, p. 333. Moreover, Isidore of Seville (560–636) cites Lucretius' poem in his book entitled preciseky *Natura Rerum.*

2. Lucretius never ascribes an origin to the motions nor does he suggest a cause for the deflections.

3. Stephen Greenblatt describes this in his Pulitzer Prize winning book *The Swerve: How the World Became Modern* (Norton 2011), Chapter 3. In the UK, the publisher (Bodley Head, 2011) changed the subtitle so the book title was *The Swerve: How*

the Renaissance Began to better reflect Greenblatt's principal thesis that the discovery in 1417 of a manuscript of Lucretius' *De Rerum Natura* dramatically changed the course of the Renaissance. Today, the largest collection of scrolls about Epicurus is found in the remains of the library of the Villa of the Papyri in Herculaneum. Following the eruption of Vesuvius in AD79, most of the manuscripts in the library were carbonized and crushed. Many of those that survived were inadvertently destroyed by the people who first discovered the library in 1752.

4. Leonardo da Vinci was born in 1452, Erasmus in 1466, Copernicus in 1473 and Michelangelo in 1475.

5. From 1309 to 1377, the papacy was based in Avignon, in the South of France. In 1378, the papacy returned to Rome and the Archbishop of Bari was elected to the position as Urban VI. However, those who had elected Urban VI soon regretted their decision and that same year held another election in Anagni, where there was a papal residence, when they elected an alternative pope. The alternate pope, Clement VII, re-established his papacy at Avignon. He had widespread support from rulers over much of Europe but was later styled as an anti-pope rather than a pope. The split papacy continued into the following century, by which time Benedict XIII was the Pope at Avignon and Gregory XII was the Pope in Rome.

6. The Council of Pisa, 1409.

7. Baldassarre Cossa has been described in the *Catholic Encyclopedia* https://www.newadvent.org/cathen/08434a.htm as an "astute financial administrator" and "skilful statesman" while at the same time being "utterly worldly minded, ambitious, crafty, unscrupulous and immoral, a good soldier but no churchman". The name 'Pope John XXIII' was taken by Cardinal Angelo Roncalli when he was elected Pope in 1958, so it is customary to speak of Baldassarre Cossa as 'antipope John XXIII'.

8. Poggio Bracciolini, formally known as Poggius Florentinus. Poggio started out as a scribe and became famous because of his invention of the font, nowadays referred to as the "Roman Font". That made writings easier to read, which is an advantage if you are a scribe, and so its use spread quickly.

9. Poggio was previously the Apostolic Scribe to the short-lived Alexander V and kept that post with John XXIII. His promotion

took place on 28 October 1414 (William Shepherd *The Life of Poggio Bracciolini*, London 1802 1st. edition, p. 54, full text available on arXiv).

10. The *Catholic Encyclopedia* (*op. cit.*) says that "the heinous crimes of which his opponents in the council accused him were certainly gravely exaggerated". After the forced abdication, Cossa was held in custody by Louis of the Palatine until he was granted liberty by the new (true) pope, Martin V, at the end of 1417. Cossa went to Florence where he died in late 1419, after the Pope had granted him the title of Cardinal Bishop of Tusculum.

11. The schism was resolved with the election of a single pope, Pope Martin V, who ruled his church from Rome and the retirement of Gregory XII and John XXIII, with the banishment of Benedict to the Kingdom of Aragon. John, now back to being Baldassare Cossa, came back to Rome in May 1419 to pay obeisance to Martin V by kissing his feet. For that he was absolved and created Cardinal Bishop of Toscolano. He died the following December.

12. Shepherd, *loc. cit.,* p. 101, who goes on to say that Poggio "was not deterred from this laudable design by the inclemency of the season".

13. Manilius described science in a classical poetic form. Among the ideas presented is that "the Sun periodically attracts Comets to itself and them lets them go, as it does with Mercury and Venus" (L. Russo. (2004). *The Forgotten Revolution.* Springer-Verlag, Berlin. p. 317).

14. Besides these four manuscripts, Poggio recovered many other significant manuscripts. See the list at https://www.britann ica.com/biography/Gian-Francesco-Poggio-Bracciolini. Shortly thereafter, in 1423, the Sicilian Giovanni Aurispa brought back a horde of manuscripts from Constantinople, some 238 of which were classified as pagan works.

15. The ancient Abbey of Hersfeld is on the river Fulda close to the Fulda Abbey and also had an important library, but that went to ruin during the 15th century. The library collection decayed and many volumes went missing. Finally, in 1761, during the Seven Years' War, the buildings were used to store gunpowder by the French. The monastery was destroyed. See https://www.newad vent.org/cathen/07296c.htm.

16. The film stars Sean Connery as the Franciscan monk William of Baskerville and Christian Slater as his acolyte, Asdo of Melk, a Benedictine novice. The film is a vastly simplified adaptation of the novel of the same name by Umberto Eco. The focus of Eco's book is the inquiry into whether the Franciscan order is guilty of heresy in arguing that the church should abandon its wealth. The meeting is held at the behest of Pope John XXII. The lost book in Eco's original version is the second part of Aristotle's *Poetics* which focused on comedy.

17. Greenblatt's book (*loc. cit*) is a beautifully written and a well-researched historical fiction, with reimagined scenes and without fictional characters. It is a thoroughly researched piece of investigative journalism, though its main thesis does not have the unanimous support of Renaissance historians. See, for example, the critiques at https://thonyc.wordpress.com/2012/05/01/the-swe rve-is-really-a-full-frontal-crash/, https://lareviewofbooks.org/article/why-stephen-greenblatt-is-wrong-and-why-it-matters/ and https://www.inthemedievalmiddle.com/2016/05/the-ethics-of-inventing-modernity.html.

18. In the 200 or so years following Poggio's discovery, there were many editions of the poem, all based on more or less accurate copies of Niccoli's version. Missing lines were arbitrarily filled in and sections of the poem transposed to other places for no specified reason other than to render the poem more readable. There were good copies and there were bad copies. In addition, there were the translations some in verse and some in prose. This inevitably reflected the author's understanding of the Latin poem, or lack thereof. Notable among these was the unpublished translation of the Greek poet and scholar Michael Marullo in which Marullo substantially modified the poem. We know about this because much of Machiavelli's own translation of the poem borrowed from Marullo.

19. See the authoritative article of D. Butterfield (2013). A sketch of the extant Lucretian manuscripts published in *The Early Textual History of Lucretius' De Rerum Natura* (Cambridge classical studies, CUP, 2013), pp. 5–45. https://doi.org/10.1017/CBO97 81139775403.003.

20. There is an incomplete manuscripts ("G"), one part of which is in Copenhagen and the other in Vienna. There are also

four fragments of another manuscript ("U") in Vienna. See G. P. Goold, "A Lost Manuscript of Lucretius", *Acta Classica*. 1957, Vol. 1, pp. 21–30 for a discussion of these four "Cardinal Manuscripts". O, Q, G and U. https://www.jstor.org/stable/24 589991.

21. D. Butterfield, *loc. cit.* and G. P. Goold, *loc. cit.*

22. It is often stated that Petrarch must have seen the Lucretius manuscript itself, but according to Passannante (*loc. cit*), he copied passages of Lucretius that had been quoted by Macrobius in his 5th century *Saturnalia.*

23. The city of Constantinople was founded in 330 CE by the Roman Emperor Constantine. The city became the seat of the Eastern Roman Empire in 395 CE, when the Empire split into two parts, until 1453 when it fell to the Ottomans. That period of just over 1,000 years is known as the Byzantine Empire. The taking of the city by the Crusaders in 1204 resulted in the division of the Byzantine Empire among the leading Crusaders. In a nutshell, the complex ensuing situation can be summarized as follows. Three-eighths of the Byzantine Empire came under control of the state of Venice and the rest divided into several states: one-quarter to the Latin Empire with Constantinople as its capital and the rest to the states of Nicaea, Trebizond and Epirus, all of which claimed the right to the Imperial Throne. This was a highly unstable situation that inevitably led to wars between the states when, in 1261, Nicea recaptured Constantinople and re-established the Byzantine Empire with Constantinople as its seat.

24. Hermes Trismegistus is supposed by some to be a contemporary of Moses and by others to be the Egyptian God Toth (Hermes being his Greek name). He is widely associated with astrology, numerology, alchemy and the occult and he is thought of as the ultimate source of Platonic Philosophy and the inventor of writing. However, that myth was laid to rest by Isaac Casaubon who in 1614 dated them as having been written in the 3rd or 4th century CE. That did nothing to stem the enthusiasm for these mystical, magical and mysterious writings: many of the founders of modern scientific thought cultivated Hermetic knowledge, notably Isaac Newton.

25. See Wouter Hanegraaf, *How Hermetic Was Renaissance Hermetism?* (Aries 2015) for a detailed discussion of the *Hermetica*. This article is also a critical repudiation of the influential thesis of Frances Yates, *Giordano Bruno and the Hermetic Tradition* (Routledge 1964) which asserted that the (flawed) Latin translation of the *Hermetica* by Ficino in 1471 sparked a Hermetic movement.

26. Robert Fludd (1574–1637) is referred to as the father of the Freemasonry. His academic interest was in "Natural Magic", the part of the occult that deals with the natural forces as discussed in alchemy and astrology. "Ceremonial Magic" is the part of occult studies that refers to the summoning of spirits.

27. While the economy of Italy was recovering during the period 1350–1450, it started to decline during the following 100 years with a fall in agricultural production and wages and a steady rise in prices. See P. Malanima, *The Italian Renaissance Economy* (1250–1600) 2008, in "Europe in the Late Middle Ages: Patterns of Economic Growth and Crisis" (see, for example, Figure 8 of his work showing wages, which is plotted on a logarithmic scale). The emergence of the crisis is often attributed to the onset of the the Little Ice Age that followed the Medieval Warm Period. That may have been a contributing factor, but the financing of the wars, in the face of a declining economy, during that period would also have been a major factor.

28. He probably died from what at the time was referred to as Gout. His father was known as Piero the Gouty. It is clear from medical records that the de Medici family suffered from bone and joint disease. A detailed medical perspective is at https://hekint.org/2019/03/11/the-gout-of-the-medici/.

29. 1494 was the start of the long series of Italian Wars, also known as the Habsburg–Valois Wars, that lasted until 1559. This marks the transition from the Medieval period to the Early Modern period of history.

30. The public burned the bank down. The Medici bank had been the biggest bank in Europe.

31. This kind of populist rhetoric has become a common phenomenon of the 21st century.

32. Leonardo's final resting place was the St. Hubert chapel of the Chateau d'Amboise on the river Loire. He was first buried in the

St. Florentine church on the Chateau grounds, but that church was destroyed by the French Revolution. What is said to have been remnants of his bones were re-entombed in 1863.

33. Albrecht Dürer (1471–1528) was very much influenced by the precision of the work of of the Flemish painter, Jan van Eyck (1390–1441). van Eyck was the inventor of oil painting and is the icon of the early Flemish Renaissance.

34. The Spanish Inquisition was formed in 1478. It had been instigated by King Ferdinand II of Aragon and Queen Isabella I of Castile and so had Royal powers to back its actions. It was a far more conservative and fundamentalist Inquisition than the later Roman Inquisition.

35. See L. Lebvre and H-J. Martin, *The Coming of the Book — the Impact of Printing 1450–1800*, published in English by NLB, 1976, and republished by Verso in 2010. p. 38. See also the later work of Elizabeth Eisenstein: *The Printing Press as an Agent of Change,* a two volume treatise on the birth of printing was released by CUP in a single volume in 1982. That and her later book on the subject, *The Printing Revolution in Early Modern Europe* (CUP 2012), raised much debate about the importance of movable-type printing as one of the drivers of the later Renaissance.

36. From Lebvre and Martin *loc. cit.*, p. 40. This is at best an informed guess since there was no listing of paper makers available from that period.

37. Copernicus wrote a first version of his ideas in 1514 in an essay called *Commentariolus.* The ideas were widely circulated in *Narratio Prima* (1540) by Georg Rheticus, who had stayed with Copernicus in 1539–41 and in 1542 carried a copy of the manuscript of *De Revolutionibus* to printer Petreius. Copernicus corrected proof sheets over some months, the last arriving, according to legend, the very day he died.

38. The outer inscription of the published map reads the following: *This orb of stars fixed infinitely up extends itself in altitude spherically, and therefore immovable the palace of felicity garnished with perpetual shining glorious lights innumerable, far excelling over [the] sun both in quantity and quality the very court of celestial angels, devoid of grief and replenished with perfect endless joy, the habitacle for the elect.*

39. By grafting endless space onto the Copernican system and scattering the stars throughout this endless space, Digges pioneered the idea of an unlimited universe filled with the mingling rays of countless stars. (See E. R. Harrison, *Masks of the Universe*, 2003 Cambridge U. Press, pp. 86 *et. seq.*) .
40. See J-P Luminet's blog: https://blogs.futura-sciences.com/e-lu minet/tag/nicholas-of-cusa/.
41. The general difficulty in thinking about the infinite is still the case. It is difficult to imagine a universe that is infinite in space. It is even more difficult to imagine a universe that is not infinite in time, hence the common question as to what happened "before" the Big Bang.
42. Cusanus, *De Docta Ignorantia* (On Learned Ignorance), Book II, #97 and #197, by which he means that the number of stars is inconceivable by mortals.
43. Newton, in his book *Opticks*, wrote the following: "There are therefore agents in nature able to make the particles of bodies stick together by strong attractions. And it is the business of experimental philosophy to find them out". The British cosmologist Edward Harrison commented on these prophetic words by Newton saying that nowadays we have high-energy particle accelerators for this purpose.
44. See, for example, Ada Palmer. (2014). *Reading Lucretius in the Renaissance*. Harvard, University Press.

Chapter 3

1. He was charged by the Venetian Inquisition in 1592 for religious heresy but not for supporting Copernicus or for stating that the Universe was infinite. The Heliocentric system was only declared a heresy in 1616 by the Inquisition. Prior to that, he was widely published and read. Bruno had traveled widely visiting academic institutions in France and Italy and spent two successful years in Oxford. See H. Gatti's. (2011). *Essays on Giordano Bruno*. Princeton, for more about the fascinating Giordano Bruno.
2. Bruno was far from being the only one condemned for heresy: the burning of heretics and hanging of witches throughout Europe continued throughout the next half century and beyond. During

the English Civil War, a puritan zealot named Matthew Hopkins described himself as the "Witch finder general" and is said to have executed between 200 and 400 women during the years 1644–1647 in the area around Manningtree — a small village 10 km northwest of Colchester in Essex.

3. The date of writing and the first performance of Shakespeare's plays are not always entirely certain. He wrote some 22 plays prior to 1600 and 16 plays during the first decade of the 17th century.

4. The current version of the King James Bible is the version edited by Benjamin Blayney and Francis Parris, published in 1769. That version corrected some 24,000 mainly typographical errors and resolved a number of ambiguities in the translation. The most popular English language version of the Bible of the 21st century is the *New International Version* (NIV), written in widely understood modern English. This was not a rewrite of the Authorized Version but a monumental international project based on all the original available sources. See https://en.wikipedia.org/wiki/Bible_translations_into_English.

5. The Siege of La Rochelle was the end of the Huguenot struggle against the Crown of Louis XIII. Previous to that, the French Wars of Religion between the Catholics and the Huguenots had raged from 1562 to 1598, a period during which it is estimated that 3 million people may have died from violence, famine and disease. Notably, in 1572, St. Bartholomew's Day Massacre resulted in the deaths of some 2,000–3,000 Huguenots in Paris. The killing of Huguenots spread throughout the country bringing the death toll to around 10,000. Some regard this number as being far too conservative, estimating numbers as high as 70,000. The massacre had been organized by the then King of France, Charles IX, a very weak and sickly monarch who was easily swayed by his advisors, among whom was his mother, Catherine de Medici. See P. Benedict's. (1978). The saint Bartholomew's massacres in the provinces. *Historical Journal*, 21.

6. It was no coincidence that the Civil War started with the end of the Thirty Years' War and the signing of the Treaty of Westphalia: the French Army was now suddenly available for other things.

7. The Battle of Cheriton took place in March 1644 near Alresford in Hampshire, southern England. It was the first time the Royalists were defeated in the south of England and left the Royalist plans in tatters. Three months later, the Battle of Marston Moor, near York in the north of England caused the Royalists to abandon their northern campaign entirely.

8. The second half of the 17th century in England saw the restitution of the Monarchy in 1652, but this was followed by three Anglo-Dutch wars and the invasion of England in 1688 by William of Orange. The King of England, James II, fled to France. His daughter Mary had been raised as protestants and had married William in 1677. She became Queen Mary II of England and by agreement reigned with William as the King. Mary had an elder brother, James, who would have been next in line to the throne after his father were he not a Catholic. James was exiled with his father and after his father's death in 1701 sought to reclaim the English throne.

9. See J. Grosslight's. Small skills, big networks: Marin Mersenne as mathematical intelligencer. *History of Science*, 51, 337. Grosslight casts Mersenne as having systematically "harvested a community" of mathematicians. That may well be, but Mersenne should also be judged in terms of his positive role as an "Influencer" rather than an "intelligencer" as the title of the article would suggest. He remarks that Mersenne's collected correspondence shows only 1896 items, of which 318 are letters that he wrote, though, it is recognized that collections of letters are inevitably dominated by those received. Sadly, the letters Mersenne wrote to Descartes are lost.

10. It is generally believed that Galileo never did the experiment since he was so sure of the result. It should be noted that Giovanni Battista Baliani (1582–1666) did perform Galileo's experiment at the Priamar Fortress at some time around 1611. He visited Galileo in 1615 and that started an intermittent correspondence between the two. Baliani wrote about his experiment in his 1638 publication *De Motu Naturali Gravium, Fluidorum et Solidorum* (*About the Motion of Bodies, Fluids and Solids*). In that, he was able to explain the important difference between *mass*

and *weight*. He also corresponded with Mersenne. For more details and references, see https://www.encyclopedia.com/scie nce/dictionaries-thesauruses-pictures-and-press-releases/baliani-giovanni-battista. Baliani was an amateur scientist and as such lived in the shadow of Galileo, one of the greatest scientists in history.

11. The experiment of dropping stones from towers probably goes back to Hipparchus (190–120 BCE) and is referred to by John Philoponus (490–570 CE) who was a Byzantine expert on the works of Aristotle and his contemporary Simplicius of Cilicia (490–560 CE). Simplicius was the first scientist we know of to use the term "kinetic energy".

12. See, for example, Richard Carrier's article *Ancient Theories of Gravity: What Was Lost?* at https://www.richardcarrier.info/ archives/14522.

13. This was the birth of the so-called Mind–Body Problem, the relationship between thought and consciousness. This is still an ongoing discussion. The philosopher John Searle in the 1980s gave the opinion that what we think of as "mind" is merely an aspect of our brain: there is no "problem" since there are different levels of description of the same thing. See https://en.wikipedi a.org/wiki/Mind%E2%80%93body_problem. Such a fundamental division between the philosophies of Descartes and Gassendi inevitably led to a degree of enmity between the two such that they could not be in the same room.

14. It is said that the writer and duellist Cyrano de Bergerac (1619–1655) was also a member of La Tétrade. He was an follower of the *Libertine* movement, a very popular movement of wealthy and influential people who chose to ignore the moral standards set by the wider society. Libertarianism was an extreme form of Hedonism that Gassendi explicitly tried to exclude in his revised version of Epicureanism. In that sense Gassendi stayed closer to Epicurus' vision of what a good life should be ('*ataraxia*'). Famous Libertines were Lord Byron, Casanova, the Marquis de Sade and Napoleon' older brother Joseph (according to Talleyrand).

15. Mersenne's Académie was the forerunner of the Royal Society and Académie des Sciences which were founded in 1660 and 1666 respectively.

16. When Anne became Regent in 1643, she seized considerable power by getting the Parlement to revoke the Will of her husband, Louis XIII. Cardinal Richelieu, the First Minister of State, had died at the end of 1642 and Anne put her strong ally Cardinal Jules Mazarin in his position. However, in 1648, a political insurrection led by the nobles with the Parlement against Queen Anne and Jules Mazarin forced Anne with her son Louis and Mazarin to flee Paris temporarily. This was a French Civil War aiming to restrict the powers of the monarchy. The war went through a number of phases, as had the ongoing English Civil War, but in the end, the King returned in 1652 and Mazarin was reinstalled the following year. Louis attained his majority in 1651 and Anne's Regency ended, but she nonetheless retained considerable power until the death of Mazarin in 1661. The outcome of the French Civil War was that the absolute power of the monarchy was re-established and reinforced.

17. Elizabeth Howard, the first wife of William Cavendish, the Earl of Newcastle, died in 1643 leaving him with five children. The First Civil War had started at the end of 1642 when William was in charge of defending the four Northern counties of England. The following year, he led his Royalist army to soundly defeat a Parliamentarians' army at the battle of Adwalton Moor in Yorkshire and he was promoted to the rank of Marquess. However, there was a reversal of fortune in 1644 when the Royalists were defeated in the battle of Marston Moor and surrendered the nearby city of York. The North of England fell into the hands of the Parliamentarians. In 1644, William Cavendish and his two sons fled England for Hamburg leaving behind all their estates and most of their money, and later, in 1645, they moved to Paris. On return to England following the restoration of the monarchy, William was made Duke of Newcastle and regained all of his estates. https://en.wikipedia.org/wiki/William_Cavendish,_1st_Duke_of_Newcastle.

18. This was detailed by William's new wife, Margaret Cavendish. According to her own accounting, William left England immediately after the battle with £100 in his pocket (£11,000 by today's standard, according to the National Archives of the UK). He

had been one of the wealthiest men in England, and when King Charles' funds were blocked by Parliament, he gave the King £10,000 (£1 million today) — which was an enormous sum, benefiting of a gift for a King with a profligate lifestyle. Over the period of exile, William's debt amounted to £941,303, a number calculated painstakingly by Margaret, a sum that would translate to a sum in the region of £110 million. See https://www.nationalarchives.gov.uk/currency-converter/.

19. See, for example, D. Norton's. (1999). *Writing the English Republic Poetry, Rhetoric and Politics 1627–1660.* Cambridge University Press.

20. As was then usual for the girls of wealthy families, Lucy did not go to school. In her *Memoir*, she says, "When I was about seven years of age, I remember I had at one time eight tutors in several qualities, languages, music, dancing, writing, and needlework, but my genius was quite averse from all but my book, ... every moment I could steal from my play I would employ in any book I could find, when my own were locked up from me."

21. There was an earlier anonymous translation into prose, dating from the early 1600s. That translation is now in the Bodleian Library, Oxford (MS. Rawl. D. 314). Lucy Hutchinson's translation was into what are known as "Heroic Couplets", a traditional form of English epic and narrative poetry consisting of pairs of rhyming lines having a particular rhythm. A translation by diarist John Evelyn of the first of the six books appeared in 1656.

22. The translation was not published until 1996 when it appeared as *Lucy Hutchinson's Translation of Lucretius: De Rerum Natura.* Duckworth (1996), an important volume edited by Hugo de Quehen of the University of Toronto.

23. The Earl of Anglesey, a friend of the Hutchinsons, intervened in the arrest of John Milton on the restoration of Charles II in 1660.

24. Marchetti (1633–1714) was not allowed to publish his translation, though the manuscript was copied and distributed informally. It was published posthumously in 1717.

25. Germaine Greer, *Horror like Thunder*, London Review of Books Vol. 23 No. 12 (21 June 2001), a brilliant review of David Norbrook's edition of Lucy Hutchinson's *Order and Disorder* (Blackwell 2001).

26. Some modern critics place Lucy in Margaret's social circle, but there is no evidence for this, and it is unlikely that the two women ever met.
27. T. Gray. (1751). *Elegy Written in a Country Churchyard.*
28. *Lucy Hutchinson's Translation of Lucretius*, p. 23.

Chapter 4

1. Fire was first used more than a million years ago by our ancestral species *Homo erectus*. Claims for the earliest evidence of use range from 1.7 to 2 Mya.
2. It is significant that this second phase of the scientific revolution also overlaps with what is dubbed the Age of Enlightenment in which all ideas could and should be freely questioned, even the monolithic authority of the Catholic Church.
3. The experiment was performed by Viviani. Viviani had been an assistant to Galileo during the period 1639–1642 and was familiar with the process of constructing experimental equipment. It seems that Viviani's role in all this has been forgotten with the passing of time. He is rarely cited today.
4. Even today, this occasions some surprise, as it did in the 17th century. The central fact in coming to terms with this is the important quantity is the *force per unit area* (which we call the *pressure*) exerted on the surface of the bowl of the mercury. This is the same regardless of the area.
5. *Noi viviamo sommersi nel fondo d'un pelago d'aria elementare.*
6. The exact date of the experiment is not well known. Some authors claim that it could be done in 1644 just before Torricelli sent the letter to Ricci. W. E. Knowles Middleton. (1963). The place of Torricelli in the history of the barometer. *Isis*, 54(1), 11–28. http://www.jstor.org/stable/228726.
7. There has been debate about whether Mersenne in fact met Torricelli when on his visit to Italy. Mersenne describes the meeting in own writings (M. Mersenne. (1647). *Novarum Observatorum Physico-Matheticarum.* Paris: Bertier). See also W. E. Knowles Middleton. (1964). *The History of the Barometer.* Baltimore: Johns Hopkins University Press and J. B. Shank (2012). *What Exactly Was Torricelli's "Barometer?"* in Ch. 7

of Ofer Gal & Raz Chen-Morris (eds.), *Science in the Age of Baroque.*

8. It was Périer who knew and recommended a suitably skilled glass blower in Rouen to make the tubes.

9. In 1662, George Sinclair, a professor of mathematics at the University of Glasgow who was apparently unaware of earlier work, obtained barometric data at Tinto Hill near Glasgow and from that deduced that the height of the atmosphere was about 11 kilometers. Sinclair is known more for his eccentric views and his book on witchcraft, *Satan's Invisible World Discovered* (c. 1685).

10. There are no surviving verifiable portraits of Hooke. See S. Inwood. (2002). *The Man Who New Too Much.* Macmillan for an excellent detailed biography of his life, though telling relatively little of his science.

11. Westminster School, which today still ranks as one of the top three academically selective private schools on many measures of ranking.

12. There had been a general exodus of academics from London during the preceding years. The successful Siege of Oxford in June 1646 by Oliver Cromwell's Parliamentarian army had put the Parliamentarians in control of the City and the strongly Royalist University. A month earlier, King Charles I had surrendered to the Scots who handed him over to the Parliamentarians, which ultimately lead to his imprisonment and execution. In 1650, Oliver Cromwell became Chancellor of Oxford University and culled the remaining academics with Royalist sympathies, replacing them with his supporters. This left numerous open positions at the university that were eagerly taken up by those with Parliamentarian sympathies.

13. On the western Mediterranean coast of modern Turkey, now the city of Izmir. In the 17th century, Smyrna was a great trading center deriving much of its great wealth from trading in Iranian Silk. Smyrna was the site of the massacre, for two weeks in 1922 when the city was set on fire resulting in the deaths of an estimated 100,000 Armenians and Greeks living in the city. http://www.genocide-museum.am/eng/online_exhibition_16.php.

14. Both the London Stock Exchange and Lloyds of London grew from coffee houses: a measure of their importance.

15. Chocolate was very expensive, which was the factor discrimi-
    nating those who frequented the chocolate and coffee houses.
    The first chocolate house in London appeared in 1652 offering
    "chocolate at reasonable rates", but it was in 1657 that choco-
    late first appeared as a luxury item for the wealthy and titled.
    See https://www.atlasobscura.com/articles/history-of-gentleme
    ns-clubs. Their wealthy clientele turned them into dens of iniq-
    uity, depicted for posterity by the artist William Hogarth in his
    1732–1734 series of eight paintings, *A Rake's Progress*, released
    as engravings in 1735. About the same time, Hogarth produced
    another series of engravings, *A Harlot's Progress*, the originals
    of which no longer exist.
16. Maintaining constant temperature was essential since there was
    no calibrated temperature scale at the time. Allowing tempera-
    ture to vary would spoil his data by adding in an unknown and
    uncontrollable variable. Boyle's Law was rediscovered indepen-
    dently by French Physicist Edme Mariotte in 1676.
17. There were several other scientists developing temperature
    scales. Rømer proposed a temperature scale, which was cali-
    brated so that water froze at 7.5° and boiled at 60°, the zero
    point being the freezing point of brine. Fahrenheit set his zero
    point as the freezing temperature of a solution of salt and ammo-
    nium chloride in water and the upper temperature at the tem-
    perature of the human body (originally thought to be 90F and
    then changed to 96F). After 1743, the year of Celsius' death,
    the Celsius scale referred to 0C for the freezing point of water
    and 100C for the boiling point (prior to that date, the numbers
    were reversed!). Today, all of these scales have more precise def-
    initions. The Celsius scale is now used throughout the world,
    with the exception of the United States of America where the
    Fahrenheit scale is still used, except in scientific work.
18. This was a variant of Philo's 3rd century BCE experiment with
    combustion. Mayow's writings were widely regarded as pretty
    incomprehensible. It is not clear that he understood what his
    nitroaerus was or that the naming was particularly appropriate.
    Like Hooke, he had performed experiments on the effect of gases
    on the blood and that the nitroaerus had an effect on the color.
    He preceded Joseph Priestley (1733–1804) and Antonie Lavoisier
    (1743–1794) in discovering the oxygen in the air by a century

and was the first to show that the purpose of breathing was not simply to "cool the heart" but rather to oxygenate the body for heat production and muscular activity.

19. In 1773, in a repeat of Mayow's experiment a 100 years before, Swedish scientist Carl Wilhelm Scheele tested combustion and respiration within an enclosed jar and got the same results as his predecessor. Scheele called the combustible gas "fire-air" and the remnant "foul-air". Scheele went on to discover chlorine and many other chemical substances. It is remarkable that Scheele, a century later, was still working within the framework of the phlogiston concept. His work was a vital step towards the eventual abandonment of Phlogiston.

20. Russo (*loc. cit.,* p. 76).

21. F. J. Moore. (1918). *History of Chemistry.* McGraw Hill, p. 30.

22. Although the strong supporters of the phlogiston theory (Joseph Priestley was one of them) nevertheless adhered to their beliefs until they died. As Max Planck (1858–1947) later said, "A new scientific truth does not triumph by convincing its opponents and making them see the light but rather because its opponents eventually die and a new generation grows up that is familiar with it." M. Planck. (1950). *Scientific Autobiography.* p. 33. A thought that is often summarized as "Science progresses one funeral at a time."

23. It was in fact merely a description of our perceptual experience of heat.

24. A. Lavoisier. (1783). *Réflexions sur le Phlogistique.*

25. The successes of the wave theory of light through the experiments of Thomas Young and Augustin-Jean Fresnel in the early 19th century generated speculation in a wave theory of heat. It was realized that heat had no weight and that it would seem possible to generate as much heat (caloric) as wanted through the action of friction, a mechanical process. Herschel had performed experiments on heat radiation and concluded that radiant heat behaved like light. It was a short leap of the imagination to view heat as a wave phenomenon: we only had to think of caloric as the ether and heat is then manifested through vibrations in the ether. Young himself had proposed this in his 1807 *Lectures on Natural Philosophy.* Strangely, although the wave theory of heat was rather popular in the 1840s, it simply faded away. See S.

Brush. (1970). The wave theory of heat. *British Society for the History of Science*, 5.

26. In that sense, Newton's work was less accessible than the theory developed by Stahl for Phlogiston. This is possibly a reason why the Phlogiston theory survived as long as it did: it was simple to grasp. There was at the time a further factor inhibiting the adoption of Newton's view: he paid considerable attention to the alchemy side of his model, by providing explanations as to why it might be possible to transmute one element into another. See B. J. T. Dobbs. (1983). *The Foundations of Newton's Alchemy.* CUP, Ch. 6 *Newton's Integration of Alchemy and Mechanism.*

27. Newton's *Opticks* was first published in 1704 and revised in 1717 when Newton added substantially to his discussions about Chemistry. The discussions of his atomic theory are in the last part of the *Opticks*, the Queries. Section 31 of the Queries is a 33-page treatise on atomism. There Newton describes a hierarchy of types of matter based on the smallest units, "indivisible by any power in Nature": the atoms. See A. Thakray. (1970). *Atoms and Powers: An Essay on Newtonian Matter-Theory and the Development of Chemistry.* Harvard (Ch. 2).

28. The names of the referees were revealed by the Royal Society in 1965: the Reverend Baden Powell who was the Savilian Professor of Geometry at Oxford and father of the Baden Powell who founded the Scout movement, and Sir John Lubbock, a banker and barrister and dilettante mathematician and astronomer, noted particularly for his work on the tides. See C. Truesdell. (1984). *An Idiot's Fugitive Essays on Science.* Springer. The story of Herepath, Waterston and others is told in detail in Chapter 33 (p. 380) of this mammoth 661-page somewhat strange collection of essays. The story of Daniel Bernoulli introduces Chapter 1 and has a 3-page chapter (Ch. 25, p. 209) to himself in a book review. This is a principal source for some of the material presented here.

Chapter 5

1. Aristotle may have met Eudoxus in Athens, and for sure he knew Eudoxus' theories of the planetary motions first-hand. See Cheng, Wei. Aristotle and Eudoxus on the Argument from

Contraries. *Archiv für Geschichte der Philosophie*, vol. 102, no. 4, 2020, pp. 588–618.

2. As stated by the historian of science Edward Grant "Whatever Aristotle may have thought about the properties of the celestial ether, there is no doubt that in *De caelo* he assumed the corporeality, and therefore the physicality, of the heavenly orbs". Grant, Edward. Celestial Orbs in the Latin Middle Ages. *Isis*, vol. 78, no. 2, 1987, pp. 153–173.

3. Here, the image is that of transparent glass or crystalline globes, Grant *loc. cit.* p. 153. The historian of science Noel M. Swerdlow (1941–2021) claimed that in Georg Peurbach's *Theoricae Novae Planetarum*, the solid spheres were assumed to be hard and rigid. See Swerdlow, Noel M. (1976). Pseudodoxia Copernicana: or, enquiries into the very many received tenets and commonly received truths, mostly concerning spheres. *Archives Internationales d'Histoire des Sciences*, xxvi, 108–158. Grant (*loc. cit.* p. 173) claims, however, that he has examined Peurbach's treatise and "have found only the usual silence on the issue of hardness or softness". Víctor Navarro Brotons and Enrique Rodríguez Galdeano refer to Michel-Pierre Lerner and his comprehensive history of the notion of the celestial spheres for clarifying these issues. See Navarro Brotons, V. and Rodríguez Galdeano, E. (1998). *Matemáticas, Cosmología y Humanismo en la España del Siglo XVI. Los Comentarios al Segundo Libro de la Historia Natural de Plinio de Jerónimo Muñoz.* CSIC-UV – Instituto de Historia de la Medicina y de la Ciencia López Piñero (IHMC). Valencia. See, also, Lerner, M.-P. (1996–1997). *Le monde des sphères.* Vols. 1 and 2.

4. This claim was used by Rocky Kolb for the title of his excellent book on the progress of our knowledge about the universe, published in 1996, *Blind Watchers of the Sky: The People and Ideas that Shaped our View of the Universe*, Helix Books.

5. The apparent retrograde motion is the motion of a planet opposite to the direction of the apparent motion of the stars revolving the Earth each 24 hours.

6. For example, Aristarchus considered the angular size of the Sun and Moon as seen from Earth to be 2°, when in fact each is half a degree.

7. Concerning elliptical orbits, Kepler's Second Law states that the speed of a planet along its orbit, with the Sun in one of the foci of the ellipse, is greater when the planet is closer to the Sun (perihelion) than when the planet is further away (aphelion).

8. In fact, as Owen Gingerich demonstrated with computational tools of the 1970s, both the Alfonsine Tables and those of Johann Stoeffler's *Ephemeridum* opus of 1532 were purely Ptolemaic, with no added entities. "Crisis" versus aesthetic in the Copernican revolution. *Vistas in Astronomy,* 17(1), 85–95.

9. Bertrand Russell in his *History of Western Philosophy*, pp. 427–433, describes the scholastic movement in some detail, referring to the scholastic period as the time during which Aristotle overtook Plato as a supreme authority. Russell goes on to say that early scholasticism may be viewed, politically, as an offshoot of the Church's struggle for power (*loc. cit.*, p. 433.)

10. Adelard of Bath (c.1080–c.1142) has long been confused with Peter Abelard (c.1079–1142) — similar names and almost identical dates. Peter Abelard was a polymath and a renowned a teacher who was widely regarded as the most important Catholic philosopher of his age. At some time around 1115, he became teacher at the cathedral school of Notre Dame. This was when he met the young intellectual Héloïse d'Argenteuil, who had already developed a reputation as a literary prodigy. An amorous relationship ensued and she became pregnant with a boy whom she named "Astrolabe". Héloïse's uncle, Canon Fulbert, disapproved of the relationship and had Abelard castrated. Part of the tragedy was that Abelard and Héloïse had decided to get married in secret. Their story became one of the world's classic love stories.

11. A. C. Crombie. (1959). *The History of Science from Augustine to Galileo.* Courier Corporation.

12. Lindberg, *loc. cit.*, Chapter 11.

13. T. C. B. McLeish *et al.* (2014). History: A medieval multiverse. *Nature*, 507, 161–163.

14. See H. E. Smithson *et al.* (2012). A three-dimensional color space from the 13th century. *Journal of the Optical Society of America A*, 29, A346–A352.

15. R. G. Bower *et al.* (2014). A medieval multiverse?: Mathematical modelling of the thirteenth century universe of Robert

Grosseteste. *Proceedings of the Royal Society A*, 470, 20140025–20140025.

16. N. M. Swerdlow. (2010). Book review of Thomas Hockey (Editor). *The Biographical Encyclopedia of Astronomers. Isis*, 101(1), 197–198.

17. The best-known astronomical instrument used during the Islamic Golden Age is the astrolabe, a device able to measure the altitude above the horizon of a celestial body. It can also be used to measure the longitude of the observer, given the local time (and vice versa) and to trace out the orbits of the planets.

18. The ecliptic is the plane of the Earth's orbit around the Sun, The term ecliptic refers to the apparent path of the Sun throughout a year. The inclination or obliquity of the ecliptic is the angle between the ecliptic and the celestial equator (the projection of the Earth's equator onto the imaginary celestial sphere). Its present value is 23.5°.

19. The celestial equator crosses the ecliptic in two points: the equinoxes. The Earth's rotational axis traces a cone around the poles of the ecliptic with a period of about 26,000 years in a movement called precession. As a consequence, the equinoxes move along the ecliptic with the same period: 1.38° per century.

20. The Tusi Couple is a special case of the geometric construct called a *Roulette*: "A roulette is the curve generated by a point which is carried by a curve which rolls on a fixed curve". See https:// en.wikipedia.org/wiki/Tusi_couple for an animated example and https://en.wikipedia.org/wiki/Roulette_(curve) for a discussion of Roulettes in general.

21. In one of the books written by al-Tusi we can read: "The organisms that can gain the new features faster are more variable. As a result, they gain advantages over other creatures. [...] The bodies are changing as a result of the internal and external interactions."

22. Theories of the oscillation of the precession of the equinoxes (called trepidation) were ruled out only at the 19th century.

23. See C. Kren. (1971). The rolling device of Nasir al-Din al-Tūsī in the De spera of Nicole Oresme? *Isis*, 62, 490–498, for a detailed discussion of Oresme's garbled description of the Tusi Couple.

24. The British historian of science Michael Hoskin and his coauthors (including Owen Gingerich on Muslim astronomy and Copernicus) were sure that Copernicus knew the work of al-Battani

(c. 850–929 CE) because it is mentioned 23 times in *De Revolutionibus* in connection with the improvement of the orbit of the Sun relative to that of Ptolemy. But they leave open the question of whether he independently discovered the Tusi couple as a way of replacing the equant (so as to get nearly uniform circular motion) or was aware of the Arabic works in Greek translation (available in Italy at the time Copernicus was there). Our private prejudice is in favor of his having known something of the Muslim work, a prejudice which perhaps derives from the Rabbinic custom of attributing at least the genesis of one's best ideas to one's teachers.

25. S. P. Blake (2016). *Astronomy and Astrology in the Islamic World*. Edinburgh University Press, p. 216.

26. The historian of science Owen Gingerich notes that the Prutenic tables showed a "notable lack of commitment" to heliocentricity. See Gingerich. (1973). *The Role of Erasmus Reinhold and the Prutenic Tables in the Dissemination of the Copernican Theory*. Poland: Studia Copernicana.

Chapter 6

1. O. Gingerich (June 2011). Galileo, the impact of the telescope, and the birth of modern astronomy. *Proceedings of the American Philosophical Society*, 155(2), 134–141.

2. Galileo Galilei published a short treatise in Latin entitled *Sidereus Nuncius (Starry Messenger)* in March 1610, a few months after he started using a telescope for his astronomical observations.

3. The names of the characters in the plot are intentional: Salviati, a brilliant sage, comes from Galileo's friend Filippo Salviati (1582–1614). Both were members of the *Accademia dei Lincei*, literally "Academy of the Lynx-Eyed". In the book, he is referred as the "Academician". Giovanni Francesco Sagredo (1571–1620), a Venetian mathematician and also Galileo's friend, provides the name for the curious, apparently neutral man. Simplicio is inspired by Simplicius of Cilicia (c. 490–c. 560 BCE), a commentator on Aristotle, but it is clear the double meaning, since *semplice* in Italian means simple.

4. For example, the drawing without explanation by the English monk and chronicler John of Worcester (died c. 1140) produced on 8 December 1128 has come down to us in the files of Corpus Christi College, Oxford. See A. van Helden. (September 1996). Galileo and Scheiner on sunspots: A case study in the visual language of astronomy. *In the Proceedings of the American Philosophical Society*, 140(3), 358–396. https://www.jstor.org/stable/987314.

5. Christoph Scheiner published his results in a book entitled *Rosa Ursina* (*The Rose of Orsini*). This treatise on sunspots was very influential for more than a century.

6. *Istoria e Dimostrazioni intorno alle Macchie Solari* (*Letters on Sunspots*) was published in Rome by the *Accademia dei Lincei*.

7. Different parts of the Sun rotate at different angular velocities, depending on the latitude: faster at the equator, slower near the poles.

8. The magnitude system used in astronomy to measure the apparent brightness of stars has its origin in the stellar classification carried out by Hipparchus of Nicaea in the 2nd century BCE. Unfortunately, Hipparchus's catalog has not survived. But almost three centuries after him, Ptolemy, crediting Hipparchus, divided the stars into six classes based on their apparent brightness. The human eye reacts to brightness logarithmically, so in reality, the stellar classification of Hipparchus and Ptolemy depends on the logarithm of brightness. First-magnitude stars (a term introduced by Ptolemy) are approximately twice as bright as second-magnitude stars, which are twice as bright as third-magnitude stars, and so on, until the sixth magnitude, which corresponds to the faintest stars observable by most people with the naked eye in a reasonable dark sky. In the 19th century, the English astronomer Norman Robert Pogson (1829–1891) established the modern system of magnitudes, in an attempt to maintain approximately the classification devised by the astronomers of the Hellenistic period (in the modern Pogson system, a first-magnitude star is 2.512 times brighter than a second-magnitude star). Note that a higher brightness corresponds to a lower magnitude on the scale and 5 magnitudes to a factor of 100 in brightness.

9. A light curve of a star is a graph that represents the variations of a star's brightness as a function of time. These variations can be the result of eclipses or transits caused by objects that pass between the stellar disc and the observer, as in the case of eclipsing binary stars, where a companion causes the brightness of the star to diminish as it passes in front of it. Exoplanet transits are another example of a slight decrease in brightness that can be detected in the host star's light curve. Variability may also be caused by stars' internal pulsations. This is the case of Cepheid variable stars, which owe their name to Delta Cephei, first studied by John Goodricke. Both the surface temperature and the radius vary periodically.

10. The dimming of the light of a star due to the transit of a planet in front of the star disc has been one of the most successful methods to detect exoplanets.

11. RV Tauri is the prototype of this kind of variable stars. The characteristic variability of this type of star shows alternating deep and shallow minima in their light curves. RV Tauri was discovered to be variable in 1905 by the Russian astronomer Lidiya Tseraskaya (1855–1931).

12. In 1950, the Dutch astronomer Jan Oort (1900–1992) proposed the existence of an immense cloud, made up of billions of icy bodies, that orbit the Sun at a distance between five thousand and fifty thousand astronomical units (an astronomical unit is roughly the average distance between the Earth and the Sun). It is known as the Oort cloud. Its total mass may be similar to the mass of five Earths. When a star or other object passes through the vicinity of the cloud, it can cause a gravitational disturbance that causes some of these ice balls to fall towards the center of the Solar System. They can become long-period comets.

Chapter 7

1. If you are standing still relative to a distant source, the light hits you right on top your head (or on your face, if you happen to be looking up). The same is true of rain if you are standing still on a windless day. If you need to walk forward into the rain, then you will stay drier if you tilt your umbrella

a bit forward. Once again, the same is true of light, if you are moving relative to the source, then you need to tip your telescope a bit forward so that the light rays can go straight down the tube. The angle by which you need to tile your telescope forward is your speed divided by the speed of light (in angular units). Bradley, studying a nearby star, discovered he needed to tilt his telescope about $20''$ forward and back through the year to keep the photons streaming down the telescope tube.

2. We aren't quite sure Brownian motion was discovered by the botanist Robert Brown (1773–1858), but it can be rediscovered by anybody with a microscope, some stray pollen, and a fluid to float it in. The pollen grains will be seen to dance about, and this was then claimed as evidence for individual atoms or molecules moving in the fluid and knocking the grains around. More recent observers, pencils and envelope backs in hand, have concluded that the individual particle knocks aren't really powerful enough and that collective fluid motion must be responsible.

3. Many thinkers have attempted to find ways around the paradox, in the process sometimes invoking some of our favorite dragons. Here is a list, more or less chronological, with names of the thinkers, years of their suggestions, and a hint of why each doesn't solve the problem:

- Starry universe not infinite (Kepler 1610). True our galaxy is finite, but other galaxies extend as far as we can see, and surely beyond.
- Universe not fully transparent, because there are holes in the ether, through which light cannot propagate (de Cheseaux 1744). Well, we killed of the ether in Chapter 1, though black holes of appropriate size could act the same way.
- Universe not homogeneous, but fractal (Immanuel Kant 1755), that is, with each larger structure and volume you consider, the average density goes down, so that sum is finite. Modern observations show that this is correct only out to structures the sizes of superclusters of galaxies and voids.

- Universe not fully transparent because of absorption by interstellar matter (Olbers himself, 1823). The catch is, with modern notions of conservation of energy, the absorbing material heats up until it also radiates as much light as it absorbs.
- Universe or contents not infinitely old, so that we see only those stars close enough for their light to have reached us (Poe 1848, Mädler 1860). True, although in an expanding, Friedmann-Robertson-Walker-Lemaître universe we see redshifted photons from very close to the beginning.
- New physics, such that photons simply lose energy on their long journeys, called "tired light" (Zwicky 1929). No evidence for such additional physics, and some against it.
- Universe not static, though otherwise as described (Bondi, Gold, and Hoyle 1948). This classic Steady State cosmology does indeed reduce the light of the night sky from infinity, but has been ruled out on other grounds (see Chapter 17). This was the topic of the announced early 1960s lecture by Raymond Arthur Lyttleton, from whom we heard the story.

4. Indeed, the ideas about an infinite universe had emerged in ancient Greece with Democritus and Epicurus, while others, such as Zeno of Citium, had proposed a finite cosmos of stars surrounded by an infinite void.
5. E. Halley. (1720). *Philosophical Transactions,* 31, 23.
6. J.-P. Loys de Cheseaux. (1744). Traité de la Comete, Bousquet et Cie. *Lausanne et Geneve*, 223–229, Appendix II.
7. There is a nice discussion of Olbers' Paradox in H. Kragh. (2007). *Conceptions of the Cosmos. From Myths to the Accelerating Universe.* Oxford University Press, pp. 83–86.
8. The first reliable measurement of a stellar distance was due to Friedrich Bessel in 1838 when he published the parallax of the star 61 Cygni and obtained a distance of about 10 light-years. Friedrich Georg Wilhelm von Struve (1793–1894) and Thomas Henderson (1798–1844) did similar measurements about the same time with Vega and Alpha Centauri, respectively. Olbers supposed that, due to absorption, we might lose one light ray in 800 in the average distance between stars. If this average distance is about one parsec (on the assumption that stars are about as bright as the Sun — no parallaxes were then known), this

amounts to something like 1.3 magnitudes of extinction per kiloparsec. This was quite close to the later estimate by Struve who in 1847 discovered evidence for interstellar obscuration not very different from the modern value (beginning with that published by the Swiss–American astronomer Robert Trumpler in 1930) of about a magnitude per kiloparsec in the galactic plane. The Swedish astronomer Carl Schalen (1902–1993) published a similar number in 1929 from completely independent considerations, but by long-standing custom receives no credit. See M. Hoskin (ed.) (1997). *The Cambridge Illustrated History of Astronomy.* Cambridge University Press, p. 225.

9. It was suggested that the medium would continue to receive heat from the stars and so the temperature of the medium would continue to rise without limit. These were the early days of thermodynamics and the concept of conservation of energy was a long way off in the future.

10. John Herschel's catalog was the go-to source for positions of known objects until it was superseded in 1888 by John Dreyer's *New General Catalogue*, "NGC", included star clusters, galaxies and other diffuse objects. It contains 7.840 entries.

11. Pecker wrote about the need to consider *les solutions alternatives, mais partielles* (alternative but partial solutions) and was a signatory, along with 33 others, of an open letter in 2004 expressing concern over scientific bias in this respect. The idea still holds sway in the 21st century, though not in most of the academic community. Not everyone is happy with the thought that our universe sprung into existence 13.8 billion years ago.

12. Lord Kelvin. (1901). On Ether and gravitational matter through infinite space. *Philosophical Magazine and Journal of Science*, 2, 161–177.

13. The night sky is indeed not completely dark in any wavelength of electromagnetic radiation. In particular, visible light from the stars of our own Galaxy, directly seen plus scattered, adds up to a measured energy density of about 1 eV per cubic centimeter, about the same as the local energy densities in magnetic fields, cosmic rays, and turbulence of the instellar gas. The visible brightness is the equivalent of about one fifth magnitude star per circle of 1° radius in the sky, or about 600 stars of magnitude 0 over the whole sky, as reported in 1901 by Simon

Newcomb, who had been observing from the Cape Breton island home of his wealthy acquaintance Alexander Graham Bell. The energy density of starlight is, of course, much smaller between the galaxies, and the same can be said of the energy density, or brightness, in X-rays, gamma rays, and so forth. There is, however, a sky brightness at radio and millimeter wavelengths whose energy density throughout the universe, at age 13.8 billion years, is also about 1 eV per cubic centimeter. And it was much larger in the past.

14. Peebles' first book on cosmology was entitled *Physical Cosmology*, published in 1971 by Princeton University Press. This was an epochal book of 282 pages setting out the future of cosmology as a branch of physics.

15. P. J. E. Peebles. (2020). *Cosmology's Century: An Inside History of Our Modern Understanding of the Universe.* Princeton University Press.

16. The Local Group is a small group of galaxies dominated by the Milky Way — our own galaxy — and Andromeda (M31), two large spiral galaxies. The third in size and luminosity is the Triangulum Galaxy (M33), another spiral galaxy, though this one is of more moderate size. The rest of the galaxies that make up the Local Group are much smaller, irregular or elliptical dwarfs. Most of them are satellites of large galaxies. The Magellanic Clouds, for example, are satellites of the Milky Way.

17. Peebles, *loc. cit.*

18. Smoot *et al.* (1992). *Astrophysical Journal Letters,* 396, L1. George F. Smoot and John C. Mather received the 2006 Nobel Prize in Physics "for their discovery of the blackbody form and anisotropy of the cosmic microwave background radiation."

Chapter 8

1. Nicolas Steno: Dissertationis prodromus, 1669.

2. Mary's father Richard, who had tuberculosis, died from a fall on the cliffs in 1810, aged 44. His death left the family seriously in debt with little possibility of paying it off. The announcement of the Ichthyosaurus discovery 2 years later was to solve the debt problem. Later, Mary herself suffered a serious fall in which her dog was killed.

3. The Jurassic Coast comprises some 150 km of coastline and is, since 2001, a World heritage site. The cliffs span almost 200 million years of geological history spanning the Triassic, Jurassic and Cretaceous eras.

4. Mary herself uncovered and cleared the entire skeleton, but it was necessary to hire workmen to dig out the block of stone containing the fossil from the surrounding rock face.

5. The name could not be made "official" until papers had been published formally describing it, but scientists nonetheless used it. The biological order "Ichthyosauria" was created in 1835.

6. It has been suggested that she had previously found another fine specimen in 1821, but there is no record of this in any of Mary's meticulous records. It is speculated that she might have sold this to Thomas Birch, but like all others, Birch never cited the provenance of his samples.

7. Mary's professional client and friend, William Buckland, was the one who coined the name "coprolite". It was he who presented the work to the Geological Society, crediting her where appropriate.

8. *The Life and Letters of the Reverend Adam Sedgwick.*

Chapter 9

1. Africa, Antarctica, Asia, Australia/Oceania, Europe, North America and South America. Sometimes, Europe and Asia are referred to as a single continent.

2. The Earth's crust, or *Lithosphere*, is a thin surface layer of rock broken up into a dozen main tectonic plates together with half a dozen smaller plates. The plates are brittle, thin and curved so as to match the curvature of the Earth's surface, and they completely cover the surface. Typically, the ocean parts of the plates are around 100 km thick while the continental parts are around twice that much. The plates are constantly moving and changing in thickness.

   Below the lithosphere is a thick region of almost solid rock, called *the Mantle*. It extends from the base of the Lithosphere to a depth of some 2,900 km where the temperature is around 4,000 K.

   Magma is hot mantle material that melts when it rises from regions with pressure about 25,000 times that of our atmosphere towards the Earth's surface. It is not surprising that magma behaves in ways quite unlike any liquid or solid that we normally experience. In simple terms, we can think of it as a solid with liquid-like properties behaving as a slow-moving viscous (thick and sticky) treacle-like layer of solid rock extending some 100 km down below the base of the lithosphere. The continents are not floating on fluid magma, but rather in hydrostatic equilibrium with the mostly-solid mantle, sticking up because they are less dense than the mantle material and thicker than the oceanic crust.

3. Most of the rocky planets and the larger satellites of planets have mantles, but the Earth is the only place in the Solar System where plate tectonics is known. It seems that Mars in the past may have gone through an early period of tectonic activity which would have ceased when the planet cooled. There is recent evidence suggesting that Venus might currently experience some form of tectonic activity. Some 4.2 billion years ago, our Moon underwent a brief period of volcanic activity. Ongoing "Moonquakes" have been discovered through instruments left on the Moon by the Apollo astronauts. Roughly half of the moonquakes originate between 600 and 1,000 km below the surface and correlate with the lunar tides. The other half originate some 20–30 km below the surface and are thought to be due to the continued cooling and associated wrinkling of the surface shell. Jupiter's satellite, Europa, is the only other Solar System body that might currently experience tectonic activity.

4. Alfred Wegener was a pioneer polar researcher and renowned meteorologist with a strong interest in astronomy. In 1906–1908, he participated in a Danish-led expedition to the uncharted northeast of Greenland. Three people died in the expedition, including the leader, Ludvig Mylius-Erichsen. Wegener's main task was the mapping of the coastline.

5. "Not Invented Here" syndrome.

6. The origins of that anti-German sentiment are found in the mid-19th century with the unification of Germany. The ill-feeling persisted up to and through World War II and into the 1950s.

7. The story of Alfred Wegener and the controversy over continental drift is told in detail in the seminal four-volume opus of Henry

Frankel, *The Continental Drift Controversy*, published in 2012 by Cambridge University Press.

8. *The Alfonsine Tables for the Use of Modern Calculators.* Much of the work concerned converting the base-60 numbers of the Arabic text to our familiar base-10 (Frenkel, *loc. cit.*, Vol. 1, p. 45).

9. Baron Alexander von Humboldt noted the current in 1846 in his book *Cosmos: A Sketch of a Physical Description of the Universe.* The current is a cold water low-salinity current that flows northwards along the west coast of South America. It is a highly productive ecosystem that is responsible for maintaining the biodiversity of the Galapagos Islands.

10. The age of the dinosaurs lasted from 252 Myr ago to 66 Myr ago when they became extinct.

11. *Die Entstehung der Kontinente und Ozeane.* Several editions were published in the early 1920s, the first English language translation appearing in 1924 was a translation of the much-revised third German edition of 1922.

12. The quotation is due to the American geologist Thomas Chamberlin who attributed it to an unnamed participant.

13. Antony Hallam has written about the history and controversy of continental drift in his books *Great Geological Controversies.* 2nd edn. Oxford, ch. 6, p. 135 (1989) and *Catastrophes and Lesser Calamities.* Oxford (2004), and in his *Scientific American* article Alfred Wegener and the hypothesis of continental drift. *Scientific American*, 232, 88–97 (1975).

14. See note 9.

15. T. Dick. (1774–1857). *Celestial Scenery; or, The Wonders of the Planetary System Displayed*, Harper, 1838. https://archive.org/details/celestialscenery00dick/page/109/mode/1up.

16. A. Snider-Pellegrini. *La Création et ses Mystères Dévoilés (The Creation and Its Mysteries Unveiled).* Ch. XXIX, p. 304 *et seq.* The before and after views are presented at the end of the chapter as two separate plates, 9 and 10. The "Proofs" of his conjecture are described in the following chapter, XXX.

17. Edward Suess was born in London. His family emigrated to Prague in 1834 and later, in 1845, moved to Vienna. In relation to the work of Suess and Taylor, see *On the Shoulders of Giants: Early Drift Theorists*, Chapter 5 of Suryakanthie Chetty's book

Africa Forms the Key — Alex Du Toit and the History of Continental Drift, Palgrave Macmillan.

18. Today, we refer to Angaraland as "Laurasia". This land mass corresponds to what is now Siberia and other old, northern hemisphere crust.

19. Presented to the Austrian Imperial Academy of Sciences, Chetty, *loc. cit.* p. 62.

20. F. Bursley Taylor. (1910). *Bulletin of the Geological Society of America*, 21, 179–226.

21. Taylor, in 1898 and 1903, had written about his early ideas in a text, *The Planetary System*, that, at the time, he considered to be incomplete. (Frankel, *loc. cit.*, Vol. 1, p. 63.)

22. F. B. Taylor, *loc. cit.*, p. 218.

23. R. Mantovani. (1889). Les Fractures de l'écorce Terrestre et la Théorie de Laplace. *Bulletin de la Société Sciences Arts Réunion*, 41–53 and his article on the splitting of the Antarctic, (1909). L'Antarctide, Je M'instruis. *La Science Pour Tous*, 38, 595–597.

24. A. W. Drayson. (1859). *The Earth We Inhabit, Its Past, Present and Future*. London: A.W. Bennett. R. A. Drayson and W. Thorp. (1859). William Thorp, on the geological evidence of the secular expansion of the crust of the Earth. *Proceeding of Geological and Polytechnic Society of the West Riding of Yorkshire*, 4, 1–27. Their work has been extensively reviewed by S. W. Hurrell. (2017). Early speculations about Earth expansion by Alfred Wilks Drayson (1827–1901) and William Thorp (1804–1860), https://www.dinox.org/.

25. With many others too numerous to mention individually, but the architect Robert Adam, the mathematician Colin Maclaurin, the poet Robert Burns, and the engineers James Watt and Thomas Telford show that it was an unusually broad intellectual community.

26. Only 100 years earlier, the English theologian Thomas Burnet had caused a stir for proposing a theory for mountain formation that deviated from the version described in the biblical *Book of Genesis*. In his book *Telluris Theoria Sacra* (*Sacred Theory of the Earth*), published in 1681, he describes the early Earth as a hollow ovoid shell containing the water that would lead to the Biblical Flood. The surface was smooth and locked in a perpetual spring — a Paradise — until the surface cracked and

crumpled leading to the formation of mountains and the release of the water to create the great flood. The deviation from Genesis was highly criticized, though Newton was inclined to support Burnet's thesis.

27. Ussher gave a rather precise date: the evening preceding 23 October, 4004 BCE. Under the old calendar, this would have been the date of the autumnal equinox. At the time, he wrote the Earth would have been 5,654 years old. Ussher was not the only person attempting this calculation. The Venerable Bede had given the date as 3952 BCE which was pretty close to Isaac Newton's 3998 BCE and so everyone could agree on roughly 4000 BCE give or take a few years.

28. There were alternative proposals. Abraham Gottlob Werner (1750–1817) suggested that the Earth started with a great all-covering ocean that receded to its current level leaving behind mountains and seas. During that process, minerals would be precipitated out to form the strata that we observe. There was no explanation of where the water went nor of how it might be renewed for the next cycle after it had gone. The strength of Hutton's discussion was that it concerned only things he had seen.

29. John Herschel (1792–1871) was the son of William Herschel who had been the first President of the Astronomical Society of London (later the Royal Astronomical Society) at the time of its foundation in 1821. He is well known for extending the work of his father and his aunt Caroline Herschel cataloging the objects of the southern sky from the Cape Colony in South Africa.

30. John Herschel graduated from Cambridge as "Senior Wrangler" in 1812 and a year later won the Smith's Prize. In 1831, he was knighted and proposed as a candidate for the position of President of the Royal Society. The election was a battle between the traditionalists and the reformers. Herschel, a reformer, narrowly lost the vote to Prince Augustus Frederick, Duke of Sussex and the sixth son of King George III of England. It is thought that this was a factor in his decision to go to the Cape Colony. The Cape Observatory had been established in 1828, and the 1835 return of Halley's Comet provided an important scientific reason to move there with his family.

31. C. Lyell. *Principles of Geology: Being an Attempt to Explain the Former Changes of the Earth's Surface, by Reference to Causes Now in Operation.* London: John Murray. The first volume of the first edition appeared in January 1830, with two other volumes appearing in 1832 and 1833. The work went through 12 editions, the last being published posthumously in 1875. Lyell's advocacy of gradualism (in contrast to catastrophism) to describe geological changes was much criticized when it first appeared, with criticism continuing even into the middle of the 20th century.

32. See Gregory Good's article John Herschel's geology: The cape of good hope in the 1830s, which is Chapter 8 in *The Romance of Science: Essays in Honor of Trevor H. Levere*, Springer (2017), edited by Jed Buchwald and Larry Stewart.

33. See A. Briggs and R. Wagner. (2016). *The Penultimate Curiosity: How Science Swims in the Slipstream of Ultimate Questions.* Oxford University Press, Ch. 36, p. 304 *et seq. The Mystery of Mysteries.* The phrase is quoted several times in Darwin's works, with acknowledgment to John Herschel.

34. Darwin also talked with the famous Scottish zoologist Andrew Smith, who spent the years 1820–1837 in South Africa. On return from the Cape Colony, Smith wrote up his five volumes, *Illustrations of the Zoology of South Africa* (1838–1850). Following their meeting at the Cape, there was extensive correspondence between Darwin and Smith.

35. James Hall (1811–1898) was a founding member of the US National Academy of Sciences and the first president of the Geological Society of America. Not to be confused with his contemporary, James Hall of Dunglass (1761–1832), the Scottish geologist who worked extensively with James Hutton and strongly supported Hutton's views. The latter James Hall became President of the Royal Society of Edinburgh.

36. James Hall was the 9th President of the AAAS, elected in 1856. That meeting of the AAAS was held in Montreal. See https://publish.illinois.edu/platetectonics/geosynclinal-theory/.

37. Joseph Henry, the secretary of the AAAS, advised Hall to introduce the theory slowly. The text of the address was not published until 1882.

38. This is a fine example of the notion that simply naming a phenomenon or object endows it with a real existence, whereas it might simply be a figment of someone's imagination.

39. See S. K. Runcorn's engaging review of Lyttleton's book *The Earth and Its Mountains.* Wiley (1982) and S. K. Runcorn. (1985). *Tectonphysics*, 111, 355–359 which is well worth reading even if the book is not. The review subtitle is *The Gospel according to St. John's*, referring to Lyttleton's college in Cambridge.

40. A. Holmes, 1931, *Radioactivity and Earth Movements*, Trans. Geological Society. Glasgow. This was a write-up of a lecture he delivered in 1929. His 1942 book, *Principles of Physical Geology*, was published in 1944 with a chapter on the movement of continents and substantially revised in 1965, the year of his death, by his widow Doris who was also a geologist. 1931 was, in a sense, too soon after Wegener's proposal: it was a time when the proposal had been firmly and dogmatically rejected by the geology community at large who, it would seem, preferred the Shrinking Earth idea. Nevertheless, Holmes had been thinking for some time about possible mechanisms that might drive continental drift.

41. By 1962, Holmes had retired after a distinguished career at the universities of London (Imperial College), Durham and Edinburgh. He was widely acknowledged as having successfully pioneered radioactive dating of rocks. It is perhaps surprising that in his seminal paper of 1962, *History of Ocean Basins*, Hess did not cite Holmes. See http://www.mantleplumes.org/WebDocuments/Hess1962.pdf.

42. A considerable fraction of the Ocean floor surveys were classified as military information during the years of the Cold War between the U.S.A and the U.S.S.R. Subduction zones had been discovered off the coast of California in the early 1950s. These were located at the continental margins where the material of one plate was diving below another. That process was ultimately recognized as the driver for continental drift and the cause of seafloor spreading. It explains why the seafloor material is relatively young while some continental crust is very old.

43. The Cantal region of Auvergne near the village of Pont Farin.

44. Stanley Keith Runcorn (1922–1995) was one of the key proponents of Plate Tectonics and one of the founders of the science of paleomagnetism. During the 1950s, his observations provided evidence that the continents had drifted across the north magnetic pole over millions of years. Tragically, he was murdered on Tuesday, 5 December 1995 in his hotel room by a person

described as a thief. He was on a visit to the Scripps Institution of Oceanography. The perpetrator was sentenced to 25 years' imprisonment after a series of three contentious trials.

45. See the short biography by C. Barton. (2002). Marie Tharp, oceanographic cartographer, and her contributions to the revolution in the earth sciences. *Geological Society of London Special Publications*, 192, 215–228. The article is based on the author's interviews with Marie Tharp.

46. Frankel (*loc. cit.*, Vol. 3, §§6.6–6.18, p. 378 *et seq.*) presents one of the best biographies of Marie Tharp (he knew her personally). It includes a detailed overview of her work and her contributions to the subject.

47. There were several important motivations for seafloor mapping, quite apart from academic interest in geology. One was to provide maps for submarine navigation. Commercial interests were also in play: it was important for the laying of transatlantic communications cables, and there was interest from the oil and gas industry.

48. S. W. Carey. (1958). The tectonic approach to continental drift, in S. W. Carey (ed.), *The 1958 Conference Proceedings Continental Drift — A Symposium*. Hobart: University of Tasmania, pp. 311–349. Later, in 1976, at a time when Plate Tectonics was gaining momentum, Carey published a monograph detailing his views: *The Expanding Earth*. Elsevier.

49. Frankel, *loc. cit.*, Vol. 4 §6.8, pp. 390–391.

50. This treatment of "research assistants", male and female, was commonplace in universities worldwide until, perhaps, the 1960s and still continues at a somewhat lower level today. Nowadays, most of the work done by such research assistants is done by a large number of postgraduate students who are working towards PhD degrees. The work of the graduate students is acknowledged in the publications arising from their PhD. However, the PhD student does not always get first-author credit, the credit going instead to the supervising professor, or even the grant holder. Of course, customs vary among disciplines and between different countries.

51. Frankel, *loc. cit.*, Vol. 4, §6.2, p. 363. The section is an insightful biography of Maurice Ewing (e.g. "Ewing himself rarely

made proposals about processes that shaped major geological features"), followed in §6.6 with a biography of Marie Tharp.

52. The Danish scientist Nicolaus Steno in his *Dissertationis Prodromus*. Use of the law was popularized by William Smith in his publications about his geological map of Britain.

53. In 1921 Henry Norris Russell argued that the age of the Earth's crust was at most 4 billion years. Russell, H. N. 1921, A superior limit to the age of the earth's crust. *Proc. R. Soc. Lond.* A99 pp. 84–86.

54. The archaean is the long time span between 4 billion years ago (when the Earth began to form continents) and 2.4 billion years ago (when the first life seems to have emerged). The atmosphere was made from volcanic gases, mostly methane and carbon dioxide, and there was no free oxygen. The appearance of life took place about the same time as the Great Oxidation Event. The Archaean was preceded by the Hadean Eon and followed by the Proterozoic Eon. The Sun was initially only 70% as bright as it is today, but the Earth was sometimes warmer than it is now.

55. Early Solar System meteorites of the kind thought to have coalesced to form planets have been found on Earth and dated to 4.5682 billion years ago (Gyr). This is currently taken to be the age of the Solar System. The age assigned to the Sun is 4.603 Gyr. Moon rocks brought back by the Apollo missions have been dated at 4.51 Gyr.

Chapter 10

1. In this chapter, we will use the nomenclature and dating shown on the *International Chronostratigraphic Chart* of the International Commission on Stratigraphy, https://stratigraphy.org/chart. "K-Pg" is the abbreviation for Cretaceous–Paleogene (previously known as the Cretaceous–Tertiary) boundary. We follow the timing of stratigraphic boundaries and the spelling of the geological names as used in this chart.

2. For convenience, we have the abbreviations Myr = millions of years and Mya = millions of years ago. We shall also use the scientific definition of a *billion* as meaning 1,000 millions, with the scientific abbreviation "Gyr" (Gigayears).

3. Traditionally, biologists have divided Life into three "Kingdoms": Plants, Animals and unicellular organisms that are collectively referred to as Protista. This simplistic view has been modified several times, and in recent decades, Bacteria, Fungi and others have been each assigned to their own Kingdoms. Each Kingdom is divided into smaller groups called "Phyla" (singular "Phylum") which are distinguished either by their "body plan" or by their evolutionary proximity. The Animal Kingdom is divided into about 31 Phyla and the Plant Kingdom into about 14 Phyla. The number depends on the precise biological definitions of the phyla.

4. Shales are rocks that were formed by the compression of various types of mud. Shales contain small fragments of minerals. Sandstones arise from sands that have been deposited and cemented together by minerals. Sandstone is generally made of quartz.

5. The word Phanerozoic derives from the Greek *phaneros* for conspicuous or visible and the Greek word *zoon* for animal.

6. This timing system and its refinements are defined and maintained as the International Chronostratigraphic Chart. See note 1.

7. The names and the time ordering of the Periods of Geological Time can be memorized with a pneumonic like "**C**amels **O**ften **S**it **D**own **C**arefully; **P**erhaps **T**heir **J**oints **C**reak? **P**lease **N**o **Q**uestions!" The only problem remaining is to recall what the initials C, O, S, D, C, P, T, J, C, P, N and Q stand for. To the astronomers, this is reminiscent of trying to recall the stellar temperature sequence O, B, A, F, G, K, M, R, N and S.

8. Here, we will quote spans of time from the start date to the end date of the interval and plot graphs so that time progresses, i.e. increases, from left to right.

9. The Precambrian Age comprises three Eons: the earliest is the Hadean, followed by the Archean and the Proterozoic.

10. This classification goes back to Linnaeus who, in his 1738 book *Systema Naturae*, produced an amazingly detailed catalog of living things, classified in a hierarchical way. This was very much like a genealogical tree of lifeforms. The basic outline of his scheme is still in use today and has remained largely unchanged even with the advent of DNA analysis.

11. Richard Dawkins and Jared Diamond have suggested that humans should be either the fifth ape or the third chimpanzee.

12. To confuse matters, there are subspecies of several of these species.

13. There are also families of Lesser Apes (Gibbons), and of Monkeys (which have tails). Gibbons are not Monkeys and nor are they "Great Apes", but Gibbons are apes. There are 18 species of extant Gibbon in four genera.

14. This statement is reported as having been made in a discussion with Niles Eldredge, in 1978. Gould and Eldredge in 1972 collaborated on a seminal paper, Punctuated equilibria: An alternative to phyletic gradualism, which, in effect, brought the theory of evolution back to Cuvier's catastrophism. It is said, somewhat provocatively, that this is the paper that won the respect of palaeobiology among the evolutionary biology community (Patricia Princehouse in *The Palaeobiological Revolution*, ed. D. Sepkoski and M. Ruse, 2009, Ch. 8).

15. "Binning" is a statistical technique for analyzing near-continuous and noisy data by grouping the data into a smaller number of "bins". Each bin is then assigned an average of the data it contains. This has the effect of smoothing the variations in the data.

16. N. D. Newell. (February 1963). Crises in the history of life. *Scientific American*, 208, 76–95.

17. D. M. Raup and J. J. Sepkoski, Jr. (1982). Mass extinctions in the marine fossil record. *Science,* 215, 1501–1503.

18. A family consists of multiple species and genera consist of many species (see Note 10). Considering the diversity at the families level rather than a species level is statistically more reliable when samples are small.

19. We can draw the following loose analogy to appreciate the relative significance of the death of a family, genus or species. Think of a *city* made up of *suburbs* that each contain *houses*. The loss of one house is sad of course, while the loss of a suburb is a tragedy. But the loss of a city is a disaster on a vast scale, having a global impact.

20. Wallace's book, *On the Law Which Has Regulated the Introduction of New Species*, presented a view of evolution quite different from Darwin's. Wallace made many important discoveries in diverse areas of biology, he wrote 22 full-length books and

over 500 scientific papers, 191 of which appeared in the journal *Nature*.

21. It is important to note that the vertical axis is a *percentage* of the genera existing at that time that became extinct. In other words, it shows the relative severity of the extinction going on at a given time. It says nothing about the *number* of genera or species that died, a high peak does not represent a higher number of extinct genera. By contrast, Figure 10.2 does show the *number* of families going extinct within Sepkoski's sample.

22. The normal rate is a bit less than one family of plant per million years going extinct.

23. Ferns are not gymnosperms because they reproduce using spores rather than seeds.

24. J. W. Valentine and E. M. Moores. (1971). Plate tectonic regulation of biotic diversity and sea level: A model. *Nature*, 288, 657.

25. J. W. Valentine. (1971). Plate tetonics and shallow marine diversity and endemism, an actualistic model. *Systematic Zoology*, 20, 253.

26. Campsie *et al.* (November 1984). Episodic volcanism and evolutionary crises, in *The Oceanography Report*, Eos, Vol. 65 and Rich *et al.* (1986). A significant correlation between fluctuations in seafloor spreading rates and evolutionary pulsations. *Paleoceanography*, 1, 85–95. The group members were John Campsie, Leonard Johnson, Janet Jones and James Rich. The story is told in the next chapter.

27. W. M. Napier and S. V. M. Clube. (1973). A theory of terrestrial catastrophism. *Nature*, 282, 455 and S. V. M. Clube and W. M. Napier. (1984). The microstructure of terrestrial catastrophism, MNRAS, 211, 953.

28. D. Whithouse. (24 July 2002). Space rock on collision course. *BBC News*. http://news.bbc.co.uk/2/hi/science/nature/214787 9.stm.

29. R. R. Britt. (29 July 2002). NASA scientists call British media's asteroid hype unethical rubbish. Space.com.

30. E. J. Öpik. (1973). Our cosmic destiny. *Irish Astronomical Journal*, 11(4), 113–124. The article was a write-up of lectures he had given in Washington D.C. and in Belfast. The *Irish Astronomical Journal* was virtually Öpik's own journal: very few other people

contributed articles. This article is listed as having had only six citations. Alvarez *et al.*, 1980 *loc. cit.* did not cite this paper.

31. H. C. Urey. (1973). Cometary collisions and geological periods. *Nature Letters*, 242, 32. This paper got 40 citations, which, by the standards of the time, was very good. It was cited by the 1980 Alvarez *et al.* paper *loc. cit.* For his back of the envelope calculations, Urey assumed an impact velocity of 45 km per second (about 100,000 miles per hour) and a mass of 1 gigatonne. The velocity of the Earth in its orbit is 30 km per second. The asteroid *Ryugu* visited by the Japanese spacecraft *Hayabusa* has a diameter of just under 1 km and a measured mass of 0.46 gigatonnes. The potentially hazardous asteroid, *Apophis*, has a diameter of 370 meters and an assumed mass of 0.06 gigatonnes.

Chapter 11

1. However, Neil deGrasse Tyson, the most famous of Astrophysics Outreach presenters, did crack a joke about the boson.
2. There are many accounts of this story by those who were participants at the time. Notably, there is Luis Alvarez's autobiography: L. W. Alvarez. (1987). *Alvarez: Adventures of a Physicist*, Basic Books. Alfred P. Sloan Foundation series, Ch. 15, Impacts and Extinctions. Walter Alvarez wrote his own story in his 1998 book *T. Rex and the Crater of Doom*. Princeton University Press. Prominent paleogeologists A. Hallam's. (1989). *Great Geological Controversies*. 2nd ed. Oxford University Press, Ch. 7 Mass Extinction, pp. 184–215, and J. L. Powell. (1998). *Night Comes to the Cretaceous: Dinosaur Extinction and the Transformation of Modern Geology*. Freeman.
3. Much has been written about this. For example, "Walter's gentlemanly style helped to offset the wrath generated among geologists by his pugnacious father", in Powell, *Night Comes to the Cretaceous*, p. xiv, *loc. cit.* Luis Alvarez was one of only three scientists who in 1954 testified to the Un-American Activities Committee against Robert Oppenheimer, the so often called "father of the atomic bomb". Oppenheimer's security clearance was revoked, leaving him a broken man.
4. A similar such statement is said to have been made by Ernest Rutherford in 1909 when he described science as "physics and

stamp collecting", which is ironic in light of the fact that he got the Chemistry Nobel Prize in 1908.

5. Powell, *loc. cit.*

6. Snowbird, Utah, October 1981, "Large Body Impacts and Terrestrial Evolution: Geological, Climatological and Biological Implications". The Alvarez presentation concluded that "...the Deccan Traps are almost certainly not the source of the anomalous iridium, the impact could have triggered the Deccan volcanism."

7. Powell, *loc. cit.*, pp. 16 and 126.

8. The word "Cretaceous" comes from the Latin "Creta", meaning "chalk" in English and "Kreide" in German. The letter "K" has been adopted as the formal abbreviation for "Cretaceous". The Cretaceous Period lasted from 145 to 66 Mya.

    In the 1970s, what are now referred to as the Paleogene (66–23.03 Mya) and Neogene (23.03–2.58 Mya), Periods were referred to as the Upper and Lower Tertiary Periods. The term Tertiary is now considered obsolete but is still used informally. So, prior to the 1980s what is now called the "K-Pg boundary" was referred to as the "K-T boundary".

9. This was also the turning point in the relationship between father and son Alvarez. In his book, Walter recalls that "Dad did not originally think that geology was an interesting science. It was my Mother, Geraldine, who got me interested in rocks when I was in high school, ... She still likes to remind me that my first rock hammer was one I borrowed from her — and lost! ... I rarely saw my father, and did not know much about him as a scientist." Powell, *loc. cit.*, p. 9 remarks that at this time Luis "had begun to worry that physics had started to leave him behind and that his career had stalled".

10. The Alvarez *et al.* paper was published in *Science* on 6 June 2008. Two weeks earlier, on the 22 May 1980, an article by J. Smit and J. Hertogen appeared in *Nature* entitled *An extraterrestrial event at the Cretaceous–Tertiary boundary*, identifying a "moment of extinction coupled with anomalous trace element enrichments, especially of iridium and osmium." They concluded that the anomaly indicated an extraterrestrial cause for the extinction

and that "The impact of an asteroid or comet 5–15 km in diameter is the most attractive." The authors acknowledged communication with Walter Alvarez, who had shared his preliminary measurements of the iridium and osmium abundances. Figure 2 of the Smit–Hertogen paper plots both sets of data. The same issue of *Nature* also had an article by the Swiss geologist, Kenneth Hsu, whose paper, Terrestrial catastrophe caused by cometary impact at the end of the Cretaceous. Walter Alvarez is thanked for comments in both papers. The Alvarez *et al.* 1980 paper does not mention Hsu, Smit or Hertogen, despite citing over 70 other papers.

11. The coastal village of Chicxulub lies almost exactly on the the center of the crater. The name is the Yucatan Maya language word meaning "The devil's flea" (*Wikipedia* 2021).

12. A lot of information was available from studies of the geological consequences of big nuclear explosions. Together with simulations of nuclear explosions made during the Cold War era, there was a substantial understanding of the geophysics and dynamics of the impact. Climactic simulations were rather crude then, but the "nuclear winter" arising in the aftermath of an all-out global nuclear war was widely recognized.

13. S. Miller, (2014), The public impact of impacts: How the media play in the mass extinction debates, in Special Paper 505 Geological Society of America. In G. Keller and A. C. Kerr (eds.), *Volcanism, Impacts, and Mass Extinctions: Causes and Effects.*

14. According to *Wikipedia* (July 2020), the dinosaur count from Hells Creek is 72 Triceratops (40%), 44 Tyrannosaurus (24%), 36 Edmontosaurus (20%), 15 Thescelosaurus (8%), 9 Ornithomimus (5%), 2 Ankylosaurus (1%) and 2 Pachycephalosaurus (1%). A big haul of important beasts, but nevertheless a small sample as compared with the smaller beasts and the plants. It is suggested that some 80% of plant taxa were wiped out in the area that is now Montana and the Dakotas.

15. See, for example, B. Cascales-Minana and C. Cleal, (2014), The plant fossil record reflects just two great extinction events. *Terra Nova,* 26, 195–200. Their data on Vascular plants reveal declines in diversity at the time of only two of the great extinctions: the Carboniferous–Permian (28%) and End-Permian (55%). However, those declines in diversity took place over a longer period

of time and are not sudden events. The authors ascribe these declines in diversity to complex environmental changes.

16. See Stephen Miller's *loc. cit.* analysis.

17. See E. Flamini, *et al.* (2019). *Encyclopedic Atlas of Terrestrial Impact Craters.* Springer, which describes almost 200 impact craters with pictures from Italian COSMO-SkyMed SAR satellite constellation and presents the science of impacts.

18. D. P. G. Bond and S. E. Grasby. (July 2017). On the causes of mass extinctions. *Paleogeography, Paleoclimatology and Paleoecology*, 478, 3–29.

19. See the study by B. M. French and C. Koeberl. (2010). The convincing identification of terrestrial meteorite impact structures: What works, what doesn't, and why. *Earth-Science Reviews*, 98, 123–170. The review concludes with the statement "The recent enthusiasm for meteorite impacts among scientists, the public, and the media, has unfortunately led to the proliferation of incorrect and questionable reports of impact events, involving both individual structures and extinction boundaries." This work sets widely accepted criteria for evaluation of candidate impact craters. This was reinforced in a subsequent review by W. Reimold and collaborators. (2014). Impact controversies: Impact recognition criteria and related issues. *Meteoritics & Planetary Science*, who say that "... one should not conclude that strange or unusual observations are necessarily impact-related just because they occur in an unusual geological setting" and advise editors to "not let enthusiasm about one possible interpretation dominate. It is necessary to carefully lay out arguments why accomplished and recognized knowledge and criteria should be abandoned in favor of so far unproven and nondiagnostic 'new evidence' for impact".

20. Bond and Grasby, *loc. cit.*, §3.6.

21. Large Igneous Provinces are also referred to as "Flood Basalts" because the lava literally floods the surrounding areas and can pile up to heights of several miles. Generally speaking, a LIP will have a volume of more than 0.1 million cubic km spread over an area of more than 0.1 million square km. They are made from magma and 75% of their volume will have been expelled by igneous pulses over a short period of 1–5 Myr. They are temporary tectonic features that typically survive less than 50 Myr

before being subducted. For details, see S. Bryan and R. Ernst. (2008). Revised definition of Large Igneous Provinces (LIPs). *Earth-Science Reviews*, 86, 175–202.

22. J. E. Rich *et al.* (1986). A significant correlation between fluctuations in seafloor spreading rates and evolutionary pulsations. *Paleoceanography*, 1, 85–95.

23. Since 2018 called *Paleoceanography and Paleoclimatology*, which ranks second in the rankings of journals on "paleontology" (https://www.scimagojr.com/journalrank.php?category=1911).

24. J. D. MacDougall. (2011). *Why Geology Matters: Decoding the Past, Anticipating the Future.* University of California Press, Ch. 10.

25. B. A. Edwards *et al.* (2020), Fifty years of volcanic mercury emission research: Knowledge gaps and future directions. *Science of the Total Environment*, 143800: "… mercury(Hg) is among the most environmentally significant [metals] due to its propensity for long-range atmospheric transport and its toxicity to many forms of life including humans". Also C. T. Driscoll *et al.* (2013). Mercury as a global pollutant: Sources, pathways, and effects. *Environmental Science Technology*, 47(10), 4967–4983.

26. This figure has been drawn from the data and figures presented in two papers: Figure 1 in E. Font and D. Bond. (2020). Volcanism and mass extinction, published in *Encyclopaedia of Geology*, 2nd edn. and Figure 11.1 in Percival *et al.* (2021). Sedimentary mercury enrichments as a tracer of large igneous province volcanism, published in R. Ernst, A. Dickinson and A. Bekker (eds.), *Large Igneous Provinces: A Driver of Global Environmental and Biotic Changes.* Geophysical Monograph 255, 1st Edn, and other and sources cited therein.

27. https://www.geolsoc.org.uk/chicxulub, "The non-smoking gun", Gerta Keller, Princeton.

28. https://www.geolsoc.org.uk/smokes, "And yet it smokes", Jan Smit, Amsterdam.

29. https://www.geolsoc.org.uk/chixculub_1, "The Chicxulub discussion part 1"; https://www.geolsoc.org.uk/chicxulub_2 "The Chicxulub discussion part 2".

30. https://www.geolsoc.org.uk/chicxulub_3, "The Chicxulub debate - What next?".

31. Participants were, in order of appearance, Alan Hildebrand, Norman MacLeod, Gerta Keller, Jan Smit, Claire Belcher, David Archibald and Wolfgang Stinnesbeck.

32. http://www.bbc.co.uk/sn/tvradio/programmes/horizon/dino_t rans.shtml.

33. B. Bosker. The nastiest feud in science. The Atlantic (updated 7 September 2008). https://www.theatlantic.com/magazine/arc hive/2018/09/dinosaur-extinction-debate/565769/.

34. At this time, in 1988, the location of the impact proposed by Alvarez was unknown.

35. When quoting from older sources, we follow the usage of the author when they use "K/T" instead of the modern "K-Pg".

36. In a 1990 paper (G. Keller and E. Barrera, (1990), *GSA Specification Paper*, 247, 563–575), the authors found that the El Kef data showed a prolonged period of ecological disruption beginning some 200,000 yr prior to the K/T Boundary (as defined by the Ir anomaly) and lasting hundreds of thousands of years afterwards.

37. Snowbird, Utah, 20–23 October 1988, *Global Catastrophes in Earth History*, sponsored by the Lunar and Planetary Institute and the National Academy of Sciences. Keller's presentations appear on pp. 88–91.

38. G. K. Gilbert. (1895). Sedimentary measurement of Cretaceous time. *Journal of Geology*, 3, 121–127. A modern review of cyclostratigraphy is presented in the review by C. Huang *et al.* (2020). Cyclostratigraphy and astrochronology: Case studies from China. *Paleogeography, Paleoclimatology and Plaeoecology*, 560, Article 110017.

39. The book was first published in German in 1941. The Serbian translation did not appear until 1998. There are two English translations in 1969 (published in Jerusalem) and 1998 (published in Belgrade).

40. J. D. Hays *et al.* (1976). Variations in the Earth's orbit: Pacemaker of the ice ages. *Science*, 194, 1121–1132.

41. Pyle and Mather, (2003), Hg is strongly enriched in volcanic emanations, and volcanoes are the only natural sources of direct Hg emission to the free troposphere and stratosphere. *Atmospheric*

Environment, 37(36), 5115–5124. Lindstrom *et al.* (2019), Volcanic mercury and mutagenesis in land plants during the end-Triassic mass extinction. *Science Advances,* 5(10).

42. In water, algae and plankton preferentially take up ^{12}C during photosynthesis. The relative change in the ratio of ^{13}C to ^{12}C thus reflects climate change, either warming or cooling. Oxygen isotopes provide similar information, and the ratio of measured ^{18}O to ^{16}O also provides information on climate. See https://timescavengers.blog/introductory-material/what-is-paleoclimatology/proxy-data/carbon-oxygen-isotopes.

43. See S. Grasby *et al.* (2019). Mercury as a proxy for volcanic emissions in the geologic record. *Earth-Science Reviews,* 196, 102880 and A. Sial *et al.* (2021). Hg isotopes and enhanced Hg concentration in the Meishan and Guryul Ravine successions: Proxies for volcanism across the Permian–Triassic boundary. *Frontiers of Earth Sciences,* Article 651224.

44. Milutin Milankovitch (1879–1958) was a Serbian geophysicist and astronomer. He worked on the problem of variations of the the Earth's motion from around 1912 until 1920 when he published his results in a book: *Théorie Mathématique des Phénomènes Thermiques Produits par la Radiation Solaire* (*Mathematical Theory of Heat Phenomena Produced by Solar Radiation*).

45. These data are independent of the dating of the impact. Supporters of the hypothesis that this was caused by a bolide impact, not surprisingly, set the date of the K-Pg boundary by the timing of the impact.

46. One of the central disputes in the schism between the impactors and the LIP groups lies in the dating of the impact. The group to which Keller is affiliated used U-Pb dating, whereas the impactors use ^{40}A-^{39}A dating (accuracy $\sim 200,000$ yr). This does not provide sufficient time resolution to distinguish the peaks in volcanic activity. There are also problems with the systematics of this timing method (it may be precise, but not accurate). See P. Renne *et al.*, 1998, Absolute ages aren't exactly. *Science* Vol. 282, pp. 1840–1841, and the section on *Standard intercalibration* at https://geoinfo.nmt.edu/labs/argon/methods/home.html.

47. The white area displays the volcanic pulses, lava megaflows, at the Deccan traps (Schoene *et al.,* 2019). The thin line represents the measurements of the abundance of mercury during this period, while the dots identify the Extreme Events of volcanic mercury emission (Keller *et al.,* 2020). The solid fat line shows the growth curve for the total amount of volcanic mercury as a function of time, relative to an adopted baseline level.

48. Condamine *et al.* in 2021 found that the dinosaurs went into diversity decline some 76 Mya (Dinosaur biodiversity declined well before the asteroid impact, influenced by ecological and environmental pressures. *Nature Communications,* 12, 3833). The decline can be attributed to climate change accompanied by an inability to adapt to a changing environment.

49. M. A. Richards *et al.* (2015). Triggering of the largest Deccan eruptions by the Chicxulub impact. *GSA Bulletin,* 127, 1507–1520. See also C. J. Sprain *et al.* (2019). The eruptive tempo of Deccan volcanism in relation to the Cretaceous-Paleogene boundary. *Science,* 363, 866–870.

50. The use of ^{40}A-^{39}A dating was developed by the English cosmochemist Grenville Turner. It gives very accurate relative ages, though must be calibrated by using the less accurate ^{40}K-^{40}A dating or orbital tuning (Milankovitch) methods. Correcting for known systematics in the methods yields a date for the Chicxulub impact at 66.0–66.1 Mya.

51. This allows for possible error of one cycle.

Chapter 12

1. *Homunculus* is a very small human or humanoid creature. In Latin, it means "little man".

2. G. Owen Gingerich. (June 2011). The impact of the telescope, and the birth of modern astronomy. *Proceedings of the American Philosophical Society,* 155(2), 134–141.

3. In N. Lane. (2015). The unseen world: Reflections on Leeuwenhoek (1677) "Concerning little animals". *Philosophical Transactions of the Royal Society,* we can read "he did not merely observe, but conducted ingenious experiments, exploring and manipulating his microscopic universe with a curiosity that

belied his lack of a map or bearings. Leeuwenhoek was a pioneer, a scientist of the highest caliber, yet his reputation suffered at the hands of those who envied his fame or scorned his unschooled origins, as well as through his own mistrustful secrecy of his methods, which opened a world that others could not comprehend." Similar arguments can be found in Ford, J. Brian. (1992). From dilettante to diligent experimenter, a reappraisal of Leeuwenhoek as microscopist and investigator. *Biology History*, 5(3), 3–21.

4. A. Roberts. (2015). *The Incredible Unlikeliness of Being: Evolution and the Making of Us.* Quercus Books.

5. The Dutch word that van Leeuwenhoek uses is *dierkens*, translated by Oldenburg as "animalcules," since *Dier* is "animal" and the ending *-ken* is a common diminutive form.

6. The ovists, on the other hand, attributed the fundamental role in reproduction to the female ovum.

7. M. Rooseboom. (1950). Leeuwenhoek, the man: A son of his nation and his time. *Bulletin of the British Society for the History of Science*, 1(4), 79–85. *JSTOR*, http://www.jstor.org/stable/4024784.

8. A spectroscope is an optical instrument in which a system of lenses and prisms allows the intensity of incident light to be analyzed at each wavelength. Fraunhofer replaced the prism with a diffraction grating thus increasing the dispersive power.

9. This story is explained in great detail in a fantastic book by B. B. Nath. (2013). *The Story of Helium and the Birth of Astrophysics.* Springer-Verlag. This book recounts the discovery of helium in a compelling way and in language suitable for a nonspecialist reader. It is a brilliant example of popular science, and in it, some of the misconceptions that have circulated for many years (and continue to do so) around the discovery of this elusive chemical element are also dismantled.

10. Nath, *loc. cit.*, p. 40.

11. Nath, *loc. cit.*, p. 171.

12. In different books and encyclopedias, Janssen is credited with discovering helium during his observations of the 1868 solar eclipse from India. Others attribute it to both Janssen and Lockyer, even emphasizing that they communicated it in letters that arrived in Paris the same day.

13. Helium is a very suitable name for this element considering that it was first discovered in the Sun (*Helios* in Greek) and not on Earth. Helium is not abundant on Earth, despite its being the second most abundant element in the universe after hydrogen. There are two reasons behind its rarity: the first is its extreme lightness, which makes it impossible for Earth's gravity to retain it, as the fact that helium-filled balloons rise demonstrates. The second is its inability to combine with other elements: it is a noble gas. By contrast, hydrogen, for example, easily combines with other elements to form heavier molecules (for example, water) that are retained by Earth's gravity, even though it is lighter than helium (by a factor 4).

14. Jargonium, nigrium, norium and asterium were postulated elements — dragons, to apply the metaphor we have used throughout this book — but they were never discovered, simply because they do not exist. The first three were postulated in 1869, 1866 and 1845, respectively, by the English scientists Henry Sorby (1826–1908) and A. H. Church (1834–1915) and by the Swede M. Swanberg through the analysis of zircon stones. Asterium (also called oronium or parhelium) was postulated as a companion of helium with similar properties by several scientists, among them were the Germans Carl Runge (1856–1927) and Friedrich Paschen (1865–1947), in the late 19th century. Its existence was also ultimately ruled out.

15. William Ramsay discovered krypton, neon and xenon in 1898, again in air (as Lord Rayleigh did earlier for argon), where these heavier gases have lingered since the formation of the Solar System, whereas helium was discovered as a disintegration product of radioactive uranium, despite of being the second more abundant element of the universe, a product of the Big Bang itself.

16. Nebulium was later found to be oxygen. Coronium, observed spectroscopically in the solar corona during the 1869 eclipse, actually originated from a highly ionized state of iron. Only helium passed all the tests and became the second element in the periodic table.

17. O. Heaviside. (1902). *Encyclopedia Britannica*. 10th Edn, 9th, No. XXXIII, 215.

18. A. E. Kennelly. (1902). *Electrical Engineering World*, p. 473.

19. E. M. Burbidge, G. R. Burbidge, W. A. Fowler and F. Hoyle. (1 October 1957). *Reviews of Modern Physics*, 29, 547.

20. (1) S. Mitton. (2005). *Fred Hoyle: A Life in Science*. London: Aurum Press; (2) C. Wickramasinghe. (2013). *Journey with Fred Hoyle*, 2nd Edn. Singapore: World Scientific; (3) K. Gregory. (2005). *Fred Hoyle's Universe*. Oxford: Oxford University Press; (4) P. Halpern. (2021). *Flashes of Creation: George Gamow, Fred Hoyle, and the Great Big Bang Debate*. New York: Basics Books.

Chapter 13

1. Alternatively, as it were, Stephen Hawking (1942–2018) said that not only does God sometimes shoot dice, sometimes he hides them under the table — and one of the Hindu "trinity" — Shiva — is said to spend his spare time playing dice with his partner Parvati (one naturally wonders whether Einstein knew about this!).

2. Tensors are mathematical objects useful for describing physical entities and processes in spaces with several dimensions. General relativity is cast in the language of tensor calculus, which is not as hard to master as it looks at first glance.

3. J. Canales. (2015). *The Physicist and the Philosopher: Einstein, Bergson, and the Debate That Changed Our Understanding of Time*. Princeton University Press.

4. The *Anschluss* is the annexation of Austria into Nazi Germany on 12 March 1938.

5. Joseph Weber designed and built the first detectors for gravitational waves.

6. Susan Jocelyn Bell was born to a sincere Quaker family in Belfast, Northern Ireland on 15 July 1943. She attended a Quaker boarding high school in York, England, and received a bachelor of physics degree with honors in 1965 from the University of Glasgow, where she was often the only woman in her classes. Bell remains an active, believing, participatory member of the Society of Friends, and some of her acceptance of events before, during, and after the events of 1974 must surely have been colored by that background. But we suspect that surviving as a female in a male environment (and at a time where getting married was deemed the pinnacle achievement of a woman!) was actually a much bigger factor.

She began graduate work at the University of Cambridge Mullard Radio Astronomy Observatory, part of the Cavendish Laboratory, in Fall 1965, working with Antony (nearly always Tony) Hewish, whose 1952 PhD thesis had been carried out under Martin Ryle and dealt with radio sources of very small angular size.

7. Radio astronomers Pilkington and Scott confirmed the reality of the first pulsar (not yet called that) by detecting it with separate equipment, thus making sure that Bell and Hewish had not been studying some horrible flaw in their own equipment Pilkington got an estimate of its distance using a dispersion technique, so confirming that it was celestial not terrestrial.

8. Martin Ryle (1918–1984) got the Nobel Prize for a very different part of radio astronomy, the development of aperture synthesis to improve angular resolution. Second-hand discussions of who got that prize for what tend to be slightly confused by the key aperture synthesis paper having been written by the pair: Ryle and Hewish in 1960.

9. The chief beneficiary of the squabble over the 1974 Nobel was surely Russell Hulse, the graduate student working with Joseph Taylor, whose very careful examination of puzzling data led to the discovery of the first binary pulsar. They shared the 1994 Physics Nobel "for the discovery of a binary pulsar". But there is always somebody left out! In this case, it was Joel M. Weisberg, who contributed hugely to the analysis of the changing orbit of PSR 1913+16. The analysis has continued down to the present, affirming every more precisely that general relativity, while it cannot be the ultimate, right theory of gravity, does an extraordinarily accurate, fail-proof job of predicting what astronomical objects and systems do. Records of the discussions leading up to choices of Nobel winners become public only 50 years after prizes are awarded, so it will be some time before we know just what was said about Joe Taylor, Russell Hulse, perhaps Joel Weisberg, and whether any memory of 1974 played a role in the process.

Chapter 14

1. John Michell (1724–1793) was a much respected Cambridge academic, holding a wide variety of positions within his College and

the University, until 1767 when he took a position as Rector of a church near the city of Leeds in northern England. Most of his research was done during this latter period of his life. See S. Schaffer's (1979). John Michell and Black Holes. *Journal for the History of Astronomy*, 10, 42.

2. Rev. J. Michell, FRS, (1783), On the means of discovering the distance, magnitude, &c. of the fixed stars, in consequence of the diminution of the velocity of their light, in case such a diminution should be found to take place in any of them, and such other data should be procured from observations, as would be farther necessary for that purpose. *Philosophical Transactions of the Royal Society*, 74, 35. Note the phrase "Consequence of the Diminution of the Velocity of Their Light" in the title. He imagined that the speed of the particles of light would diminish as they climbed out of the gravitational force field of the star. More than 50 years prior to this work, Bradley had already measured the speed of light with considerable accuracy. This perhaps reflects a general disbelief in the notion that the speed of light is finite.

3. The escape velocity of a ball does not depend on the mass of the ball. The escape velocity from Earth's surface is 11.2 km/sec (kilometers per second) or 25,000 miles per hour. This means that all bullets fired into the air must come back down to Earth. The escape velocity from the Sun's surface is 615 km/sec. If the Sun were the size of the Earth, the escape velocity rise would be 6,455 km/sec. If the Sun was 3 kilometers in radius, the escape velocity from its surface would be 300,000 km/sec, the speed of light. There is an on-line calculator for this at https://www.om nicalculator.com/physics/schwarzschild-radius.

4. These two papers by Laplace were not well known until they were discussed in Appendix A of S. Hawking and G. Ellis'. (1973). *The Large Scale Structure of Space–Time*. CUP. P.-S. Laplace. *Exposition of the System of the World*, part II, p. 305. It was the Institute of Theoretical Astronomy librarian, David W. Dewhirst, who provided the reference to Laplace's calculation: *Allgemeine Geographische Ephemeriden, Verfasset von Einer Gesellschaft Gelehrten, Weimer, IV, Bd 1 St. 1799. Ed. F.X. von Zach.* Appendix A of Hawking and Ellis (*loc. cit.*) gives a full English translation of that article.

5. J. G. Sodlner. *On the Deflection of a Light Ray from Its Rectilinear Motion, by the Attraction of a Celestial Body at Which It Nearly Passes by*, Berlioner Astronomisches Jahrbuch, pp. 161–172. See S. Jaki. (1978). Johann Georg von Soldner and the gravitational bending of light. *Foundations of Physics*, 8, 927–950 for a detailed discussion.

6. *Ceteris paribus* translates as "Other things being equal".

7. The reason for such diverse values being derived from Rømer's data is that neither the distances nor the distorted Keplerian elliptical shapes of the orbits of the Earth, Jupiter, and its moons were well known.

8. A triumph it may have been, but it was not without considerable controversy. For details, see the beautiful article by P. Coles. (2019). A revolution in science: The eclipse expeditions of 1919. *Contemporary Physics*, 60, 45–59. This also gives a fine didactic explanation of the physical process as understood within the framework of the General Theory of Relativity. For more on the controversy, see G. Gilmore and G. Tausch-Pebody. (October 2020). The 1919 eclipse results that verified general relativity and their later detractors: A story re-told. *Royal Society Journal of the History of Science*.

9. The Event Horizon Telescope, "EHT", is an international collaboration of about 60 institutions which was set up in 2009 to image two super-massive black holes with the largest apparent size: the one at the center of our Galaxy, and the other at the center of the giant elliptical galaxy M87. Radio telescopes from around the world are linked using highly accurate clocks to create a giant almost Earth-sized telescope with very high angular resolution and sensitivity.

10. See the paper, The black hole fifty years after: Genesis of the name by C. A. R Herdeiro and J. P. S. Lemos at https://arxiv.org/abs/1811.06587.

11. The endpoint of stellar evolution can be one of three objects: a white dwarf if the final mass is less than 1.4 solar masses (the Chandrasekhar limit), a neutron star if its mass is less than about 2.16 solar masses (the Tolman–Oppenheimer–Volkoff limit) or a black hole. Which of these fates a star suffers depends on the mass of the core that is left after nuclear burning and mass loss.

12. Kip Thorne had been one of John Wheeler's graduate students. Thorne went on to win the Nobel prize for his part in the discovery of gravitational waves. John Wheeler had another PhD student who went on to win the Nobel Prize for Physics: Richard Feynman.

13. From "black-hole" in https://www.dictionary.com/browse/black-hole. The term had already been used before the writings of Michell and Laplace when referring to the 24-square meter Fort William jail known informally by British soldiers serving in the British East India Company in Bengal as the "Black Hole of Calcutta". The jail became infamous following the incarceration there, in June 1756, of 64 prisoners taken in the siege of the Fort by the Nawab of Bengal: 43 died overnight from suffocation and heat exhaustion (these are the modern numbers; earlier accounts suggested that 120 out of 146 prisoners died. It is hard to imagine even 64 people squeezed into that 24 m^2 space, let alone 146). See https://en.wikipedia.org/wiki/Black_Hole_of_Calcutta.

14. There are a couple of small but important details about the region just outside of the event horizon. First, at one and a half times the event horizon radius light can travel in a closed circular orbit around the black hole. All circular orbits at smaller distances are unstable. For a massive particle or body, three times the event horizon radius is critical: there are no stable circular orbits closer to the black hole. This has the physical consequence that matter orbiting around a black hole and getting ever closer to the hole will, at this critical radius, find itself in an unstable orbit careering towards the singularity. The three Schwarzschild radius sphere is thought to mark the inner edge of an accretion disk. Andrew Hamilton has created video clips showing the view someone traveling around a black hole would get when traveling in a diverse selection of orbits: https://jila.colorado.edu/~ajsh/insidebh/schw.html.

15. Roy Kerr was working at the University of Austin in Texas at the time the paper was published: R. Kerr. (1963). Gravitational field of a spinning mass. *Physical Review Letters*, 11, 237. In February 1963, Maarten Schmidt *et al.* announced the discovery of quasars, very distant and very luminous for the time. Fowler and Hoyle had proposed gravitational collapse as the energy source for radio galaxies. The discovery and the idea led to the first of the series of

the Texas Symposia on Relativistic Astrophysics, held in Dallas Texas in December, 1963. All four of them spoke there.

16. We are familiar with this phenomenon when we see spinning ice skaters pull in their arms: the rate of spin increases.

17. Objects that were suspected to host supermassive black holes had been discovered in the early 1960s, e.g. the quasars, but the capability of verifying that did not come until the appearance of a new generation of instrumentation and photon detectors.

18. W. L. W. Sargent *et al.* (1978). Dynamical evidence for a central mass concentration in the galaxy M87. *Astrophysical Journal*, 221, 731. This seminal paper was an exploratory study using the first two-dimensional image photon counting system ("IPCS") developed by the British astronomer Alexander Boksenberg. The IPCS was a significant driver of UK extragalactic astronomy in the pre-CCD era.

19. M87 is big enough and bright enough to be seen in an amateur telescope or binoculars. The galaxy covers an area in the sky about ¼ the diameter of the Moon, though only the central ½ of that can be seen without a large telescope. The jet is only ¹⁄₁₀₀ the Moon's diameter (i.e. 0.3 arcmin) long and lies buried in the very brightest part of the galaxy. Amateur astronomers with 20″ telescopes and appropriate filters have made disputed claims to have seen it.

20. See the list at https://en.wikipedia.org/wiki/List_of_most_massive_black_holes. There are a number of objects, galaxies and quasars that are hosting black holes that are 10 times more massive than the one found in M87.

21. There is an excellent overview of this by J.-P. Luminet. *An Illustrated History of Black Hole Imaging: Personal Recollections (1972–2002)*, https://arxiv.org/abs/1902.11196. See also J.-A. Marck's animation posted by Luminet at https://www.youtube.com/watch?v=5Oqop50ltrM.

22. The EHT group chose to use the word "shadow" while others argued that this was not the correct description and suggested "silhouette". In light of what the models reveal, neither is particularly satisfactory. See the article by Matt Strassler at https://profmattstrassler.com/2019/04/09/a-non-experts-guide-to-a-black-holes-silhouette/ with clarification at https://profmattstra

ssler.com/2019/06/14/a-ring-of-controversy-around-a-black-hole-photo/.

23. Centaurus A, also known as NGC 5128, is the fifth brightest galaxy in the sky and so has been the subject of many studies at southern hemisphere observatories.

24. Andrea Ghez was also working in the GRAVITY collaboration.

25. *Observational Evidence For Black Holes in the Universe*, Proceedings of a Conference held in Calcutta, India, 10–17 January 1998, Springer 1999, ed. Sandip K. Chakrabarti.

26. For a comprehensive overview of this discovery, see Ethan Seigel's *Starts with a Bang* blog: https://www.forbes.com/sites/startswithabang/2019/07/25/general-relativity-rules-einstein-victorious-in-unprecedented-gravitational-redshift-test/.

27. The Nobel Committee for Physics released a more technical and detailed presentation of the scientific background for the 2020 prize: https://www.nobelprize.org/uploads/2020/10/advanced-physicsprize2020.pdf.

28. The distance of 1.4 ± 0.6 billion light-years is estimated from the total power output of the event as computed from models of the black hole merger constrained by the details of the chirp merger, and ringdown signal received.

29. Many of the experts in the field, including Einstein himself, supposed that the gravitational waves were a coordinate artifact, a consequence of the selection of a particular set of coordinates.

30. In the case of the carpet you could view it from two different positions. We do not have that option in the universe!

31. Much of this early work was done at King's College in London by the mathematical physicists Hermann Bondi, Felix Pirani, Ivor Robinson, Rainer Sachs and Andrzej Trautman. They exploited new coordinate-free methods of analysis, thereby sidestepping the question of whether or not the waves could be a mere consequence of coordinate choices. This gave physicists the "Green Light" to embrace gravitational waves as a part of physics. See the article of C. D. Hill and P. Nurowski. (2016). *How the Green Light was given for Gravitational Wave search*, https://arxiv.org/pdf/1608.08673.pdf. In 1957, Richard Feynman described a thought experiment for detecting gravitational waves.

32. What is so special about the curvature? The curvature is a property of the carpet — it is just a number telling us how curved the surface is. If the carpet is flat, the answer is zero curvature everywhere. The height of the carpet above the floor, however, depends on external references: in this case the floor, which might also be bumpy.

33. Joseph Weber (1919–2000).

34. J. Weber and J. A. Wheeler. (1957). Reality of the cylindrical gravitational waves of Einstein and Rosen. *Reviews of Modern Physics,* 29, 509–515.

35. Virginia Trimble, Joe Weber's wife "for the last 28½ years of his life", wrote a short, engaging and highly informative biography of Joe's scientific life, Wired by Weber: The story of the first searcher and searches for gravitational waves, published in the *European Physical Journal,* 42H (2017). This open access article can be downloaded from https://link.springer.com/article/10.11 40/epjh/e2016-70060-5.

36. The story of the gravitational wave discovery is very well summarized in M. Bartusiak. (2016). The long road to detecting gravity waves. *Science News,* 189, 24–27.

37. The annual Gravity Essay Prize was set up in 1949 by Roger Babson's Gravity Research Foundation, an organization at first created to encourage research into gravity and, in particular, anti-gravity. Many notable relativists have since won it, including Stephen Hawking in 1971. The reward for winning was prestige and the first prize of $1,000 when the prize was first set up. It has since risen to $4,000.

38. J. Weber. (1969). Evidence for discovery of gravitational radiation. *Physical Review Letters,* 22, 1320.

39. Kip Thorne has estimated to good accuracy that the Weber detector was not sensitive enough to detect the expected signal (the amplitude of the signal being several order of magnitudes smaller than the Weber bars limit).

40. J. Weber. (1970). Anisotropy and polarization in the gravitational-radiation experiments. *Physical Review Letters,* 25, 180.

41. In 1973, other failures to verify Weber's claims were reported by Anthony Tyson of Bell Laboratories (Murray Hill, New Jersey) and by James Levine and Richard Garwin of the IBM Thomas

Watson Research Center (Yorktown Heights, NY). Their story is told in some detail in Chapter 5 of Janna Levin's (2016). *Black Hole Blues and Other Songs from Outer Space*. Knopf Publishing.

42. J. Weber and B. Radak. (1996). Search for correlations of gamma ray bursts with gravitational radiation antenna pulses. *Nuovo Cimento B*, 111, 687–692.

43. M. Bartusiak. (2017). *Einstein's Unfinished Symphony. The Story of a Gamble, Two Black Holes, and a New Age of Astronomy*. New Haven: Yale University Press. The first edition of this book was published in 2000 by Joseph Henry Press, Washington, DC.

44. The story is in fact a little more complex. The LIGO team was testing with rare "fake" signals added to the data stream without any prior warning. After the event, these were logged. Marco Drago did what all good scientists do and checked to see whether anyone knew of any test signal that had been sent that night. That started a long period of checking the degree of confidence in the observation and the 1,000-strong LIGO group managed not to talk about the details publicly before the official announcement on 11 February 2016. In many ways, their collective silence was impressive and an example of how science should be done avoiding premature press releases. Marco modestly said that anyone in the team might have been on duty and been the first to see the event. The story is told by Adrian Cho in his February 11th news item for Science, *Here's the first person to spot those gravitational waves*.

45. The design was more specific for detecting collisions of binary neutron stars that were known to exist, like the Hulse–Taylor binary pulsar. The detection of gravitational waves from the merging of black holes was a nice "surprise".

46. At a more technical level, interferometers measure the phase difference between the light paths in the two arms. LIGO needed to do this to one part in 10^{10} in order to achieve the required sensitivity. So, the more light the better. There is an upper limit on the arm length, however, because the gravitational wave cancels itself if it goes through a complete wave cycle during the time it takes for the gravitational wave to pass. There is a fine albeit somewhat technical overview of the LIGO interferometer

in a lecture given by Alan Weinstein. *Physics of LIGO Lecture 2*, https://dcc.ligo.org/public/0033/G000164/000/G000164-00.pdf.

47. With a 4-km arm length it was necessary to allow for the curvature of the Earth when building the arms. Following the surface of the Earth, the center of the arm would be more than 1 meter above the ends. Building the base for the straight arms was one of the technical challenges.

48. Traditionally, speculum consisted of two parts copper to one part tin with a small admixture of arsenic, lead and silver. Bronze was a mixture of 7 parts copper with one part tin and used in ancient mirrors. Speculum expanded and contracted with changes in temperature, so William Herschel had to allow several hours each night to let his 40-inch speculum mirror come to a stable equilibrium before he could use it. His 20-inch was much more useful.

49. See https://www.ligo.caltech.edu/page/facts for gee-whiz facts about the LIGO technology. It should be emphasized that LIGO only measures the change in the length of the light path in one arm relative to the other and not its absolute length which might be impossible.

50. The early part of the LIGO story is told in great detail in Harry Collins' 2004 book, *Gravity's Shadow: The Search for Gravitational Waves*. University of Chicago Press. See in particular Chapters 31 and 32. Collins in 2017, then wrote *Gravity's Kiss: The Detection of gravitational Waves*. MIT Press. Collins is convinced that Barry Barish and his project manager Gary Sanders rescued "LIGO, and gravitational wave astronomy, from the brink of the grave".

51. Until 1993, Barish had been working on the design of the "GEM" (Gammas, Electrons and Muons) experiment for the NSF funded Superconducting Super-Collider (SSC) being built in Texas. When the SSC Super-Collider project was finally canceled in 1993, Barish was offered the position at LIGO.

52. See https://www.ligo.org/news/index.php#nobel for a brief running history of the LIGO consortium. The name "Virgo" came from the Virgo cluster of galaxies (at about 20 Mpc) because its sensitivity was designed to detect neutron star binaries at this distance. The Virgo project started in 1993 and has been working in parallel with LIGO since 1 August 2017.

53. The memorandum of agreement between LIGO, Virgo and KAGRA was signed on 4 October 2019.

54. At the time of writing, the LIGO–Virgo–KAGRA consortium will start its first joint operational season in May 2023.

55. We are familiar with pulsars, which are rapidly spinning neutron stars. A teaspoon of neutron star material would weigh around a billion tons.

56. There is some uncertainty in these numbers which are derived using complex models for binary neutron star mergers. The level of uncertainty is, however, not so significant that it would change our picture of the event.

57. Both signals traveled for 140 million years on their way to Earth and arrived within two seconds of each other: their speeds must have been almost identical. Think of two marathon runners who cross the finishing line one second apart after running for over two hours: their average speeds would have been almost identical, differing by at most one part in 7,200.

58. The mass estimates for neutron stars encountering black holes are somewhat uncertain. For example, the black hole mass assigned to GW200105 was somewhere in the range of 7.4–10.1 solar masses, while the neutron star mass was in the range of 1.7–2.2 solar masses. See https://www.ligo.caltech.edu/LA/news/ligo20210629 Fact Sheet.

59. Simulations of these black hole and neutron star mergers by the Max Planck Institute for Gravitational Physics are shown at https://www.aei.mpg.de/726542/gw200105-gw200115.

60. See Janna Levin *loc. cit.*, at the end of Chapter 9, Weber and Trimble.

Chapter 15

1. M. Bartusiak. (2009). *The Day We Found the Universe*. New York: Vintage Books.

2. V. Trimble. (1995). The 1920 Shapley-Curtis discussion: Background, issues, and aftermath. *Publications of the Astronomical Society of the Pacific*, 107, 1133–1144.

3. Readers interested in this topic should read Martin Rees's book *On the Future: Prospects for Humanity*. In this extraordinary

and provocative text, Lord Rees affirms that multi-verses are "speculative science, not just metaphysics".

4. D. Sobel. (2017). *The Glass Universe: How the Ladies of the Harvard Observatory Took the Measure of the Stars.* New York: Penguin Books.

5. H. Shapley. (1969). *Through Rugged Ways to the Stars.* New York: Charles Scribner's Sons.

6. D. Sobel, *loc. cit.*, p. 191.

7. O. Gingerich. (1990). *Shapley, Hubble, and Cosmology.* ASP Conference Series, Volume 10, Evolution of the Universe of Galaxies; Proceedings of the Edwin Hubble Centennial Symposium (San Francisco: ASP), ed. by Richard G. Kron, p. 19.

8. Otto Struve, director of Yerkes Observatory, described Cecilia Payne's PhD Thesis as "the most brilliant PhD thesis ever written in astronomy", as explains Patrick A Wayman in his paper, Cecilia Payne-Gaposchkin: Astronomer extraordinaire. *Astronomy & Geophysics*, 2002, 43, 1.27–1.29.

9. E. Hubble. (1929). A relation between distance and radial velocity among extra-galactic Nebulae. *Proceedings of the National Academy of Sciences of the United States of America*, 15(3), 168–173.

10. N. Aghanim *et al.* (2020). Planck 2018 results VI. Cosmological parameters. *Astronomy and Astrophysics,* 641, A6.

11. A. G. Riess *et al.* (2022). Comprehensive measurement of the local value of the hubble constant with 1 km s^{-1} Mpc^{-1} uncertainty from the Hubble Space Telescope and the SH0ES team. *The Astrophysical Journal Letters*, 934, L7.

12. Allan Sandage explained Hubble's views on the cosmic expansion: "Hubble, to the very end of his writings maintained this position, favouring (or at the very least keeping open) the model where no true expansion exists, and therefore that the redshift represents a hitherto unrecognized principle of nature". In A. Sandage. (1989). Edwin Hubble 1889–1953. *The Journal of the Royal Astronomical Society of Canada*, 83(6).

13. Features in the spectra of astronomical objects can appear shifted to longer wavelengths (redshifted) for three reasons: (1) the light has had a tough climb out of a gravitational potential, (2) the source and observer are moving away from one another (this is

called a Doppler shift, was first detected for sound in air, is correctly described in special relativity, and can also produce blue shifts to shorter wavelengths, if source and observer are moving towards each other), and (3) expansion of space-time itself (adequately described in general relativity). At various times each of the three has been invoked as the cause of Hubble's law (or redshift-distance relation, linear only for small distances; curved various ways at large distances in various cosmological models (that is, various values of cosmic density, pressure, and so forth). Decades of observations since Hubble's time have ruled out (1) and (2). Only (3) remains and is almost universally regarded as the right answer.

Chapter 16

1. H. N. Russell, R. S. Dugan and J. Q. Stewart. (1926). *Astronomy.* Boston: Ginn & Co.
2. J. C. Kapteyn. (1922). First attempt at a theory of the arrangement and motion of the sidereal system. *The Astrophysical Journal,* 55, 302.
3. F. Zwicky. (1933). Die Rotverschiebung von extragalaktischen Nebeln. *Helvetica Physica Acta,* 6, 110–127. Translation provided by Robert H. Sanders in his book *The Dark Matter Problem. A Historical Perspective.* Cambridge University Press, p. 14 (2010).
4. M. Schwarzschild. (1954). Mass distribution and mass-luminosity ratio in galaxies. *The Astronomical Journal,* 59, 273.
5. J. H. Oort. (1940). Some problems concerning the structure and dynamics of the galactic system and the elliptical Nebulae NGC 3115 and 4494. *The Astrophysical Journal,* 91, 273.
6. H. W. Babcock. (1939). *Lick Observatory Bulletin,* 19, 41. Asked in 1986 by one of your authors why neither he nor anyone else had followed up on this at the time, he responded that he had almost been laughed off the stage when he presented the results at a conference. And then, of course, there was a war, and in due course he followed his father, H. D. Babcock, into solar astronomy.
7. K. Lundmark. (1939). Über die bestimmung der entfernungen dimensionen, massen, und dichtigkeiten für die

nächstgelegenen anagalaktischen sternsysteme. *Meddelande från Lunds Astronomiska Observatorium*, No. 125.

8. G. Münch. (1959). The mass-luminosity ratio in stellar systems. *Publications of the Astronomical Society of the Pacific*, 71, 101. He found the lack of correlation with anything so discouraging that he never touched the subject again.

9. V. C. Rubin and W. Kent Ford II. (1970). Rotation of the Andromeda Nebula from a spectroscopic survey of emission regions. *The Astrophysical Journal*, 159, 379. It is probably worth mentioning that, when Rubin started out to track the rotation curve, she had hopes of finding different sorts for spirals of different morphologies and was initially somewhat annoyed, according to her own autobiography, when they all looked the same and flat.

10. M. S. Roberts. (1966). Manuscript for Volume IX, *Galaxies and the Universe*, A. Sandage, M. Sandage, and J. Kristian (eds). University of Chicago Press. Though the manuscript was prepared then, the volume finally appeared in 1971. W. K. Ford, V. C. Rubin, and M.S. Roberts. (1971). A comparison of 21 cm radial velocities and optical velocities of galaxies. *The Astronomical Journal*, 76, 22 is also of interest in these connections.

11. D. H. Rogstad. (1967). Rotation and masses of galaxies as determined by single spacing interferometry of 21 cm hydrogen emission. PhD thesis, submitted in partial fulfillment of the requirements for a PhD in astronomy at the California Institute of Technology. Work done under Gordon Stanley. Much of the data also appear in D. H. Rogstad, G. W. Rougoor, and J. B. Whiteoak. (1967). *The Astrophysical Journal*, 150, 9.

12. L. Smolin. (2006). *The Trouble with Physics: The Rise of String Theory, the Fall of a Science, and What Comes Next*. Boston: Houghton Mifflin.

13. This is a type of radioactive decay in which a fast energetic electron or positron and a neutrino are emitted from an atomic nucleus and a neutron in the nucleus becomes a proton, or conversely a proton becomes a neutron.

14. The next paragraphs are excerpted from the script of the video *Hágase la Masa* (2013) in Spanish (*Let there be mass*), on the

Higgs boson. https://www.youtube.com/watch?v=trN6IeMLv3 o&t=64s, directed by Javier Díez, Vicent J. Martinez was the scientific director and the authors of the script were Fernando Ballesteros, Vicent J. Martínez and Amelia Ortiz.

Chapter 17

1. The history of astronomical observation is a story of an ever-expanding view of "the Universe" and an ever-growing ability to acquire higher-quality data. Galileo's telescope is an outstanding example of a revolution in expanding our view of the universe, as was the recent development of the LIGO interferometer that detected gravitational waves for the first time. Cosmology is driven by developments in instrumentation for data gathering and computers for modeling and analysis.

2. The basic building blocks of atoms are neutrons, protons and electrons. Helium is made up from a nucleus of two neutron and two protons orbited by two electrons. With a very simple model for the cosmic expansion starting from an ultra-hot Big Bang, we find that roughly 10% of atoms in the universe will be helium atoms making up some 25% of the total mass in the universe.

3. Cosmologists nevertheless enjoy speculating about the universe, as we shall see later on in this chapter. It might even be argued that such speculation is an important driver of new ideas to pursue, especially when there is a possibility of experimental verification of the idea.

4. The Antikythera mechanism is perhaps the outstanding example of an ancient astronomical calculator. See Chapter 1 and Section 1.4.

5. By "plausible" here, it is meant that such an experiment is conceivable in terms of known or incipient technology and, importantly, of cost. Many of the great astronomical experiments of the 20th century, such as the Large Hadron Collider and the LIGO project to detect gravitational waves, took decades to build and cost billions of euros/dollars. The cost of Tycho Brahe's Uraniborg Observatory was on the order of 1% of the Danish GDP. See https://thonyc.wordpress.com/2014/10/30/financing-tychos-little-piece-of-heaven/. Since 2000 the US government spends around 0.5% of its annual fiscal budget on NASA.

6. An example of this in modern times was the so-called "Helium Problem" which arose when astronomers in the early 1960s estimated the amount of helium in the universe. They estimated that 10% of the atoms in the universe were helium atoms. The question was as follows: Where did those atoms come from? The hot Big Bang theory had a ready explanation: they were "cooked" in the first few minutes after the Big Bang. The alternative steady-state theory had no solution to offer other than to propose that the Universe was populated by as yet unseen helium producing stars. In other words, they sought to solve the problem by introducing an entity outside of the current thoughts on stellar evolution and whose only reason for existing was to solve the problem. The correct Big Bang prediction by itself was not sufficient to cause a mass exodus of steady-state supporters to the Big Bang camp.

7. Obviously, other groups were involved, particularly those involving astrophysicists who had been working on theoretical cosmology prior to the discovery. Cambridge UK (Sciama and Rees), Harvard (Layzer and Silk), Berkeley (Sachs), Massachusetts (Harrison), Paris (Audouze) and Hiroshima (Nariai and Tomita).

8. To get a sense of this, we can think of looking at the Pacific Ocean from the Moon: it looks much the same all over. When we get closer, or zoom in on the map, islands emerge and even closer still we see ships and ultimately people moving around.

9. There had been a serious controversy involving Einstein and Hilbert. Both had almost simultaneously published papers presenting what were later to be called "the Einstein Field Equations". There had been allegations that Hilbert had plagiarized Einstein's work, but Hilbert took the initiative of referring to them as Einstein's, despite the fact that he was undoubtedly the first to publish them.

10. The novelty of Einstein's geometric view of space–time was the great insight and also the first hurdle in getting to grips with the theory. In the early days when Marcel Grossman introduced him to the geometry of curved spaces, Einstein expressed exasperation with its complexity in his *Zurich Notebook*. Even after surmounting that difficulty, there was, for a considerable time (decades), confusion when making the link between the outcome of a relativistic calculation and the real world. The confusion

was a consequence of the freedom with the theory to choose coordinate systems that might not have a direct link with the physics. Many people, including de Sitter himself, had deduced an expansion law for de Sitter's model that the velocity would be proportional to the square of the distance, as opposed to Hubble's discovery of a linear relationship. This was only resolved in the early 1930s by Howard Robertson. But even after that, many errors were made in interpreting results.

11. Droste had been working under the direction of Lorentz looking for simple solutions to an early version of General Relativity published in 1913. His thesis produced the same solution as was later found from the 1916 final version of the theory.

12. Einstein had realized that his method of deriving his equations allowed for an extra constant energy term to be arbitrarily included. He had no prior reason to include this term except that it allowed him to find this solution of his equations in which the term would act like a pressure term balancing the pull of gravity. If that pressure was too large, it would overcome the force of gravity and cause the model to fly apart, which was almost de Sitter's solution.

13. When Einstein was asked how it felt to be the smartest man on Earth, he responded that "you should ask Nikola Tesla about that".

14. Slipher had earlier presented a list of 15 of those velocities to the 1915 17th meeting of the American Astronomical Society. Later, in 1922, he sent a list of 41 velocities to Arthur Eddington in Cambridge for inclusion in his book *The Mathematical Theory of Relativity* (CUP 1923).

15. There was another fundamental division in the thinking of the two groups which persists to the present day. The astronomers would describe the system of galaxies as "rushing away from one another", whereas the relativist would see this as the expansion of the geometry of the fabric of space–time in which the galaxies are embedded. In the relativistic point of view, the galaxies inherit their observed velocity from the change with time of the space–time geometry and may additionally have "peculiar" velocities due to their mutual gravitational interactions.

16. Eddington, *loc. cit.* §69, p. 162.

17. Knut Lundmark had done something similar in 1924, but only for the de Sitter empty universe model. The article concentrated on using a local approximation to de Sitter's model for the system of globular clusters. Nevertheless, he was able to plot a "Hubble diagram" for Slipher's nebulae using brightness and diameter as distance indicators. The large scatter, due to greatly uncertain distances, made it impossible to say that there was a linear distance–velocity relationship.

18. Members of the Zeldovich group, Andrei Doroshkevich and Igor Novikov, concluded in 1964 that the radiation was detectable.

19. In 1955, Émile le Roux had measured a background "noise" in his microwave antenna at the Nançay Radio Observatory to which he assigned a temperature of 3 ± 2 kelvins (degrees above absolute zero) and by Tigran Shmaonov who in 1957 reported a background noise of 4 ± 3 kelvins. In 1961, Edward Ohm, using also the Holmdel horn antenna reported a couple of degrees of "inallocable antenna temperature" in the *Bell System Technical Journal.*

20. At the same time, and quite independently since this was the Cold War period, the Moscow group of the great Soviet physicist, Yakov Zeldovich, pursued the same goals of defining physical cosmology and in particular the problems of the origin of cosmological structure and the properties of the background radiation field.

21. Peebles has recently published an excellent account about this story: *Cosmology's Century. An Inside History of Our Modern Understanding of the Universe.* Princeton (2020).

22. The temperature is proportional to the frequency: As the temperature drops so does the frequency, and obviously the wavelength increases.

23. In the case of singing, it is possible to understand the shape of this sound spectrum by modeling the human vocal tract. Likewise, in the universe, we can extract the properties of the universe 380,000 years ago using a physical model of the process of recombination.

24. The spectrum of microwave background radiation matches (almost perfectly) what physicists call a black body: An ideal object that absorbs all the radiation that hits it without reflecting anything. Radiation emitted by a black body depends only on its temperature.

25. Temperatures are given on the Kelvin scale in which the zero-point is the coldest temperature that the laws of physics allow, $-273.16C$ on the Celsius scale. As of 2019, the size of a kelvin was redefined in terms of seven more fundamental constants.

26. Many previous ground-based and high-altitude balloon-based experiments had previously put limits on the anisotropy on various angular scales. The balloon-based experiments provided the first maps and collectively these experiments provided the first view of the expected signature peak of the CMB angular spectrum.

27. Both stressed that the results were the outcome of the work of the COBE team of more than 1,000 scientists and engineers.

28. The Moon subtends an angle of half a degree on the sky.

29. Balloon Observations Of Millimetric Extragalactic Radiation and Geophysics. This was one of many end-of-the-century balloon-borne experiments given exotic acronymic names like MAXIMA, Saskatoon, TOCO and others that saw the transition from COBE to the next generation of space experiments, WMAP and then Planck.

30. See https://www.nytimes.com/2000/04/27/us/clearest-picture-of-infant-universe-sees-it-all-and-questions-it-too.html for the reactions of a number of prominent inflation scientists.

31. However, the analysis of the background radiation could not have been done without the prior knowledge of what the universe looked like and prior knowledge of what the universe was made of.

32. From the figures on the graph, the fraction of *baryonic matter*, seen and unseen, is (total baryonic) /(total matter) = 4.9/(4.9 + 26.1) = 0.158.

33. Loosely put, the dark energy does not change throughout the expansion, unlike the density of matter which decreases as the universe expands. So, in the past, the matter dominated the first 10 billion years or so of the cosmic expansion. When the matter level fell below the dark energy level, the dark energy took over the expansion. It has dominated the past 4 billion years of cosmic history. Our Sun and Earth were born only slightly before that time.

34. What has been called the causal horizon here is more commonly called the "particle horizon", formally defined as the distance

that light can travel between two times. "Causal horizon" seems more *apropos* in the current context.

35. All magnets have a North Pole and a South Pole; no experiment has ever demonstrated the existence of a bare pole without raising considerable controversy. The magnetic monopoles arise from attempts to unify all the forces between elementary particles, the so-called Grand Unified Theories.

Chapter 18

1. The process of fertilization takes place in the fallopian tube and takes about 30 hours when a zygote is produced.

2. In this section and the following sections, we are going to see some extremely large numbers and also some extremely small ones. The scientific notation for these numbers is such that 10^{12} means a one followed by 12 zeros. Thus, a thousand is 10^3, a million is 10^6, a billion is 10^9, a trillion is 10^{12} and a quadrillion is 10^{15}. Note that for the numbers that are small than one, we write one millionth as 10^{-6}: there are *five* zeros followed by a one. A billionth is 10^{-9} and so on. When we speak of a pico-second. See also footnote 13.

3. There are *macroscopic* level, empirical and laws for determining the boiling point or freezing point of a solvent. These involve empirically determined constants. The physics underlying such a law, and the values of the constants, is an extremely complex problem in molecular dynamics.

4. This is an aspect of the "particle-wave duality" of quantum mechanics in which, for example, light can be viewed either as a wave or as a stream of particles (photons).

5. The following papers can be said to chart the origin of the inflationary universe idea:

   - A. D. Linde. (1979). Phase transitions in gauge theories and cosmology. *Reports on Progress in Physics*, 42, 379.
   - D. Kazanas. (1980). Dynamics of the universe and spontaneous symmetry breaking. *Astrophysical Journal Letters*, 241, L59–L63.

- K. Sato. (1981). Cosmological Baryon-number domain structure and the first order phase transition of a vacuum. *Physics Letters*, 99B, 66–70.
- K. Sato. (1981). First-order phase transition of a vacuum and the expansion of the universe. *Monthly Notices of the Royal Astronomical Society*, 195, 467–469.
- A. H. Guth. (1981). Inflationary universe: A possible solution to the horizon and flatness problems. *Physical Review D*, 23, 347–356.
- A. H. Guth and S.-H. H. Tye. (1980). Phase transitions and magnetic monopole production in the very early universe. *Physical Review Letters*, 44, 631.

6. In the world of quantum physics, we regard the action of the forces as being mediated by particles through which the particles influence one another. This is an aspect the famous wave-particle duality of quantum theory in which the fundamental components of matter can be considered either as a particle or a wave depending on the circumstances.

7. Therefore uncharged particles have no antiparticle. Formally, an electrically neutral particle is its own antiparticle.

8. The two, W bosons and Z bosons, are anomalous since they have mass which has been acquired through their interaction with the Higgs. The two W's, W^+ and W^-, are also the only charged force particles.

9. Richard Feynman objected to using a "flavor" on the grounds that it added nothing to our understanding and was at best misleading.

10. The mass of the gluon is zero, but like the photon, it carries energy.

11. The hadrons get their mass from the gluons they contain.

12. François Englert's collaborator Robert Brout died in 2011 and so could not be awarded the Prize. There were suggestions that Tom Kibble of Imperial College, London, should have shared this prize (Kibble died in 2016). Carl Hagen (University of Rochester) and Gerald Guralink (Brown University) also wrote key papers on the same subject.

13. Yotta is a decimal prefix of the metric system denoting a factor of 10^{24}, a "1" followed by 24 zeros. The mass of the Earth is about 6,000 yottagrams (Yg). See https://en.wikipedia.org/wiki/Metric_prefix.
14. Formally, in physics, a gas exhibits negative pressure when an increase in volume causes a decrease in entropy.
15. A colloidal solution is a suspension of particles in a fluid, as opposed to a solution of one substance in another. Examples of colloids are blood, mayonnaise, milk, muddy water and paint.
16. The usual way of explaining this is to view the vacuum between the mirrors as a collection of waves. Wavelengths that are longer than the separation between the mirrors are excluded from the gap (they don't fit), and so there is less quantum vacuum in the gap than there would otherwise be. Put another way, there is more energy outside the gap than inside (fewer waves). So, the plates move towards one another.
17. The quantum version of the theory of electromagnetism.
18. We are easily fooled by nonflat geometry. The shortest flight between Singapore and New York is via Alaska and over Hudson Bay (18 hr 45 m, the longest nonstop commercial flight). On the standard Mercator Map, this seems quite extraordinary: it looks far longer than flying the straight line over Honolulu and Los Angeles.
19. E. Kolb and M. Turner. (1990). *The Early Universe*. Addison Wesley, p. 86.
20. Cross-disciplinary science has never been easy, and despite the Internet and the wider dissemination of research work, it is still not easy.
21. NASA ADS search on "abs:"inflation" year: 1970–2021".
22. According to the article, A cosmic controversy, *Scientific American*, 10 May 2017, signed by 33 leading scientists.
23. This is a lightweight requirement: there may be an extra parameter in the theory that can be tweaked or an added epicycle that gives the observed result.
24. The announcement was made by what was then the LIGO–Virgo collaboration. The Virgo detector was not online at that time; it was commissioned in 2016 and made its first detection in August 2017 (GW170814).

25. Sound waves are waves in air that are compressed in the *direction* of travel of the wave: they are longitudinal waves and they are not polarized. Light is a transverse wave oscillating in a plane perpendicular to the direction of travel, like a wave on a rope that is fixed at one end.

26. Cheap sunglasses have the laminate on the surface of the lens; expensive ones have the laminate between two layers of lens material.

27. Not only is the altitude and air clarity at the South Pole especially favorable, but the lack of a 12-hour alternation between night and day makes the environment very stable over long periods of time.

28. We stress the pattern around the maxima of the temperature fluctuation map because the polarization pattern around the minima is almost indistinguishable from the B-mode.

29. See B. Keating. (2018). *Losing the Nobel Prize: A Story of Cosmology, Ambition, and the Perils of Science's Highest Honor.* W. W. Norton & Company. The book is not only a fine introduction to cosmology but also an autobiographical account of the story of the BICEP program from the point of view of the instigator of the project.

30. 150 GHz corresponds to wavelength = 2 mm. For comparison, terrestrial 5G communications work in the range of 3.4–4.0 GHz depending on service provider. Light of this frequency is focused using special lenses.

31. To quote from the abstract of the paper, "...tensor-to-scalar ratio r with $r = 0$ disfavored at 7.0 σ."

32. https://www.nytimes.com/2014/03/18/science/space/detection-of-waves-in-space-buttresses-landmark-theory-of-big-bang.html, written by Dennis Overbye.

33. The Planck experiment had published maps of the sky at multiple frequencies. Astonishingly, the BICEP team digitized the published versions of the maps and used that to remove the contribution of the Galaxy to their data.

34. Strassler discussed a figure in the paper which worried him.

35. Pronounced "archive" (the middle letter is a Greek chi), the arXiv is the main e-print repository for astronomy, physics and other sciences, https://arxiv.org/. Most arXiv papers posted are about to be or are still under peer review and many appear in

several revisions. This Planck paper is available at https://arxiv.
org/pdf/1405.0871v1.pdf.

36. http://resonaances.blogspot.com/2014/05/is-bicep-wrong.html.
37. Keating, *loc. cit.*
38. J. Dunkley. (2020). *Our Universe: An Astronomer's Guide.* Penguin Random House, UK: Pelican Books, pp. 216–217.
39. A. Ijjas, P. J. Steinhardt and A. Loeb. (2014). Inflationary paradigm in trouble after Planck 2013. *Physics Letters B*, 723, 261.
40. A. Ijjas, P. J. Steinhardt and A. Loeb. (2014). Scale-free primordial cosmology. *Physical Review D*, 89, 023525.
41. This was based on far more complex earlier work published in 2001: J. Khoury, B. A. Ovrut, P. J. Steinhardt and N. Turok. (2001). Ekpyrotic universe: Colliding branes and the origin of the hot big bang. *Physical Review D*, 64, 123522. This was followed by a cyclic universe version: P. J. Steinhardt and N. Turok. (2002). A cyclic model of the universe. *Science*, 296, 1436–1439.

Chapter 19

1. A valuable collection of documents and programs associated with Babbage's Analytic Engine is available at https://www.fou rmilab.ch/babbage/contents.html. This was the second of the two machines that Babbage developed. The first was the *Difference Engine* that was a mechanical calculator about the size of a piano. It could do elementary arithmetic but it required too much human intervention to be economical and was abandoned in 1822. Revisions followed until 1833 when the project was stopped, largely because investors saw that the project was going nowhere. Undaunted by failure, work started on a completely new concept in 1837: the *Analytic Engine*.
2. The unification of Italy was a process spanning most of the 19th century. The capture of Rome and the fall of the Papal States in 1871 marks the completion of the process, although it was not until 1918 that Italy as we know it today emerged. Menabrea served as prime minister from 1867 to 1869.

3. L. F. Menabrea. (1842). *Bibliothèque Universelle de Genève*, 42, 352–376. The journal, published by a group of Genevan Scholars, had a wide audience in French-speaking Europe.

4. English titled society has a Byzantine system of inheritance of titles and names. Ada's father was the poet George Byron, the 6th Baron Byron. So prior to her marriage, Ada was known as the Hon. Augusta Byron. Ada became "Lady King" on her marriage (July 1835) to William, 8th Baron King. At that point, she became known as Lady Ada King. Ada was a descendant of the extinct Baron Lovelace family and in 1838 her husband was made the 1st. Earl of Lovelace and Viscount Okham, whereupon Ada became known as Ada King, Countess of Lovelace.

5. The Lovelace translation and commentary are published as L. F. Menabrea. (1843). Sketch of the analytical engine invented by Charles Babbage. *Scientific Memoirs*, 3, 666–731. If this were published today, it would, with Menabrea's agreement, be referred to as the Lovelace and Menabrea paper. Menabrea later wrote expressing his appreciation of her contribution.

6. Like many titled young girls, she made her formal societal debut at the age of 17, after which her tutor, Mary Somerville, took her to the Saturday salon of Charles Babbage. (Mary Somerville was a distinguished Scottish scientist who, alongside Caroline Herschel, became the first female Honorary Members of the Royal Astronomical Society.) Babbage was evidently impressed with Lovelace's knowledge: that meeting was the start of a long and fruitful platonic friendship.

7. Lovelace's work was revived 101 years after her death when it became a centerpiece for the "History and Theory of Computing" section of the 1952 Manchester Symposium on Digital Computing machines, the Proceedings of which were edited by B. V. Bowden (1953). *Faster than Thought*. Pitman. Bowden presents a strong focus on the parallels between the Babbage-Lovelace vision and the digital computers being built during and after the Second World War.

8. A program to compute the "Bernoulli Numbers", a sequence of rational numbers that appear frequently in the representation of functions by power series.

9. The Jacquard loom can be regarded as a programmable computer that drives a machine and, in that sense, the Jacquard

card decks represent a computer program. The machine does not have an internal data store, which is why each row of material has its own card, even when the pattern is repeated. The consequence of which is that a woven pattern may take tens of thousands of cards to create. Lovelace comments extensively on this in her Note C, asserting that an internal data store is a necessary feature of a computing machine.

10. Turing completeness is a deep mathematical concept. In practice it means that a Turing complete computer can, given sufficient time and memory, run any logically correct program, including an emulation of another Turing complete machine. The simple implication of this is that it is unnecessary to build a new special-purpose machine for each task to be computed: a Turing complete machine can run them all. We see this in the modern world where the same "apps" (programs) can in principle run on all brands and species of mobile phone, or on desktops or super-computers. A desktop computer can emulate a super-computer. The next Turing complete machine to be designed was Ludgate's 1909 Analytic Engine, which, like Babbage's, was mechanical and never built. The first built and working Turing Complete machine, according to https://en.wikipedia.org/wiki/Ana lytical_Engine#Comparison_to_other_early_computers, was the electronic "E.N.I.A.C.", designed in 1945 by John Mauchly and J. Presper Eckert of the University of Pennsylvania for the US Army Ordnance Corps.

11. The prolific and celebrated statistician Karl Pearson, famous among other things for introducing the Pearson Correlation Function (1896) and Chi-square test (1900), was one of the pioneer users and an influential advocate for the use of calculating machines. He saw them as a great time-saver and when asked how he wrote so many papers famously answered that "I never answer a telephone or attend a committee meeting".

12. She did not write about other nonmathematical uses such as word and document processing which 140 years later both Apple and Microsoft saw as the key to expanding their markets beyond the 1980s.

13. This is easiest to achieve with data that has been pre-classified ("face" or "no face"). However with millions of pictures super-computers can learn what a face is without pre-classification: it learns what is or is not a face.

14. A recent, highly technical, investigation into this question has been attempted by the Princeton researcher H. Qin. (2020). Machine learning and serving of discrete field theories. *Science Reports*, Art. 19329.

15. J. Jumper *et al.* (2021). Highly accurate protein structure prediction with AlphaFold. 596, 583–589. The work was first made public at an online workshop in May–July 2020 when, under the name *AlphaFold2*, it won the CASP14 assessment of methods for predicting protein structure by a large margin. (CASP is the acronym for the *Critical Assessment of Structure Prediction*, a community-wide experiment to determine and advance the state of the art in modelling protein structure from amino acid sequences.)

16. At the time of writing, it would be over-enthusiastic to assert that the problem had been "solved", though we should nevertheless agree that this is an extraordinary success. What is missing is verification by other groups using their own databases. To this end, Google and DeepMind have put all of their work into the public domain. See the discussion at https://en.wikipedia.org/wiki/Talk:AlphaFold.

17. AlphaFold is one of several developments of the *DeepMind* research group who were famous for building *AlphaGo* that played the game of Go and *AlphaZero* that played Chess, both achieving levels of play beyond those achieved by any human player. Both games learned the rules simply by watching the game. DeepMind merged with Google in 2014, and in 2015 Google restructured itself, creating Alphabet Inc. which owns both Google and DeepMind.

18. Some 100 million distinct proteins have been identified, a small fraction of which have exactly known 3-D structures.

19. These 20 amino acids are found in the genetic code. Nine of these are classified as "essential"; they cannot be made by the body and come from what we eat. A further two can be added using special mechanisms.

20. See, for example, the excellent *The Structure of Life* booklet issued by the U.S. National Institute of General Medical Science: https://www.nigms.nih.gov/education/Booklets/The-Structures-of-Life/Pages/Home.aspx.

21. A virus alone does not have the means to replicate itself. It needs the resources of a *living* cell, and the process may kill the cell.

22. Proteins called "antibodies" are the "search and destroy mechanism" for defense against invasion of our bodies by alien invaders, viruses, bacteria, fungi and so on. They have a specific "Y"-shape, the top branches of which connect to the invader, while the stem connects to resources that can kill the invader. The tests for the antibodies of a given type allows us to identify whether or not a person has hosted a bacterium of another infecting agent.

23. Protein shape may be affected by ambient variations in acidity and temperature. The misfolded proteins clump together to produce large unwanted polymers which interfere with normal processes. The amyloid and tau proteins are among the proteins thought to interfere with the communication between brain cells. Disorders due to misfolding are also associated with non-neurological disease.

24. "AI", interpreted as "Artificial Intelligence", is a widely used term that is poorly defined and frequently misunderstood. The alternative interpretation as "Augmented Intelligence" would provide a somewhat better description of what AI is about.

25. See Davies *et al.* (2021). Advancing mathematics by guiding human intuition with AI. *Nature,* 600, 70–74.

26. Failure to properly curate a database is one of the most error-prone stages in designing this kind of AI. One difficulty is that some form of exploratory analysis has always been a first step when trying to rationalize what our sense of perception tells us, despite the unconscious biases that this may involve.

27. There was much discussion that poorly understood, human-designed, "algorithms" might have played a part in causing the Global Financial Crisis of 2008. There is little concrete evidence for that; see https://en.wikipedia.org/wiki/Subprime_mortgage_crisis.

28. See M. Ossendrijver. (2016). Ancient Babylonian astronomers calculated Jupiter's position from the area under a time-velocity graph. *Nature,* 351, 482–484 and a popular account

at https://www.scientificamerican.com/article/babylonians-tra cked-jupiter-with-fancy-math-tablet-reveals/. Ancient Babylonian astrologers made long-term predictions of the position of the planet Jupiter simply from the existing data, with no underlying theory for the motion of the planet. The court astrologers, who were well versed in mathematics and geometry, discovered a mathematical method, an "algorithm" in modern parlance, with which they could use previously observed positions of the planet to make long-term predictions of its position in the sky.

29. The formal designation of this comet is C/1680 VI. The discovery in November 1680 by Gottfried Kirsch was the first telescopic discovery of a comet, and hence this was the first comet to be discovered before it was visible to the naked eye.

30. D. Seargent. (2009). *The Greatest Comets in History*. Springer, Ch. 7 lists 52 Daylight Comets from 185 BCE to 2009 CE.

31. The Sun's radius is about 700,000 km.

32. On 28 December, Robert Hooke reported that the tail was 90° long and 2° wide.

33. These four all developed methods for fitting straight-line paths to observations of comet positions. Discussed in C. A. Wilson. (1970). From Kepler's laws, so-called, to universal gravitation: Empirical factors. *Archive for History of Exact Sciences*, 6, 89–170, see p. 152.

34. See D. K. Yeomans. (1991). Comets: A chronological history of observations. *Science, Myth, and Folklore*. Also T. Heidarzadeh. (2008). *A History of Physical Theories of Comets, From Aristotle to Whipple*, Springer, Ch. 4. Wiley Science Editions, in particular Ch. 3.

35. Kepler had long before determined the orbit of Mars without reference to gravitation and seen that it was not circular.

36. The source of the data was Hevelius' *Cometographia*. He selected mainly those comets that had seen both before and after their closest approach to the Sun.

37. Ruffner, J. A. (2000). Newton's propositions on comets: Steps in transition, 1681–84. *Archive for History of Exact Sciences*, 54, 259–277.

38. Both Halley and Newton reported on this conversation in their reminiscences many years after the event.

39. Newton acknowledged Boulliau. Boulliau had argued, using the analogy with the intensity of light falling off inversely as the square of the distance, that an inverse square force law was necessary for Kepler's Laws of planetary motion to hold, albeit for quite different reasons. In the modern sense of field theory, Boulliau was correct: in Einstein's general relativity the inverse square law arises from what is called Gauss' Law for conservation of flux. See http://hyperphysics.phy-astr.gsu.edu/hbase/electric/gaulaw.html.

40. As put by Edward Dolnik in his 2011 book, *The Clockwork Universe: Isaac Newton, the Royal Society, and the Birth of the Modern World.* Harper-Collins. See Ch. 52, "In search of God", final two paragraphs.

41. Newton had tried unsuccessfully to prove that the comets of 1680 and 1681 were the same observed twice: before and after its orbit had taken it close by the Sun.

42. The main source here is D. Grier. (2006). *When Computers Were Human.* Princeton. Grier describes the project and those involved in considerable detail in Chapter 1, The first anticipated return: Halley's Comet 1758, pp. 11–25. Other sources are G. Bernardi. (2016). *The Unforgotten Sisters: Female Astronomers and Scientists Before Caroline Herschel.* Springer, Ch. 19, pp. 121–127 and the fun-to-read story told in C. Sagan and A. Druyan's. (1997). *Comet.* Ballantine Books, Ch. 4, The time of the return.

43. This was the year the prize was set up.

44. Grier, *loc. cit.*, p. 21.

45. Uranus and Neptune in fact contributed 6 days. The four inner planets contributed another 6 days and errors in the masses for Jupiter and Saturn contributed 4 days. See P. Broughton. (1985). The first predicted return of comet Halley. *Journal for the History of Astronomy*, 16, 123–133.

46. There was no simple mathematical solution to Newton's laws for a system of more than two bodies. However, given sufficient computing power, human or otherwise, we can achieve numerical solutions. Using modern supercomputers, we can follow the evolution of gravitating systems consisting of billions of point masses.

47. So called because Stevin developed a notation based on units of $0.1, 0.01, 0.001$, *etc.* Thus, he could write 3.14 as 314 in units 0.01.

By extending to arbitrarily small "tenths", he could represent any number with as much precision as was required. This gave rise to the idea of *infinitesimals* which in turn led to the calculus.

48. For the full history, see F. Cajori. (1828). *A History of Mathematical Notation*, Open Court, Vol. 1 *Notations in Elementary Mathematics* and Vol. 2 *Notations Mainly in Higher Mathematics*.

49. This is the description of the quadratic equation $(10 - x)^2 = 81x$. From https://en.wikipedia.org/wiki/Muhammad_ibn_Musa _al-Khwarizmi.

50. See https://en.wikipedia.org/wiki/Table_of_mathematical_symb ols_by_introduction_date for an extensive list.

51. In 1615, Henry Briggs (1561–1630) visited Napier and recommended rescaling Napier's logarithms to better represent the fact that we count in tens. Briggs' rescaled logarithms became the most widely used. Napier went on to design a device for doing arithmetic, "Napier's bones" which could be used to extract square roots. This was followed by the forerunner of the Slide Rule by the English mathematician William Oughtred (1574–1660). Slide rules and log tables were used up until the 1970s when the first pocket scientific electronic calculator arrived on the scene, the Hewlett Packard HP-35. The 1970s saw the advent of desktop programmable computers like the Wang 2200 (1973 — US$7,400), the Altair 8800 (1975, sold as a kit for US$621) and the Apple-1 (1976 — US$666.66). In the UK, we saw the BBC Micro (1981 — £235) and the ZX Spectrum (1982 — £125), both of which used the owner's TV for display.

52. See John Norton's A peek into Einstein's Zurich Notebook at https://sites.pitt.edu/~jdnorton/Goodies/Zurich_Notebook/. Einstein's frustration with the mathematical complexity of the what he is trying to do is shown at the end of the section *The Riemann tensor at last* where, in frustration, he scribbled "zu umstaendlich", "too involved" across the bottom of the page. The notebook is one of the most fascinating documents in science.

53. K. Popper. (1959). *The Logic of Scientific Discovery*. Hutchinson (first English-language translation). As Popper puts his thesis, "all the statements of empirical science (or all "meaningful" statements) must be capable of being finally decided, with

respect to their truth and falsity" (*loc. cit.*, p. 17). The question this raises is as follows: Who is to be the judge of truth or falsity whose verdict will be equally acceptable by both sides? Astrology offers a case in point.

54. See, for example, S. A. McLeod. (2020). Karl Popper — Theory of falsification, www.simplypsychology.org/Karl-Popper.html, where it says "For Popper the scientist should attempt to disprove his/her theory rather than attempt to continually prove it."

55. Adding a further epicycle is, in modern terms, just like adding an additional Fourier component to the representation of a function in order to improve resolution. That is normally regarded as an acceptable procedure.

56. The story is engagingly told in David Bodanis' 2005 book *Electric Universe — How Electricity Switched on the Modern World*. Crown.

57. At the *Google Zeitgeist Conference*, England, May 2011.

58. See also the philosopher Graham Hartman's commentary on this in his 2012 publication *Concerning Stephen Hawking's Claim that Philosophy Is Dead*, published in Filozofski Vestnik, Vol. 33, pp. 11–22. This can be downloaded from https://ojs.zrc-sazu.si/filozofski-vestnik/article/download/3240/2957.

Acknowledgments

Our first, greatest gratitude clearly goes to our parents, who started us on paths of learning, and then to the educators who taught us science, history and philosophy or at least tried to.

Moving a little closer to the present, student Bernard Jones first encountered post-doc Virginia Trimble in 1969 at Fred Hoyle's Institute of Theoretical Astronomy in Cambridge (UK). Student Vicent Martínez then encountered Professor Jones who became his advisor in 1986 at NORDITA and the Niels Bohr Institute in Copenhagen. And Professor Trimble encountered Dr. Martínez as his junior co-organizer of a September 2000 summer school on Historical Development of Modern Cosmology in Valencia, Spain. Enn Saar of Tartu Observatory (Estonia) joined our team in 2004 for a review article on "Scaling Laws in the Distribution of Galaxies" and was, early on, involved in the project represented by this volume.

Fast forward another decade or so, we encountered both the Alexander Pope's quotation "Be not the first by whom the new is tried, nor yet the last to lay the old aside" and the idea that some astronomical entities had once been thought essential but turned out not to exist, stir vigorously, and you get Trimble, Martínez and Thomas Hockey (of the University of Northern Iowa), 15 pages in a mostly Polish-language calendar, published in 2011, including a tribute to the late Konrad Rudnicki (a philosopher–cosmologist–priest). Previously, Martínez and Trimble had published the contribution *Cosmologists in the Dark* in the proceedings of the meeting "Cosmology Across Cultures" held in Granada, Spain, in 2008, and edited by J. A. Rubiño-Martín, J. A. Belmonte, F. Prada, J. A. and A. Alberdi.

A year earlier, in 2007, Martínez had published in the University of Valencia Press (PUV) the small treatise *Marineros que Surcan Los Cielos*. A few excerpts from these publications have been adapted for inclusion in this book. We are grateful to the editors.

There seemed to be some bits left over. Hmmm, said Martínez; maybe this should be a short book, Trimble thinking that some of those entities had acted like the semi-mythological dragons of old maps, keeping intellectual navigators from venturing as far from Aristotelian shores as they could.

Re-enter Bernard Jones, saying firmly that yes, it should be a book, and there are similar entities in very different branches of science that we should also address as best we can. Here is the product, and we won't say "final product" because we each and all still have lots of ideas about other things that might be said!

Turning now to the almost-finished product, there are at least eight groups of people and institutions to whom we are deeply grateful. And they should all come first. On two-dimensional papers or VDU screens, this is not possible, so we have provided NASA bullets rather than a numbered list:

- An Army, said Napoleon, marches on its stomach. Correspondingly, any scientific project marches on euros, dollars and other fungibles, so our hearty thanks to the Kapteyn Astronomical Institute in Groningen, the Astronomical Observatory and the Department of Astronomy and Astrophysics at the University of Valencia, the Department of Physics and Astronomy of the University of California Irvine and, at the early stages, the Department of Astronomy of the University of Maryland. This work has been funded by the project PID2019-109592GB-100/AEI/10.13039/501100011033 from the Spanish Ministerio de Ciencia e Innovación — Agencia Estatal de Investigación, by the Project of Excellence Prometeo/2020/085 from the Conselleria d'Innovació, Universitats, Ciència i Societat Digital de la Generalitat Valenciana and by the Acción Especial UV-INV-AE19-1199364 from the Vicerrectorado de Investigación de la Universitat de València.
- Authors and publishers need to be brought into contact, one way or another. Special contacts were provided by our encouraging literary agent Renata Kasprzak, who thought that this project could

be guided by Keith Mansfield, editor at World Scientific at that time. Keith's first reading and annotation of the manuscript led to considerable re-writing and reorganization of significant chunks of text. This book would not be what it is had it not been for that insightful, forthright and constructive critique. For that, we are deeply grateful.

- Lots of folks have skills we don't and have provided them for this volume, including both the technical folks at World Scientific (Soundararajan Raghuraman, Michael Beale, Adam Binie, Nandha Kumar, Shezmin Miah and Shanette Quao) and Esther Ibáñez and Sofía Fuentes at the Valencia Astronomical Observatory and for VT a succession of administrative assistants, Carrollanne Simmons, Alison Lara and Jan Strudwick. This also includes the artists responsible for some of the illustrations and infographics: Fernando Ballesteros, Isabel Gálvez and Javier Pérez Belmonte.
- To put it as mildly as possible, we were not the first to think of all the ways all our topics might be shown, so special thanks to the authors, publishers and institutions who have given permission for reproduction of their images and conceptions: Lucio Russo, Alice Roberts, the British Museum and Elizabeth Bray (account manager of British Museum Images), Tony Freeth from Images First Ltd, the History of Science Collections, University of Oklahoma Libraries and JoAnn Palmeri (librarian and curator), the Houghton Library, Harvard University and Zoë Hill (reference librarian), the National Army Museum and Chris Andrews (Archives Assistant), the Natural History Museum and Stephen Atkinson, the Carnegie Institution for Science and Kit Whitten (librarian and archivist), the Harvard University Archives and Edward Copenhagen (reference archivist), the photographer Denyse Applewhite from Princeton University, Brian Greene from Columbia University, History of Science Museum (University of Oxford), *Daily Herald*, the University of Maryland Libraries, the Geological Society of Glasgow, Lamont Doherty Earth Observatory and of course the Wikipedia commons.
- It is a sad confession to make, but we three don't know everything, so we took the precaution of running individual sections, and in a few cases, the entire ensemble, past experts on many of the topics and their histories. Comments, corrections, additions, subtractions and occasional encouraging remarks came from the following to

whom we are both collectively and individually grateful: Pablo Arnalte Mur, José Adolfo de Azcárraga, Fernando Ballesteros, Marcia Bartusiak, Jocelyn Bell-Burnell, Óscar Brevià, Peter Coles, David DeVorkin, Alberto Fernández-Soto, Juan Fabregat, Tony Freeth, José Antonio Font, Beatriz Gallardo Paúls, José Carlos Guirado, Alan Heavens, Raúl Jimenez, Gerta Keller, Avi Loeb, Jean-Pierre Luminet, Enrique Martínez-González, Jordi Miralda Escudé, Iván Martí-Vidal, Lorena Nieves Seoane, Miguel Ángel Sanchis Lozano, Víctor Navarro Brotons, Amelia Ortiz-Gil, Carlos Peña Garay, María Jesús Pons-Bordería, Jim Peebles, Juli Peretó, Vicent Peris, Martin Rees, Enn Saar, José Luis Sanz, Dava Sobel, Peter Tallack, Luigi Toffolatti, Licia Verde, Patricio Vielva, and Rien van de Weygaert.

- We are particularly grateful to Janet Jones (whom author Trimble first met when she was Fred Hoyle's student Janet Clinch) who has been acting as an editor, organizer, minor contributor and arbiter of good grammatical style for much of the revised text. Janet also proposed additional topics for inclusion, as in the part of the book devoted to volcanoes, asteroid impacts and dinosaur extinction.
- VJM is grateful to his wife Laura, their sons Albert and Jordi and their daughter Clara, for inspiring after-dinner discussions on the topics the book deals with.
- We should also acknowledge the use of Wikipedia and the work of all those expert authors who have written erudite books and papers on subjects we can only touch upon. BJ benefited from the clarity of the blogs of Ethan Siegel and Paul Sutter, both of whom make science seem so simple and enjoyable. VJM is grateful for the translation of texts from Spanish to English carried out by Tom Corkett for *Tecnolingüística SL*.

So, hearty thanks to everyone mentioned here (none of whom should be blamed for our residual errors — but feedback from readers would be much appreciated) and also to anyone and everyone whom we should have credited and did not.

Cover

Designed by Javier Pérez Belmonte based on a drawing made by Isabel Gálvez. Following a suggestion by Albert Martínez Artero, the main theme of the cover is the image of Ouroboros, often interpreted as a symbol for eternal cyclic renewal. Images inside the body of Ouroboros are: the sky temperature of the cosmic microwave background radiation (credit: Damien P. George, ESA/Planck collaboration), the galaxy NGC 7331 (courtesy of Vicent Peris and Gilles Bergon, Calar Alto Observatory), the homunculus depicted by N. Hartsoeker (courtesy of History of Science Collections, University of Oklahoma Libraries), an original diagram of the Michelson's 1881 interferometer (Javier Pérez Belmonte), an original drawing of a dinosaur skull (Isabel Gálvez), an engraving of the Torricelli's barometer in a book by Camille Flammarion (public domain, Wikipedia Commons), and the spheres of the geocentric system from the Sky Atlas by Andreas Cellarius (*Orbium Planetarum Terram Complectentium Scenographia.* Courtesy of the Barry Lawrence Ruderman Map Collection, David Rumsey Map Center, Stanford Libraries). Isabel Gálvez is also the author of the illustrations appearing at the beginning of the different parts of the book, on pages 1, 63, 123, 199, 257, 295 and 337.

About the Authors

Bernard J. T. Jones is Emeritus Professor at the Kapteyn Astronomical Institute of the University of Groningen. His research has covered many areas of astrophysics, both theoretical and observational, with a strong emphasis on cosmology, where he is widely published and cited. He is the author of the treatise *Precision Cosmology* (Cambridge University Press, 2017) describing the advances in physics and statistics on which our understanding of the universe is based. He has also developed sophisticated computer imaging systems for diverse mission-critical industrial and surveillance markets.

Vicent J. Martínez is Professor of Astronomy and Astrophysics at the University of Valencia, where he was the Director of its Astronomical Observatory for 11 years. His research is focused on the large-scale structure of the universe. He is the author with Enn Saar of the book *Statistics of the Galaxy Distribution* published by CRC Press in 2002. He received the José María Savirón Award on Popular Science 2013 and the Award for Teaching and Disseminating Physics from the Spanish Royal Physics Society and the BBVA Foundation.

Virginia L. Trimble is Professor of Physics and Astronomy at the University of California, Irvine. She has been working on the structure and evolution of stars, galaxies and the universe, and of the communities of scientists who study them, and the history of science, especially astronomy and physics. She received the 2019 Andrew Gemant Award of the American Institute of Physics and was elected as a Legacy Fellow of the American Astronomical Society (2020) and a Fellow of the Scientific Research Honor Society Sigma Xi (2021). She was President of the Division of Galaxies and the Universe and of the Division of Union-Wide Activities of the International Astronomical Union. She is Doctor *Honoris Causa* by the University of Valencia (2010).

Index of People

Index of Subjects

Printed in the USA
CPSIA information can be obtained
at www.ICGtesting.com
LVHW020618010224
770403LV00001B/1